Introduction to Mathematical Elasticity

Introduction to
Mathematical Elasticity

Leonid P. Lebedev
National University of Colombia

Michael J. Cloud
Lawrence Technological University, USA

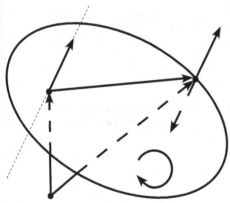

World Scientific

NEW JERSEY · LONDON · SINGAPORE · BEIJING · SHANGHAI · HONG KONG · TAIPEI · CHENNAI

Published by

World Scientific Publishing Co. Pte. Ltd.
5 Toh Tuck Link, Singapore 596224
USA office: 27 Warren Street, Suite 401-402, Hackensack, NJ 07601
UK office: 57 Shelton Street, Covent Garden, London WC2H 9HE

British Library Cataloguing-in-Publication Data
A catalogue record for this book is available from the British Library.

INTRODUCTION TO MATHEMATICAL ELASTICITY
Copyright © 2009 by World Scientific Publishing Co. Pte. Ltd.

All rights reserved. This book, or parts thereof, may not be reproduced in any form or by any means, electronic or mechanical, including photocopying, recording or any information storage and retrieval system now known or to be invented, without written permission from the Publisher.

For photocopying of material in this volume, please pay a copying fee through the Copyright Clearance Center, Inc., 222 Rosewood Drive, Danvers, MA 01923, USA. In this case permission to photocopy is not required from the publisher.

ISBN-13 978-981-4273-72-5
ISBN-10 981-4273-72-4

Desk Editor: Tjan Kwang Wei

Printed in Singapore.

Foreword

In the twenty first century much of Applied Science is concerned with problems of great subtlety and complexity, and the greater the difficulty the more sophisticated are the mathematical and computational techniques needed in order to make significant progress. It is therefore particularly unfortunate that the gap between applied scientists and mathematicians is growing, indeed one is tempted to say has grown, so large that these groups are usually unable to understand each other and so to communicate. The training of undergraduate applied scientists frequently fails to cover even the basic level of the powerful mathematical tools available. On the other hand, mathematical students often learn little about the problems which originally motivated the very mathematics they are studying. Thus communication between the various groups at research level is often exceedingly difficult, and as a consequence joint research, essential now in many areas, faces major obstacles. It is thus timely that the present volume, which is aimed at bridging the gap, should appear.

The book is aimed, on the one hand, at those who study that part of Applied Science dealing with the properties and dynamics of materials, that is Continuum Mechanics. "Materials" in this context may cover a wide range of substances, from fluids to steel to exotic plastics for example with memory; such a range is of obvious importance to almost all types of engineering disciplines, including aeronautics, metallurgy, and civil engineering to mention just a few. Less obvious perhaps, but just as important, is the central role of Continuum Mechanics in several other branches of Applied Science, for example Earth Science, where the properties and movements of the core and mantel of the earth are much studied in traditional contexts such as Volcanology, but also have recently become headline issues with their connection with the safety of storage of nuclear waste.

On the other hand, the book will be of interest to the many mathematicians who are unaware of the wide range of problems to which mathematics may be applied in an interesting and effective manner. They may also be unfamiliar with the extension of standard results which are needed in applications. Thus for example, although functional analysts will know a great deal about Banach spaces and linear operators on them, they may not be familiar with how solving practical problems necessitated the introduction of Sobolev Spaces and of generalised solutions for a whole class of differential equations.

A brief outline of the approach of the authors is as follows. The aim is to start with classical Newtonian Mechanics and then to proceed to show carefully how the basic structure of the governing equations of continuum mechanics are developed. Of course it would be impossible in a single volume to cover the ramifications of this vast subject, however, the methods the authors put in place form the basis for extension to rather general settings. In the development, some key tools of mathematics are introduced naturally and it is shown how they are instrumental in gaining insight into the applications. The first is the Calculus of Variations. The great power of this theory is perhaps unfamiliar to most people without a considerable mathematical background, and the authors outline how effective its methods are in deriving the basic equations as well as in devising powerful solution techniques including numerical methods.

The second main mathematical area introduced is Functional Analysis, the basic theory of which is developed here. Several of its wide-ranging implications are covered in the book. In particular, the authors describe with great elegance how the idea of a generalised solution, which may at first sound obscure, actually much simplifies the analysis of the differential equations associated with continuum mechanics, and in particular is important in understanding the meaning and accuracy of numerical methods, particularly when they are applied to common engineering problems with, for example, discontinuous loads.

The book is written jointly by an engineer and a mathematician, and is sensitive to the background of both applied scientists and mathematicians. The authors take great care to introduce the material in such a way that it is readily available and interesting to each group. I believe that many students and researchers in both categories would benefit a great deal from a study of this excellent volume.

University of Sheffield   V. Hutson
England

Preface

In the early days of mechanics, it was not uncommon for investigators to develop mathematical methods to fit their immediate needs. Modern mathematicians and mechanicists, on the other hand, are often too specialized to understand each other at all. This book is an attempt to relate continuum mechanics with some methods of contemporary mathematics, and to present the latter in a sound mechanical context.

Of course, each topic we treat (e.g., functional analysis, the calculus of variations) has its own vast literature, and we naturally restrict ourselves to those portions needed for our purposes. But even a mathematically prepared reader can benefit from seeing how abstract notions arose from applications. For example, generalized solutions to boundary value problems are closely related to solutions obtained via the extremal principles of mechanics. S.L. Sobolev, who pioneered the use of such solutions and introduced the functional spaces that now bear his name, understood this link because he discovered generalized solutions while solving a hydromechanics problem. But later mathematicians who developed the theory did not maintain this relationship with mechanics. The same thing happened throughout mathematics; ultimately the phenomenon can be attributed to overspecialization of researchers working in the area.

Narrow specialization has led us to the point where experts in different areas cannot communicate. Even those working in closely-related subareas can fail to perceive common points between their disciplines. These points exist precisely because they were inherited from older versions of scientific theory, of course, but many specialists lack the historical perspective necessary to see this. They regard the methods they employ as essentially independent of other areas. This partially explains why we decided to start with classical mechanics, although many readers will be familiar with much of it already. Our plan is to examine continuum mechanics as a growing discipline in which one model arises on the basis of pre-existing models, and therefore inherits certain properties while disregarding others. Size constraints prevent us from covering classical mechanics in detail, so we merely stress those points that show how continuum mechanics developed. We take a somewhat unified view of mechanics as a science rooted in the simple models of a particle and rigid body, and then move to more complicated models. We hope this will assist readers coming from various areas in mechanics, engineering, and mathematics.

Plan of the Book. The book is meant to partially bridge the gap between mathematics and mechanics. It presents a set of topics that were developed in mechanics as mathematical tools.

The purpose of Chapter 1 is to review some elementary ideas of classical mechanics. Although the objects of classical mechanics — particles and rigid bodies — are elementary, the mathematical techniques needed to describe them are not. Hence we also introduce some older portions of mathematics that were developed for application in mechanics. These include the calculus of variations, the elements of Lagrangian mechanics (which we will not use in the sequel but which is important in its own right in many applications), and basic functional analysis. Thus the first chapter is preparation for the introduction and discussion of continuum mechanics.

Equipped with the necessary tools, we can begin working with the objects of continuum mechanics in Chapters 2 and 3. Some — like the spring, string, and bar — are quite elementary. Others — like the beam, membrane, and linearly elastic body — are more advanced. Each time we start by deriving the equations that describe the object. Then we introduce well-posed formulations for practical problems. Finally, we study these formulations using mathematical methods. From Chapter 2 onward, the models considered are not those of classical mechanics. We use results and equations from classical mechanics, but apply them only after introducing certain assumptions that may be regarded as additional axioms for the deformable objects. The two most important hypotheses are the continuum assumption and the solidification principle. These permit us to apply the tools of calculus and the equations of mechanics to deformable objects. Next, we introduce other assumptions such as the smallness of deformations and linearity in the form of Hooke's law. In this way we encounter a few of the models mentioned above. While it might seem that the model of a three-dimensional elastic body should encompass all the other models, this is not the case. The string and membrane models are independent of the model of a three-dimensional body. And the bar and beam models are based on additional hypothesis, so these can be regarded as independent of the three-dimensional model as well.

Unlike classical mechanics where the focus is on initial value problems, in continuum mechanics we encounter boundary value problems for statics and initial-boundary value problems for dynamics. The statics problems for the bar, beam, and string are described by ordinary differential equations and associated boundary conditions. Although the resulting boundary value problems for these objects are relatively simple and can be solved using elementary tools, we take the opportunity to introduce a more powerful approach that also applies to problems involving membranes and three-dimensional bodies. This is the notion of generalized solution as the point of minimum total energy for the "body-force system." Using this mechanical principle, we convert each boundary value problem for an equation or system of equations to a problem of minimizing the total energy. In the process, we come to adopt a very different point of view. Indeed, the mechanics of a problem can dictate certain conditions that are necessary for solution but that may not be evident from a purely mathematical point of view. Moreover, questions such as solvability or uniqueness of a generalized solution will naturally lead us into the realm of functional analysis with its generalized derivatives, Sobolev spaces, and many abstract results.

Despite the abstractness of these studies, they lead to very practical outcomes. The various finite element methods, along with many other numerical methods upon which engineers have come to rely, have these generalized solutions as limits of the resulting numerical approximations. So to understand how practical numerical methods work, one should study the questions presented in this book.

For each mechanical object we cover in this book, we present a derivation of the governing equations that permit engineers to solve practical problems. We also present mathematical tools necessary to study the related boundary value problems. Many of the objects under consideration require only the calculus of functions in one or two variables. The linearly elastic body, however, is presented in the context of tensor analysis. The tensor apparatus, while not strictly required in this relatively simple case, is needed to develop the theories for the corresponding nonlinear and inelastic models. Moreover, the tensorial notation is of value in its own right because of the compact and vivid representation it provides.

In Chapter 3 we touch on some questions concerning vibration. Here the reader can see how the abstract theorems of functional analysis yield results basic in engineering such as the discrete nature of the eigenfrequencies of certain systems while also providing a firm background for topics such as Fourier analysis and the method of separation of variables.

In this book we attempt to present only a minimal set of mathematical tools needed for the qualitative investigation of mechanics problems. We demonstrate that a knowledge of mechanics can assist the pure mathematician, while at the same time a knowledge of mathematics can lead to a deeper understanding of purely mechanical questions. We accomplish these things in an unconventional but organized manner, progressing from simple models to three-dimensional linear elasticity.

Clearly, a great many questions in continuum mechanics are of mutual interest to mathematicians and engineers. A large number of these lie outside the scope of this small book, but are important nevertheless. We ask the reader to bear in mind that the present book is only a brief introduction to the interaction between mechanics and pure mathematics. As such, it cannot be complete.

Acknowledgements. We are deeply indebted to Dr. Ricardo Oscar Grossi (Facultad de Ingeniería, Universidad Nacional de Salta, Argentina) for pointing out errors in the manuscript and suggesting a significant number of pedagogical enhancements. Vivian Hutson and Victor Eremeyev also provided many useful comments. Finally, we are grateful to the staff of World Scientific, including our editor Kwang Wei Tjan and production supervisor Yolande Koh.

The book is dedicated to our wives, Natasha Lebedeva and Beth Lannon-Cloud.

Department of Mechanics and Mathematics   L.P. Lebedev
Rostov State University, Russia
&
Department of Mathematics
National University of Colombia, Colombia

Department of Electrical and Computer Engineering   M.J. Cloud
Lawrence Technological University, USA

Some Notation

$\{x: P(x)\}$	set of all x such that proposition $P(x)$ holds
$x \in A$	x belongs to the set A
u_x or $\partial u/\partial x$	partial derivative with respect to single variable
D^α	multi-index notation for partial derivative
$\dot{u}$	time derivative of u
$\mathbb{R}^n$	n-dimensional Euclidean space
$\mathbf{x}$	vector in Euclidean space
$\mathbf{0}$	zero vector in Euclidean space
$\mathbf{x}^2$	$\mathbf{x} \cdot \mathbf{x}$
x	abstract vector
δ_{ij}, δ^{ij}, or δ_i^j	Kronecker delta
$d(\cdot, \cdot)$	metric, distance function
(S, d)	metric space
$\|\cdot\|$	norm
$\|\cdot\|_X$	norm on space X
$(S, \|\cdot\|)$	normed space
$(\cdot, \cdot)$, $\langle \cdot, \cdot \rangle$	inner products
H	Hilbert space
X	normed space, Banach space
E	energy space
Ω	compact domain
$L^p(\Omega)$	Lebesgue space; space of elements absolutely integrable with degree p over Ω
$W^{m,n}(\Omega)$	Sobolev space
$C(\Omega)$	functions continuous on Ω

$C^{(n)}(\Omega)$	functions continuous and having continuous derivatives up to order n on Ω
$\{x_n\}$	sequence
$\{x_{n1}\}$	subsequence of sequence $\{x_n\}$
$x_n \to x$	(strong) convergence of $\{x_n\}$ to x
$x_n \rightharpoonup x$	weak convergence of $\{x_n\}$ to x
$A \hookrightarrow B$	embedding
$\square$	end of proof
$\sup S$	supremum (least upper bound) of set S
$\inf S$	infimum (greatest lower bound) of set S
$\max S$	maximum of set S
$\min S$	minimum of set S
$\lvert \cdot \rvert$	absolute value, magnitude
$\det(\cdot)$ or $\lvert \cdot \rvert$	determinant
δu	virtual displacement
$\mathcal{E}$	energy functional
$\overline{\alpha}$	complex conjugate of α
$A \dotplus B$	direct sum
$M^\perp$	orthogonal complement of M

Contents

Foreword		v
Preface		vii
Some Notation		xi
1.	Models and Ideas of Classical Mechanics	1
	1.1 Orientation	1
	1.2 Some Words on the Fundamentals of Our Subject	2
	1.3 Metric Spaces and Spaces of Particles	4
	1.4 Vectors and Vector Spaces	8
	1.5 Normed Spaces and Inner Product Spaces	11
	1.6 Forces	16
	1.7 Equilibrium and Motion of a Rigid Body	21
	1.8 D'Alembert's Principle	23
	1.9 The Motion of a System of Particles	25
	1.10 The Rigid Body	31
	1.11 Motion of a System of Particles; Comparison of Trajectories; Notion of Operator	33
	1.12 Matrix Operators and Matrix Equations	40
	1.13 Complete Spaces	44
	1.14 Completion Theorem	48
	1.15 Lebesgue Integration and the L^p Spaces	54
	1.16 Orthogonal Decomposition of Hilbert Space	60
	1.17 Work and Energy	63
	1.18 Virtual Work Principle	67
	1.19 Lagrange's Equations of the Second Kind	70

1.20	Problem of Minimum of a Functional	74
1.21	Hamilton's Principle	83
1.22	Energy Conservation Revisited	85

2. Simple Elastic Models — 89

2.1	Introduction	89
2.2	Two Main Principles of Equilibrium and Motion for Bodies in Continuum Mechanics	89
2.3	Equilibrium of a Spring	91
2.4	Equilibrium of a String	95
2.5	Equilibrium Boundary Value Problems for a String	100
2.6	Generalized Formulation of the Equilibrium Problem for a String	105
2.7	Virtual Work Principle for a String	108
2.8	Riesz Representation Theorem	112
2.9	Generalized Setup of the Dirichlet Problem for a String	115
2.10	First Theorems of Imbedding	116
2.11	Generalized Setup of the Dirichlet Problem for a String, Continued	120
2.12	Neumann Problem for the String	122
2.13	The Generalized Solution of Linear Mechanical Problems and the Principle of Minimum Total Energy	126
2.14	Nonlinear Model of a Membrane	128
2.15	Linear Membrane Theory: Poisson's Equation	131
2.16	Generalized Setup of the Dirichlet Problem for a Linear Membrane	132
2.17	Other Membrane Equilibrium Problems	145
2.18	Banach's Contraction Mapping Principle	151

3. Theory of Elasticity: Statics and Dynamics — 157

3.1	Introduction	157
3.2	An Elastic Bar Under Stretching	158
3.3	Bending of a beam	168
3.4	Generalized Solutions to the Equilibrium Problem for a Beam	175
3.5	Generalized Setup: Rough Qualitative Discussion	179
3.6	Pressure and Stresses	181
3.7	Vectors and Tensors	188

3.8	The Cauchy Stress Tensor, Continued	196
3.9	Basic Tensor Calculus in Curvilinear Coordinates	202
3.10	Euler and Lagrange Descriptions of Continua	207
3.11	Strain Tensors	208
3.12	The Virtual Work Principle	214
3.13	Hooke's Law in Three Dimensions	218
3.14	The Equilibrium Equations of Linear Elasticity in Displacements	221
3.15	Virtual Work Principle in Linear Elasticity	224
3.16	Generalized Setup of Elasticity Problems	227
3.17	Existence Theorem for an Elastic Body	231
3.18	Equilibrium of a Free Elastic Body	232
3.19	Variational Methods for Equilibrium Problems	235
3.20	A Brief but Important Remark	243
3.21	Countable Sets and Separable Spaces	243
3.22	Fourier Series	245
3.23	Problem of Vibration for Elastic Structures	249
3.24	Self-Adjointness of A and Its Consequences	252
3.25	Compactness of A	255
3.26	Riesz–Fredholm Theory for a Linear, Self-Adjoint, Compact Operator in a Hilbert Space	262
3.27	Weak Convergence in Hilbert Space	267
3.28	Completeness of the System of Eigenvectors of a Self-Adjoint, Compact, Strictly Positive Linear Operator	272
3.29	Other Standard Models of Elasticity	277

Appendix A Hints for Selected Exercises	281
Bibliography	293
Index	295

Chapter 1

Models and Ideas of Classical Mechanics

1.1 Orientation

An introductory section is usually written to convince readers that their lives will be incomplete unless they buy the book. At some risk, the present authors would like to state that a reader will profit most from this book if he or she seeks a clearer view of continuum mechanics from two concurrent viewpoints: that of the engineer, and that of the mathematician.

Continuum mechanics, which began with simple problems in hydromechanics and the strength of materials, spans multiple theories including elasticity, plasticity, and viscoelasticity. It employs models describing not only how objects deform under load, but their thermal, electric, and magnetic properties. The various sub-theories within continuum mechanics can reach high degrees of complexity. We will ultimately focus on the linearized theory of elasticity, a departure point for many extensions of basic continuum mechanics.

Any engineer will know at least some elements of continuum mechanics. (It is worth noting that James Clerk Maxwell utilized notions from hydrodynamics when formulating his famous equations of electromagnetism.) The understanding of a typical applied mathematician is, however, quite different. To a mathematician working in the theory of shells, say, the whole subject may commence with a statement of the form "The following system of partial differential equations describes a shell in equilibrium. Supplementing these with the boundary conditions, we arrive at the boundary value problem considered in the next 300 pages." Then the mathematician can forget what was denoted by u or F: he or she can begin to play with equations in a manner completely divorced from physical considerations.

Engineers and mathematicians are therefore unlikely to understand each other, even when discussing the same problem. The commonality between their worlds is minimal. The purpose of the present book is to address this unfortunate gap. Only a better mutual understanding can further the collaborative efforts of these two technical communities. Mathematicians should understand that the physical models they view as axioms are actually derived under rather crude assumptions. A mathematician can run into trouble by oversimplifying the behavior of a real object or unwittingly introducing highly artificial features into a model. Engineers, on the other hand, should start to understand why mathematicians spend so much time talking about things like the ill-posedness of a problem, or the weak convergence of a sequence of approximations.

Our ideal reader will possess a wealth of curiosity and a desire to apply it to the fascinating gulf that still exists between real physics and rigorous mathematics. With that thought firmly in mind, let us begin.

1.2 Some Words on the Fundamentals of Our Subject

Among the most primitive of the sciences that treat the behavior of bodies in space, the *theory of elasticity* is simple in some respects and complex in others. It describes the motion and deformation of real bodies, but does so by dealing with idealizations. Neglecting atomic structure, the subject treats the motion and deformation of geometrical figures; unlike geometry, however, it attributes the properties of mass and elasticity to the parts of such figures. We might say that the theory of elasticity deals with spatial transformations of geometrical figures having these mechanical properties.

In this book we shall consider some principal models and mathematical questions in the theory of elasticity. Just as pure mathematics had its roots in the ideas of arithmetic and elementary geometry, and developed these so far that a novice may not see connections between the former and the more advanced parts of modern mathematics, elasticity was based on the ideas of *classical mechanics*. Classical mechanics also treats real natural objects, but using highly simplified models. The set of all the models in continuum mechanics constitutes a hierarchy entailing increasing complexity but still resting on the laws of motion and equilibrium of real bodies (which continuum mechanics inherited from classical mechanics). So before proceeding to the theory of elasticity, we should touch on a few essential points from classical mechanics.

First, we should point out that mathematicians are not alone in having to deal with abstractions. The objects of mechanics may have more elaborate properties than those of pure mathematics, but they are *still* abstractions. Such properties were assigned only after long experience with the behavior of real bodies. They were also influenced by the mathematical tools available for their study. Essential portions of mathematics, in turn, were developed to meet the needs of mechanics, and the interplay between these subjects is still strong. Although various viewpoints are possible, it can be argued that classical mechanics is now a branch of mathematics. It is common for sciences to branch out into separate areas initially, only to reunite after reaching a more mature stage. Indeed, the ultimate aim of any natural science is the study of one thing: Nature. Some mathematicians believe they study an ideal world, but this latter "world" is an attempt to describe Nature in a certain way.

So what are the objects of classical mechanics? We shall not delve into the notions of space and time here. These may seem simple and evident to students today, but classical mechanicists up to and including Newton did not regard them as such. Although many of the ideas in mechanics were elaborated long before Newton, we frequently refer to him as the founder of classical mechanics. In fact, Newton collected known results and created a general approach to modern mechanics just as Euclid did for geometry.

The most elementary object that exists in equilibrium or moves through the space and time of classical mechanics is the *material point* or *mass point*. We shall refer to it loosely as a *particle*. As with a geometric point, a mass point has no spatial dimension — although it does have finite mass. The notion of *mass* is regarded as primitive (and undefinable) in mechanics. In elementary books we encounter statements to the effect that mass is "the measure of inertia" of a body. But such "definitions" are meaningless (despite the comfort we often take in them).

Next come collections of mass points and, after that, *rigid bodies* (i.e., bodies that cannot be deformed). Newton used the term "corpuscle" instead of "mass point". He avoided the term "rigid body" as well. But time changes everything and we are discussing the present form of classical mechanics.

In many mechanics books, a rigid body is defined as a collection of mass points whose relative positions are fixed. Even for a body that could realistically be considered as a finite collection of particles, however, the definition is not complete until we specify *how* the particles can interact. But the practical necessity of considering bodies that appear as geometric figures having continuous mass distributions basically forces us to employ limit passages from finite sets of mass points to continuous bodies. To justify such passages we must bring in additional assumptions which, from a mathematical viewpoint, could be regarded as new axioms of classical mechanics. Overall, however, it is more convenient to take the rigid body itself as a primary notion. We then formulate as *axioms* for a *continuous* rigid body the properties *derived* for a rigid body composed of a finite number of particles. This is implicitly done in almost any book on theoretical mechanics. None of these constructs — point mass, rigid body, etc. — exist in Nature, but all can serve as good approximations to real bodies.

Classical mechanics studies the motion and equilibrium of its objects under the action of *forces*. Forces likewise may not truly exist in Nature, but the force concept gives us a way to describe the effects of bodies or fields on the motion of a given body. Force is another primitive notion in mechanics; it is left undefined, but certain properties are attributed to it. In that sense it is similar to the primitive notions of pure geometry, such as those of point, line, and plane.

With the advent of relativity, classical mechanics lost much of its status as an exact science. This hardly affected its usefulness as an engineering tool. We could maintain, furthermore, that classical mechanics *still is* an exact science in the same sense as mathematics is. Its structure, in fact, is similar to that of a branch of mathematics: it has a set of primitive notions (space, time, particle, force, etc.), as well as a set of axioms. Unlike the axioms of mathematics, however, the axioms of mechanics are sometimes left unstated.

Before embarking on the theory of elasticity, we shall provide an overview of the conceptual base on which it rests. This includes a collection of topics from classical mechanics, along with certain tools of the theory of elasticity that happened to arise in the context of classical mechanics.

1.3 Metric Spaces and Spaces of Particles

Newtonian mechanics considers the motion of mass points and rigid bodies in an absolute space. Of course, this implies that the latter exists and has properties like those of the space of ordinary Euclidean geometry. We call such a space a *(Newtonian) reference frame*. If we consider one absolute space in which a system of particles moves, then there exist (infinitely many) other absolute spaces in which this system can be taken as moving. The spaces themselves translate with respect to one another at a constant velocity. We call any two such frames *inertial reference frames*. Note that one inertial frame cannot rotate with respect to another; rotation about some axis implies that the velocity of various points is proportional to their distances from the axis, and this is obviously not the same for all points.

Newton's first law implies that there is no *preferred* absolute space: no experiment can distinguish one inertial reference frame from another. In particular, it is impossible to determine which reference frame might be "stationary" in an absolute sense. Nonetheless, it is conventional to construct a reference frame that is "stationary with respect to the distant stars" (or even "stationary with respect to the Earth's surface," although any point of that surface executes a complex motion produced by the Earth's rotation about its axis, its revolution about the Sun, etc.). In Newton's time, a good number of stars appeared to be fixed in position, so they were used to mark out a reference frame. Today we know that all stars are in motion, but the idea is still convenient for ordinary calculations.

When an ideal particle has a fixed position in an absolute space, it coincides with a point of the space. The space itself is *isotropic* and *homogeneous* — its properties are the same in all directions at all points — so the only meaningful relation between any two of its points is one of separation distance. We wish to apply the notion of distance to other objects not necessarily related to geometrical space, so for the mass points we will generalize it as follows. Suppose that to any pair of points A and B we assign a nonnegative finite number denoted by $d(A, B)$. In this way we get a function in two variables that is defined for each pair of points in the space. It is called a distance function or, in mathematics, a *metric*, if it satisfies three axioms of the usual distance employed in geometry:

M1. $d(A, B) \geq 0$, with $d(A, B) = 0$ if and only if A and B coincide;
M2. $d(B, A) = d(A, B)$;
M3. $d(A, B) \leq d(A, C) + d(C, B)$, where C is any other point of the space.

Exercise 1.3.1. *Demonstrate that M1 can be changed to "$d(A, B) = 0$ if and only if the points A and B coincide." So $d(A, B) \geq 0$ is a consequence of the altered system of axioms.*

The absolute space is physically empty, composed only of fictitious points. Material points are always associated with material objects; the reference frame is a mental construction for the sake of expediency. Let us take a fixed time instant. Now we can consider only the set S of mass points, which could be finite or infinite, and pair this set with a metric d

that was defined in the absolute space (but is now applied only to mass points). In mathematics, we denote such a pair by the symbol (S, d) and call it a *metric space*. When there is no chance for confusion (i.e., when only one metric d is employed in a given discussion), we loosely refer to S itself as a "metric space".

The notion of metric space is general and can be used with sets S that do not consist of spatial or mass points. The elements of S can be of any nature if an appropriate metric d can be defined. In the term "metric space", the word "space" is simply a synonym for "set". Hence, by definition, even a set consisting of just one mass point is a metric space. Indeed, labeling this point A, we could define the metric by setting $d(A, A) = 0$. The reader can verify that M1–M3 hold in this simple case.

When dealing with metric spaces we often borrow mental pictures from elementary geometry. For example, we can define a *ball* having center x_0 and radius $r > 0$ as the set of points x of the metric space that fall within distance r of x_0. The ball is *open* if the inequality $d(x, x_0) < r$ is used; it is *closed* if the inequality $d(x, x_0) \leq r$ is used. Sometimes we require the notion of a *neighborhood* of x_0. By this we mean a subset of the space that contains some open ball with center x_0 and nonzero radius.

We have said that a metric can take any form satisfying the necessary axioms. Consider, for instance, a set of mass points whose motion is confined to the surface of a sphere. In this case it is natural to measure distance along the great circle that connects any two points (of the two possible arcs along the great circle, we must take the shorter one in order to satisfy the metric axioms). This is essentially how we measure ordinary distances between points on the Earth's surface.

As another example we could consider how distances should be measured in a town where the streets form a uniform rectangular grid. A metric can be defined as the minimal distance between any two points *when measured along the grid lines*. If we introduce Cartesian coordinates in the plane and identify points with these coordinates, e.g.,

$$A = (a_1, a_2), \qquad B = (b_1, b_2),$$

then we can represent the "taxicab metric" by the expression

$$d_1(A, B) = |b_1 - a_1| + |b_2 - a_2|. \qquad (1.3.1)$$

The reader can also verify that the function

$$d_p(A, B) = (|b_1 - a_1|^p + |b_2 - a_2|^p)^{1/p} \qquad (1.3.2)$$

is a valid metric for any fixed $p \geq 1$. When $p = 2$ we get the Euclidean distance. The reader should guard against a tendency to accept such statements without actually checking for satisfaction of the axioms. For the p-metric above, the only nontrivial axiom to check is the "triangle inequality" M3. Satisfaction of this follows from Minkowski's inequality

$$\left(\sum_{i=1}^{m} |a_i + b_i|^p\right)^{1/p} \leq \left(\sum_{i=1}^{m} |a_i|^p\right)^{1/p} + \left(\sum_{i=1}^{m} |b_i|^p\right)^{1/p} \quad (1.3.3)$$

which holds for any $p \geq 1$ and any two sets of real numbers $a_1, \ldots, a_m$ and $b_1, \ldots, b_m$.

Exercise 1.3.2. *Demonstrate that for $0 < p < 1$, the function $d_p(A, B)$ cannot serve as a metric for points on the plane.*

To introduce Cartesian coordinates as we have done above, we must appoint an origin. We noted previously, however, that all points in an absolute Newtonian space stand on an equal footing. So the choice of coordinate origin is arbitrary and has no ultimate physical significance. When we write $A = (a_1, a_2, a_3)$ we in reality introduce a directed line segment extending from the frame origin to the point A. We can denote this segment by $a_1\mathbf{e}_1 + a_2\mathbf{e}_2 + a_3\mathbf{e}_3$ where $\mathbf{e}_1$, $\mathbf{e}_2$, and $\mathbf{e}_3$ are unit vectors along the orthogonal frame axes. We can even draw this vector in the geometrical space, where the mass points are, but must keep in mind that it merely symbolizes a correspondence between a certain vector and the position of a point as mentioned above; in particular it does not belong to our initial set of mass points in the space. The reader has surely made use of "position vectors" in solving mechanics problems. The concept is useful because it allows us to impose all the machinery of vector algebra on a space that really possesses only the metric property. In order to make the best possible use of this correspondence between mass points and vectors, we should introduce it in such a way that it is one-to-one and preserves the distance (metric) between pairs of respective elements. A one-to-one correspondence between two metric spaces in which distance is preserved is said to be an *isometric* correspondence.

We started with a simple metric space having no algebraic structure and arrived at another metric space with algebraic structure. We shall continue to work with the latter space, loosely regarding position vectors as points. Again, this is permissible only because of the isometric correspondence mentioned above. Just as there are no position vectors in the space of mass points, there are no mass points in the space of position vectors. It is natural to employ a mixture of the two different kinds of objects only because we are used to drawing points and vectors on the same plane. But from a mathematical viewpoint, the vectors and points are objects of different natures and are treated using very different tools.

Let us quickly summarize. The principal objects of classical mechanics are mass points, which are described using a reference frame that has been imposed on an idealized absolute space. In this space, a free mass point (i.e., one that is not experiencing forces or collisions with other bodies) maintains both its speed and its direction of motion. From a mathematical viewpoint, however, classical mechanics deals with the images of those mass points under a one-to-one correspondence with a space of position vectors. The rules for working with these vectors are taken from the theory of vector spaces, but are supplemented by the rules of mechanics itself.

1.4 Vectors and Vector Spaces

Many of us were exposed to the vector concept in high school mathematics. Unfortunately, beginning students are prone to assign the term "vector" to any arrow drawn on the chalkboard. A nice demonstration that such an arrow need not represent a vector was given by A.P. Minakov. Sketching the perpendicular intersection of two one-way streets with a shop standing on one corner, Minakov had his students imagine traffic flows of 30 cars per minute down one street and 40 cars per minute down the other. He noted that nothing would prevent anyone from labeling these flows with appropriately sized arrows. He was quick to point out, however, that if these arrows represented vectors then a resultant flow of $(30^2+40^2)^{1/2} = 50$ cars per minute would be entering the doors of the shop! Clearly we cannot apply vector addition to just any quantities that happen to be represented by directed line segments. Quantities have a vectorial nature only when we can carry out vectorial operations with them. These include vector addition, subtraction, and multiplication by a scalar.

The lesson here is that one cannot perform mathematical operations on objects without first verifying that these objects share all properties required for validity of the operations. This holds for the formation of a metric and for the treatment of quantities as vectors. So what is a vector — or, more precisely, a linear space of vectors? In mathematics, an element of an n-dimensional Euclidean space is a special object denoted variously by symbols such as $\mathbf{x}$, $\bar{x}$, $\underline{x}$, or $\vec{x}$. But the mere use of notation does not automatically make something a vector. With vectors, we must be able to carry out two principal operations: vector addition and scalar multiplication. These operations are, in turn, subject to the axioms of a vector space. Let $V \neq \emptyset$ be a set along with suitably defined operations of addition and scalar multiplication. That is,

(a) to each pair $x, y \in V$ there corresponds a unique vector $x + y$, and
(b) to each $x \in V$ and each scalar λ there corresponds a unique vector λx.

This structure — consisting of the elements with two operations of addition and multiplication by scalars (real or complex) — can be a vector space only if the following hold:

(1) V is algebraically closed with respect to the two operations. That is, $x + y$ and αx both belong to V for any $x, y \in V$ and any scalar α.
(2) Addition is both commutative and associative; that is, we have
$$x + y = y + x, \qquad x + (y + z) = (x + y) + z,$$
for any $x, y, z \in V$.
(3) There is an additive identity element in V. This unique element is called the *zero vector* and is denoted by 0; it has the property that $x + 0 = x$ for any $x \in V$.
(4) Each $x \in V$ has a unique additive inverse in V. This vector is denoted by $-x$ and has the property that $x + (-x) = 0$.
(5) If $x, y \in V$ and α, β are any scalars, then
  (1) $\alpha(x + y) = \alpha x + \alpha y$,
  (2) $(\alpha + \beta)x = \alpha x + \beta x$,
  (3) $(\alpha \beta)x = \alpha(\beta x)$.

  Moreover, we have $1x = x$.

Of course, these axioms are so simple that they obviously hold for ordinary vectors in two or three dimensions. But such formalization allows us to apply them to more abstract sets. Indeed, the notion of vector space applies not only to sets of forces or position vectors, but also to finite (or infinite) sets of trigonometric polynomials of the form
$$\sum_k (a_k \sin kx + b_k \cos kx).$$

When considered on some interval $a \leq x \leq b$ (finite or infinite), a set of these polynomials can constitute a vector space (of dimension $2n$ if we sum over k from 1 to n only). Here, however, the use of arrows to represent vectors would be fruitless. We urge the reader to verify the axioms of a vector space for this example and thereby justify labeling trigonometric polynomials as vectors (even though in many situations it would not be advisable to do so).

We have referred to vector space dimension. This notion relates, as the reader knows, to that of linear independence. A set of vectors $\{x_1, \ldots, x_m\}$ is said to be *linearly independent* if from the equation

$$c_1 x_1 + \cdots + c_n x_m = 0$$

with scalar coefficients c_k it follows that $c_1 = \cdots = c_m = 0$. The *dimension* n of a vector space is the maximal number linearly independent vectors in the space. A set of n linearly independent vectors is called a *basis* of the n-dimensional space; any vector from the space can be uniquely represented as a linear sum of the basis vectors.

If we cannot find a finite n for the dimension of the space, we call the space *infinite dimensional*. Here the problem of basis is not simple, however. Above we considered the $2n$-dimensional space of trigonometric polynomials. For some problems this space is of great interest; the trigonometric polynomials are used to represent solutions to differential equations of the hyperbolic or parabolic type (and not only these, of course, but this is where interest in such polynomials originated). But infinite polynomials

$$b_0 + \sum_{k=1}^{\infty} (a_k \sin kx + b_k \cos kx),$$

called *Fourier series*, are also employed. In calculus, these series are considered apart from differential equations. Instead, they are used to represent a 2π-periodic continuous function, and it is shown that the Fourier coefficients a_k and b_k are defined uniquely. The set of continuous 2π-periodic functions is obviously a vector space. Furthermore, it has infinite dimension since a finite set of functions $1, \sin x, \sin 2x, \ldots, \sin rx, \cos x, \cos 2x, \ldots, \cos rx$ is linearly independent. Thus we have found an infinite set of linearly independent "vectors" (i.e., continuous functions) in the space.

Exercise 1.4.1. *Propose a few metrics over spaces of trigonometric polynomials.*

1.5 Normed Spaces and Inner Product Spaces

It is clear that the space of functions continuous on a segment $[a,b]$ is also an infinite-dimensional vector space. An extension of this idea is the space of vector functions — functions taking values from a vector space such as $\mathbb{R}^3$ or $\mathbb{R}^2$ — that depend continuously on some parameter. If the parameter is time t, such a space can be regarded as the set of all continuous trajectories of a point in space for t in some segment such as $[0,T]$. The space of continuous vector functions on $[a,b]$ is important in mechanics. We often must characterize the difference between two trajectories $\mathbf{f} = \mathbf{f}(t)$ and $\mathbf{g} = \mathbf{g}(t)$, not only at a given time instant (which could be done with the metrics we have considered for the mass points), but "in total" on the segment. We could accomplish this with the metric space notion and introduce, say,

$$d(\mathbf{f},\mathbf{g}) = \max_{t \in [a,b]} |\mathbf{f}(t) - \mathbf{g}(t)|. \tag{1.5.1}$$

Here, instead of the absolute value, we could use any metric on $\mathbb{R}^n$ to characterize the distance between points on the trajectories at the instant t. But the vectorial structure of the space leads us to use a particular kind of metric, one based on the norm that appears in linear algebra.

A *norm* on a vector space is a function that assigns to every element x in the space a finite nonnegative number $\|x\|$. This function must satisfy the following axioms:

N1. $\|x\| \geq 0$, with $\|x\| = 0$ if and only if $x = 0$;
N2. $\|\lambda x\| = |\lambda|\,\|x\|$ for any scalar λ;
N3. $\|x + y\| \leq \|x\| + \|y\|$ for any two vectors x, y in the space.

A vector space V, when paired with a norm $\|\cdot\|$, is called a *normed space*. From a mechanical viewpoint, the notion of norm brings in the idea of the *homogeneity* of space. First, if two pairs of elements have equal differences then the norms of these differences will be equal, regardless of the regions of space from which we take the elements:

$$\|(x+z) - (y+z)\| = \|x + z - y - z\| = \|x - y\|.$$

Second, axiom N2 guarantees homogeneity with respect to multiplication by a scalar. The triangle inequality N3 extends the usual triangle axiom of Euclidean space. Note that every normed space is automatically a metric space. Indeed, the function

$$d(x,y) = \|x - y\| \tag{1.5.2}$$

is easily seen to satisfy M1–M3 on page 5.

Exercise 1.5.1. *Prove that N1 can be changed to "$\|x\| = 0$ if and only if $x = 0$". That is, $\|x\| \geq 0$ is a consequence of the new set of axioms.*

Exercise 1.5.2. *Show that the inequality*

$$\big| \|x\| - \|y\| \big| \leq \|x - y\| \tag{1.5.3}$$

holds for any $x, y \in V$.

It is clear that the metric (1.5.1) is induced by the norm

$$\|\mathbf{f}\|_{C(a,b)} = \max_{t \in [a,b]} |\mathbf{f}(t)|, \tag{1.5.4}$$

and therefore the space of continuous vector functions on $[a, b]$ with this norm is a normed space. It is usually denoted by $C(a, b)$. We stress that this notation indicates not only that a set of continuous vector functions is under consideration, but that the norm (1.5.4) is assumed as well. Perhaps it would be more reasonable to write $C[a, b]$, but our notation is traditional and in this book we deal exclusively with compact domains such as closed and bounded intervals. The subscript $C(a, b)$ is appended to the norm symbol because we shall introduce other norms on the same set of vector functions. For example, the norm

$$\|\mathbf{f}\|_{L^2(a,b)} = \left(\int_a^b |\mathbf{f}(t)|^2 \, dt \right)^{1/2} \tag{1.5.5}$$

characterizes the difference between two continuous vector functions in an integral rather than a pointwise sense. Another difference between (1.5.5) and (1.5.4) will become apparent when we consider the results of performing limit passages for sequences of elements.

Exercise 1.5.3. *On the set of all functions continuous on $[0, 1]$, introduce*

$$\|f\| = \sup_{x \in [0,1]} \frac{|f(x)|}{x}.$$

Is the result a normed space?

Remark 1.5.1. When we say that some quantity like a norm is "defined on" a space, we mean that it must be defined (hence finite) at every point of the space. □

There is another integral quantity whose relationship to the norm (1.5.5) is analogous to that between the ordinary dot product in $\mathbb{R}^n$ and the ordinary length of a vector. It is given by

$$(\mathbf{f}, \mathbf{g})_{L^2(a,b)} = \int_a^b \mathbf{f}(t) \cdot \mathbf{g}(t)\, dt,$$

and we have

$$(\mathbf{f}, \mathbf{f})_{L^2(a,b)} = \|\mathbf{f}\|^2_{L^2(a,b)}.$$

This, along with the linearity of $(\mathbf{f}, \mathbf{g})_{L^2(a,b)}$ with respect to the arguments $\mathbf{f}$ and $\mathbf{g}$, suggests that we could use this integral form in the same way we use a dot product in $\mathbb{R}^n$.

Exercise 1.5.4. *Using a uniform Riemann sum approximation to the integral, confirm that the analogy between the dot product and the norm in $\mathbb{R}^n$ really corresponds to the relation between $(f,g)_{L^2(a,b)}$ and $\|f\|_{L^2(a,b)}$ for ordinary functions continuous on $[a,b]$.*

Let us introduce the general case covering such analogies to the dot product. An *inner product space* is a vector space V together with a function (f, g) defined for any pair of elements $f, g \in V$; this function, termed an *inner product*, satisfies the following axioms:

I1. $(f, f) \geq 0$ for all $f \in V$, with $(f, f) = 0$ if and only if $f = 0$;
I2. $(g, f) = (f, g)$ for all $f, g \in V$;
I3. $(\lambda f + \mu g, h) = \lambda(f, h) + \mu(g, h)$ for all $f, g, h \in V$ and real scalars λ, μ.

In much of this book we employ real spaces. It is worth mentioning, however, that for a complex space we need only change axiom I2 to read

I2'. $(g, f) = \overline{(f, g)}$ for all $f, g \in V$

and then change the real scalars in I3 to complex scalars.

The inner product structure lets us introduce the idea of orthogonality between elements of a vector space. We say that f and g are *orthogonal* if

$$(f, g) = 0. \qquad (1.5.6)$$

This extends the familiar condition $\mathbf{f} \cdot \mathbf{g} = 0$ in $\mathbb{R}^3$. In $\mathbb{R}^3$, of course, we can go further and introduce the full notion of angle. In general this is not possible; however, the orthogonality idea deserves special mention because it lets us carry out orthogonal projections even in an abstract space.

Exercise 1.5.5. *Let e be a unit vector: $(e,e) = 1$. Prove that $f - (f,e)e$ is orthogonal to e. This shows that the operation $(f,e)e$ is analogous to orthogonal projection onto an axis in $\mathbb{R}^3$.*

The next thing to notice is that an inner product space is automatically a normed space under the *natural norm*

$$\|f\| = (f,f)^{1/2} \tag{1.5.7}$$

where the positive square root is taken. Hence it is also a metric space under the induced metric

$$d(f,g) = \|f - g\| = (f-g, f-g)^{1/2}. \tag{1.5.8}$$

As always, we cannot simply state that $(f,f)^{1/2}$ is a norm; we must prove that the axioms hold. Verification of N1 and N2 is trivial, but this is not the case for N3. The triangle inequality is equivalent to

$$\|x+y\|^2 \leq \|x\|^2 + 2\|x\|\|y\| + \|y\|^2,$$

which, in terms of the inner product, can be rewritten as

$$(x+y, x+y) \leq (x,x) + 2\|x\|\|y\| + (y,y).$$

By I1–I3 we have

$$(x+y, x+y) = (x,x) + 2(x,y) + (y,y).$$

We see that N3 is satisfied if we can establish the

Cauchy–Buniakowski–Schwarz inequality. *We have*

$$|(x,y)| \leq \|x\|\|y\|. \tag{1.5.9}$$

Equality holds if x or y is zero, or if there is a constant c such that $y = cx$.

Proof. Clearly (1.5.9) holds as an equality for $x = 0$. Now assume $x \neq 0$ and consider the vector

$$z = y - \frac{(y,x)}{\|x\|^2}x$$

$$= (y,e)e \text{ where } e = \frac{x}{\|x\|}.$$

By Exercise 1.5.5, we have $(z,x) = 0$. By I1,

$$0 \leq \|z\|^2 = \left(y - \frac{(y,x)}{\|x\|^2}x, \; y - \frac{(y,x)}{\|x\|^2}x\right) = (y,y) - \frac{(y,x)(x,y)}{\|x\|^2},$$

which is equivalent to (1.5.9) squared. Equality in (1.5.9) holds only when $\|z\| = 0$; this means that $z = 0$, hence $y = cx$ with $c = (y,x)/\|x\|^2$. $\square$

Exercise 1.5.6. *Prove* (1.5.9) *for a complex space* V.

Some equalities from elementary geometry extend to an inner product space. One is the *parallelogram equality*: the sum of the squares of the diagonals of a parallelogram is twice the sum of the squares of its sides. The reader should prove this in abstract form.

Exercise 1.5.7. *Show that*

$$\|x+y\|^2 + \|x-y\|^2 = 2(\|x\|^2 + \|y\|^2). \tag{1.5.10}$$

Because in a real inner product space

$$4(x,y) = \|x+y\|^2 - \|x-y\|^2, \tag{1.5.11}$$

we can represent an inner product using only its norm. Clearly the form on the right-hand side of (1.5.11) does not satisfy the axioms of the inner product in just any normed space, hence not every normed space is an inner product space.

It is useful to present one special inner product space.

The space l^2

This space consists of all the real infinite sequences $X = (x_1, x_2, x_3, \ldots)$ for which the series

$$\sum_{k=1}^{\infty} |x_k|^2 \tag{1.5.12}$$

converges. The inner product of X with the sequence $Y = (y_1, y_2, y_3, \ldots)$ is given by

$$(X, Y) = \sum_{k=1}^{\infty} x_k y_k. \tag{1.5.13}$$

We emphasize that from the set of all infinite sequences we select *only* those for which (1.5.12) is convergent.

The space l^2 is quite special. Let us explain why, using the results and terminology that will appear later in the book. In a Hilbert space H with an orthonormal basis $(e_1, e_2, e_3, \ldots)$ (i.e., a separable Hilbert space), any $x \in H$ can be represented in the form

$$x = \sum_{k=1}^{\infty} x_k e_k \tag{1.5.14}$$

with uniquely defined Fourier coefficients $x_k = (x, e_k)$. Moreover

$$\|x\|^2 = \sum_{k=1}^{\infty} |x_k|^2 \tag{1.5.15}$$

and so we obtain a one-to-one correspondence between H and l^2. This means we could present the entire theory of separable Hilbert spaces using only l^2.

In a similar fashion we can introduce normed spaces l^p, $p \geq 1$, starting with the same set of sequences X but under the condition that the series $\sum_{k=1}^{\infty} |x_k|^p$ must converge. The norm is given by

$$\|X\|_{l^p} = \left(\sum_{k=1}^{\infty} |x_k|^p \right) \qquad (p \geq 1). \tag{1.5.16}$$

The inner product of l^2 can be used with a space l^p for $p > 2$, but such a space is incomplete under the induced (i.e., l^2) norm. We will understand what this means after introducing Banach and Hilbert spaces.

Exercise 1.5.8. *Show that for any integer k the elements 1, $\sin kx$, and $\cos kx$ are mutually orthogonal in $L^2(-\pi, \pi)$. Find the unit basis vectors of this space and calculate the projections of the above elements along the directions defined by the basis. In this way we obtain the Fourier coefficients of a function given on $(-\pi, \pi)$. The present general viewpoint (with the inner product) allows us to consider other expansions of continuous and discontinuous functions. Such expansions (in particular, involving the use of orthogonal polynomials) are widely used in analysis.*

1.6 Forces

The term "force" lacks a rigorous definition. But we think of force as the quantity that effects the motions or deformations of bodies and characterizes their mutual interaction. That forces have a vectorial nature is also well known (in fact, force was the prototype for the general notion of a vector). By this we mean that the *resultant* of several forces acting on the same particle can be found by vector addition. The original force system can be replaced by its resultant, and the motion of the particle will be unchanged.

Since forces are applied to certain points, the addition of two forces that act on different mass points would be senseless. Therefore the sets of forces acting on different particles constitute different vector spaces.

In mathematics, vector quantities do not carry physical units. Hence their norms, which can be regarded as characterizing their sizes or intensities, are dimensionless numbers. But many mechanical quantities do carry units and this is the case with force. The SI base unit of force is the Newton (N). The need to perform extensive numerical calculations now requires engineers to convert many of their equations and relations to *dimensionless form*: a form in which everything is expressed in relative figures in such a way that variable quantities typically lie near unity. This has its advantages for calculation, but can obscure what happens with real objects. Of course, some routine calculations can be performed entirely by a computer — even to the plotting of final graphs. But the analysis of intermediate and final results is often easier using quantities whose physical meanings are clear.

We will frequently use energy norms. These inherit the dimensional units of the corresponding energy quantities. We will establish various inequalities among the energy norms, and constants will appear in these relations. It is important to understand that such constants often carry dimensional units and would therefore have different values in other unit systems. And, of course, we cannot directly compare the values of constants that have different dimensions.

We will not stop to review the familiar processes of adding or subtracting forces that act on the same particle. However, we should mention some issues concerning rigid bodies. To say that force is a vector is really an oversimplification. With rigid bodies, we run into various modifications of the vector concept: we must distinguish between sliding vectors, free vectors, etc. This is due to the peculiarities inherent in the effects produced by forces acting on rigid bodies (in particular, the possibility of inducing rotation). Before discussing this further, let us recall that when we depict a force vector acting on a body, we in fact superpose pictures for two different spaces: an absolute space, and a space of force vectors. From a logical standpoint, this particular combination of pictures is even "worse" than that of points and their position vectors. In applications, however, convenience always triumphs over formal requirements, so for mathematicians there is no recourse other than attempting to justify such "illegal" actions. Engineers often make use of objects or tools that are imperfect from a mathematical viewpoint. In the more extreme cases, entirely new branches of mathematics have been created in response to this. A good example was the δ-function, used so intensively in physics that it gave rise to the theory of *distributions* or *generalized functions.*

For a system of separate mass points, it is forbidden to shift a force from one point to another. With a rigid body, because of the constraints connecting the points, it is possible but not completely straightforward: clearly an *arbitrary* shift in the point of application of a force on a rigid body can introduce a rotational tendency that was not present originally. This leads us to consider another characteristic of force: the *moment* it produces about a point. In elementary physics we learn that moment equals "force times lever arm." This definition suffices for planar structures where clockwise or counterclockwise rotation are the only two possibilities. In general, however, the moment of a force $\mathbf{F}$ about a point O is the vector quantity

$$\mathbf{M} = \mathbf{r} \times \mathbf{F}. \qquad (1.6.1)$$

Here $\mathbf{r}$ locates the point of application of $\mathbf{F}$ with respect to O. The reference point O is arbitrary but typically placed at the coordinate origin.

When forces are applied to a particle, the resultant force is the simple vector sum of all forces acting. We cannot distinguish whether the particle moves under the action of some number of forces, or under the action of their resultant. What is the simplest force complex to which we can reduce the action of some set of forces acting on a rigid body, in such a way that the resulting motion (i.e., the acceleration of all points of the body) is indistinguishable from that produced by the original force set? The answer, it turns out, consists of a resultant force and a resultant "couple."

Let us mention, first, that long experimentation brought physicists to the idea that a force acting on a rigid body can be shifted along its own *line of action*. That is, the point of force application can be moved along this line without affecting the resulting motion.[1] A vector that can be "attached" at any point of its line of action without affecting other characteristics of a problem is called a *sliding vector*. A good deal more is required if we wish to move the point of application off the original line of action. Suppose a force $\mathbf{F}$ acts at a point A on a body and we want to shift this force in a parallel fashion to some other point B (see Fig. 1.1). We begin by introducing a pair of forces $\pm\mathbf{F}$ at B. This is certainly permissible since the effects of these additional forces completely cancel. But now we can regard the pair of forces consisting of $\mathbf{F}$ acting at A and $-\mathbf{F}$ acting at B as a *couple* $\mathcal{C}$. So when we transfer the point of application of the original force $\mathbf{F}$ from A to B, we must, in effect, compensate through the introduction of $\mathcal{C}$. Continuing to refer to Fig. 1.1, we can see that force couples possess some very important properties. First, the resultant force associated with

---
[1] This is not true for a deformable body.

$\mathcal{C}$ is zero. Second, the moment of $\mathcal{C}$ is given by

$$\mathbf{M} = -\overline{AB} \times \mathbf{F}.$$

We can obtain this by considering the sum of the moments of each of the forces introduced above. Indeed, the moment of $\mathbf{F}$ acting at A is $\overline{OA} \times \mathbf{F}$, and the moment of $-\mathbf{F}$ acting at B is $\overline{OB} \times (-\mathbf{F})$; hence the sum of these moments is $\overline{OA} \times \mathbf{F} + \overline{OB} \times (-\mathbf{F}) = -\overline{AB} \times \mathbf{F}$.

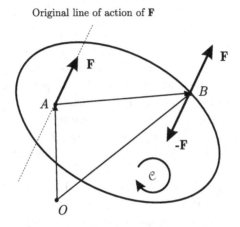

Fig. 1.1 Shifting a force off its line of action on a rigid body. When $\mathbf{F}$ is transferred from point A to point B, the couple $\mathcal{C}$ appears.

The motion of the rigid body under the force $\mathbf{F}$ applied at A is exactly the same as that under $\mathbf{F}$ applied at B in the presence of the couple $\mathcal{C}$. It turns out, however, that $\mathcal{C}$ is completely characterized by its moment $\mathbf{M}$. This means we can replace $\mathcal{C}$ by any other force couple (i.e., by any other pair of forces that have equal magnitudes, opposite directions, and non-coincident lines of action) that happens to produce the same moment $\mathbf{M}$, and the motion of the body will remain the same. Furthermore, we can attach the moment $\mathbf{M}$ to any desired point of the body. Whereas force is an example of a sliding vector, the moment of a couple is a *free vector*. A free vector is one that we can attach to any point of a body without changing the other characteristics of the problem.

Several moments $\mathbf{M}_i$ can be added according to the usual rules of vector addition. This means that any set of forces acting on a rigid body can be replaced by a single resultant force $\mathbf{F}_R$ and a single couple having moment $\mathbf{M}_R$. Indeed, we can transfer all the given forces to any fixed point D.

Each time a force is moved off its line of action, a couple appears. The force vectors can be summed at D to obtain $\mathbf{F}_R$, and the moments of the couples can be summed to obtain $\mathbf{M}_R$. Again, the original set of forces and the pair ($\mathbf{F}_R$ at D, $\mathbf{M}_R$) will be equivalent in the sense that each will have precisely the same effect on the rigid body. We should add that, whereas the magnitude and direction of $\mathbf{F}_R$ do not depend on the point to which it is attached, the corresponding characteristics of $\mathbf{M}_R$ do depend on the line of action of $\mathbf{F}_R$.

Consideration of the set (space) of forces acting on a rigid body has led us to a "vector addition" different from that used in pure mathematics. In particular, the notion of couple enters the picture as we transfer all force vectors to a common point in order to facilitate their combination. So a force $\mathbf{F}$ applied to a rigid body at some point A possesses characteristics beyond those of an ordinary vector. It has magnitude, a line of action along which it can slide without affecting the resulting motion, and a moment $\mathbf{M} = \overline{OA} \times \mathbf{F}$ about any arbitrary reference point O. From a mathematical viewpoint then, the force vector is a distinctly different object from the vector of formal linear algebra. It does obey a set of well-defined rules, however, and it is somewhat surprising that mathematicians have not seized the opportunity to use these rules as the basis for a new abstract formal system. Possibly this happened because these rules appear to hold only for forces, hence they were left for mechanicists to consider and employ.

Exercise 1.6.1. *Two forces of different magnitude and opposite direction are applied to a rigid body. Can their action be equivalent to only the resultant force (without a couple)? If so, when?*

We have said that mechanics could be based on an explicit set of axioms. Apparently, there have been no serious attempts to select the minimal set of axioms. However, for the equivalence of sets of forces acting on a rigid body, an attempt at axiomatization was made by the Polish mathematician Stephan Banach. Banach initially received an engineering diploma and only afterwards became a mathematician. He lectured on mechanics and even wrote a book (*Mechanics*) on the subject. Although much of this book was written from a traditional mechanical viewpoint, it also contains interesting glimpses of a purely mathematical approach. One passage, concerning the equivalence of sets of forces acting on a rigid body, is worth quoting here:

> "In order to deduce the conditions for the equilibrium of a rigid body, we shall assume the following hypotheses:

I. To a system of forces acting on a rigid body which is in equilibrium we can add (or remove from the system) without disturbing equilibrium:
   (a) two forces equal in magnitude and acting along the same line, but oppositely directed;
   (b) several forces having a common point of application and whose sum is zero.
II. Zero forces balance one another; in other words: if no forces act on a rigid body, then the body can remain in equilibrium.

These hypothesis can be verified experimentally. We shall deduce from them the necessary and sufficient conditions for the equilibrium of forces."

1.7 Equilibrium and Motion of a Rigid Body

It is impossible to tell whether a rigid body moves under the action of some set of forces (and couples), or only under the action of the pair discussed above (resultant force and resultant couple). If no resultant force acts on a body, it remains in a state of "uniform motion". But the meaning of this phrase is not as simple as it is in the case of a particle. If the resultant for a mass point is zero, it moves along a straight line at constant speed with respect to an inertial frame (or is in equilibrium when this speed is zero: we can treat both cases in a unified manner and speak only about equilibrium). This is Newton's first law. But the situation is different for a rigid body. A vanishing resultant is not enough for a rigid body to be in equilibrium with respect to an inertial frame, since the body can rotate.

Suppose we can neglect the size of a rigid body and consider it as a particle. In this case we neglect all couples (their moments become zero), and apply all forces to the particle. Thus we can replace the forces with a single resultant. The mass of the particle should be taken as the whole mass of the rigid body. This is

$$M = \int_V \rho(\mathbf{r})\, dV, \qquad (1.7.1)$$

where $\rho = \rho(\mathbf{r})$ is the mass density of the body as a function of position, and V is the volume occupied by the body. In this case we obtain the motion of the body in an "integral sense" where rotation is neglected. In actuality the body may rotate, of course, but there is one point whose motion coincides precisely with that of the "equivalent" mass point under the action of the resultant. This is the *center of mass*, given by

$$\mathbf{r}_M = \frac{1}{M} \int_V \rho(\mathbf{r}) \, \mathbf{r} \, dV. \tag{1.7.2}$$

Newton's first law is traditionally formulated for a mass point. To formulate it for a rigid body, we should say that the center of mass moves along a straight line with constant velocity if the resultant force acting on the body is zero. (Again, however, the body may rotate about its center of mass.) In standard textbooks on classical mechanics we find that the linear momentum of the body now remains constant during the motion.

The center of mass is a particularly convenient point to attach the resultant force. The resultant couple is then called the *principal couple*. In classical mechanics it is shown that if the force resultant is zero, then the magnitude and direction of the resultant couple do not depend on the point to which all the forces were referred to obtain the resultant.

Our present goal is to formulate the conditions for equilibrium of a rigid body. Suppose the body is stationary with respect to a stationary frame of the absolute space and that no forces act on it. Since there are no forces, the body remains in equilibrium. But we have said that it is impossible to distinguish whether a rigid body is under the action of some set of forces or under the action of their resultant force and couple. Thus if these latter quantities are together zero, it is equivalent to the case of the absence of any forces; hence a body in equilibrium will remain in equilibrium if the resultant force and couple vanish. The condition that the resultant force and couple both vanish is equivalent to the statement that the body remains in equilibrium. Forces acting on a rigid body that satisfy this condition are said to be *forces in equilibrium*.

We mentioned that if the resultant is zero, the resultant couple does not depend on the point of reduction of the forces. Transferring all the forces so that their lines of action pass through the frame origin, we find that the moment of the corresponding couples equals the moment of the forces with respect to the origin. So the condition for equilibrium of a rigid body can be written as

$$\sum_i \mathbf{F}_i = \mathbf{0}, \qquad \sum_i (\mathbf{r}_i \times \mathbf{F}_i) = \mathbf{0}, \tag{1.7.3}$$

where $\mathbf{r}_i$ locates the point of application of $\mathbf{F}_i$ for each i. In component form this gives six equations. Equations (1.7.3) are used for three-dimensional objects. For two-dimensional problems we get three scalar equations: two for the components of the resultant force, and one for the couple.

Equilibrium problems for rigid bodies normally involve geometrical constraints as well as applied forces. Usually the effects of constraints are translated into reaction forces and moments, which are then found using the equilibrium equations (1.7.3). Thus, for a three-dimensional problem involving a rigid body, we can find up to six unknown components of reaction forces or moments, and for a two-dimensional problem only up to three. If a structure consists of several joined parts, we should also introduce reactions in the joints and consider each body as separate under the actions of all forces and reactions. If the number of equations is equal to the number of unknown components and the system of equations is linearly independent, it can be solved and the reactions determined. This is a typical problem of classical mechanics; such problems are common in both civil and mechanical engineering. However, problems where the number of equations is less than the number of reaction components are even more frequent in practice. For such a problem we cannot find the reaction by solving the system of equilibrium equations; instead we must bring in the laws of deformation of the structure and use a model of a deformable body. Among the simplest of such models (though still not simple) is that of a linearly elastic body.

1.8 D'Alembert's Principle

It is clear that the derivatives of the position vectors of the same mass point in two inertial frames differ by the velocity of their relative motion. However, the acceleration of the mass point is the same in both frames and appears in the mathematical formulation of Newton's second law for the motion of a particle having mass m:

$$\mathbf{F} = m\frac{d^2\mathbf{r}}{dt^2}. \tag{1.8.1}$$

The position vector $\mathbf{r}$ is often called the *radius vector*; it can be drawn in the absolute space as a directed segment starting at the origin of the immovable frame and ending on the mass point. See Fig. 1.2.

The derivative of a vector function $\mathbf{f} = \mathbf{f}(t)$ is defined by analogy with that for a scalar function:

$$\frac{d\mathbf{f}(t)}{dt} = \lim_{\Delta t \to 0} \frac{\mathbf{f}(t + \Delta t) - \mathbf{f}(t)}{\Delta t}. \tag{1.8.2}$$

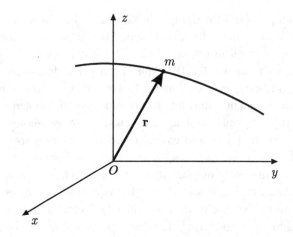

Fig. 1.2 Trajectory of a moving particle.

If we express the function in component form as

$$\mathbf{f} = \sum_{k=1}^{3} f^k \mathbf{e}_k \qquad (1.8.3)$$

and the basis vectors $\mathbf{e}_k$ do not depend on t, then

$$\frac{d\mathbf{f}(t)}{dt} = \sum_{k=1}^{3} \frac{df^k(t)}{dt} \mathbf{e}_k. \qquad (1.8.4)$$

We shall now invoke Einstein's summation rule: when we see repeating sub- and superscripts in a term involving components of vectors, such as $a^i b_i$, we should perform a sum over i, with i taking values from 1 to the dimension of the space in which the vectors are considered. We therefore write

$$\frac{d\mathbf{f}(t)}{dt} = \frac{df^k(t)}{dt} \mathbf{e}_k. \qquad (1.8.5)$$

Exercise 1.8.1. *Write out $d^2\mathbf{f}(t)/dt^2$ in component form when the frame basis is (a) time independent, and (b) time dependent.*

Newton's second law for a particle can be rewritten in the form

$$\mathbf{F} - m\frac{d^2\mathbf{r}}{dt^2} = \mathbf{0}. \qquad (1.8.6)$$

This simple transformation, after introduction of the notation

$$\mathbf{F}_I = -m\frac{d^2\mathbf{r}}{dt^2}, \qquad (1.8.7)$$

brings us to

$$\mathbf{F} + \mathbf{F}_I = \mathbf{0}, \tag{1.8.8}$$

which looks precisely like the equilibrium equation for the particle. This transformation was proposed by d'Alembert. The expression $\mathbf{F}_I$ is called the *inertia force*. Equation (1.8.8) can be regarded as a statement of

d'Alembert's principle. *During the motion, the set of all forces (including the inertia force) forms a system of forces in equilibrium.*

Thus, the problem of finding the acceleration of a particle reduces to the problem of equilibrium of the system of all forces acting on the particle.

Of course, there is little point in using this principle for a free mass point. It does offer advantages, however, when constraints are present (e.g., when points form a rigid body). Let us consider this possibility further.

1.9 The Motion of a System of Particles

Consider the motion of a more complex object: a finite system of n particles. Of course, for each particle we could think of simply writing down Newton's second law. However, we wish to focus on the interaction between particles and the properties that result from this.

The theory for a finite system should inherit some features from the theory for a single mass point. In particular, when we consider the system as a mass unit without extent, as is done for the distant stars, the equations should reduce to those for a mass point. Indeed, we shall see that the center of mass of the system moves exactly as a material point having mass equal to the total mass of the system and acted upon by the resultant force. There is a hierarchy in the theories of classical mechanics.

Let the ith particle have mass m_i and position vector $\mathbf{r}_i$ with respect to the origin of an inertial frame. Consider the forces acting on this particle, including those produced by the actions of the other particles in the system. If the distances between pairs of particles are all preserved during the motion (because of massless constraints), we call the system a rigid body; we shall not, however, limit ourselves to this case. To characterize the system as a whole at a time instant t, let us introduce the total mass

$$M = \sum_{i=1}^{n} m_i \tag{1.9.1}$$

and the position $\mathbf{r}_C$ of the center of mass defined by

$$M\mathbf{r}_C = \sum_{i=1}^{n} m_i \mathbf{r}_i. \tag{1.9.2}$$

The position of the ith particle relative to the center of mass is given by the vector

$$\boldsymbol{\rho}_i = \mathbf{r}_i - \mathbf{r}_C. \tag{1.9.3}$$

Then, since (1.9.2) can be written as

$$\sum_{i=1}^{n} m_i \mathbf{r}_i - \sum_{i=1}^{n} m_i \mathbf{r}_C = \mathbf{0}, \tag{1.9.4}$$

we have

$$\sum_{i=1}^{n} m_i \boldsymbol{\rho}_i = \mathbf{0}. \tag{1.9.5}$$

Exercise 1.9.1. *Putting $\boldsymbol{\rho}_i = (\xi_i, \eta_i, \zeta_i)$, expand (1.9.5) in a Cartesian frame. Note that in this frame the center of mass is the origin.*

Henceforth we shall use an overdot notation for time derivatives. Equation (1.9.2) holds at any instant. Differentiating it with respect to time t, we get

$$M\dot{\mathbf{r}}_C = \sum_{i=1}^{n} m_i \dot{\mathbf{r}}_i.$$

Consequently,

$$\sum_{i=1}^{n} m_i \dot{\boldsymbol{\rho}}_i = \mathbf{0}. \tag{1.9.6}$$

A second differentiation gives

$$M\ddot{\mathbf{r}}_C = \sum_{i=1}^{n} m_i \ddot{\mathbf{r}}_i$$

and

$$\sum_{i=1}^{n} m_i \ddot{\boldsymbol{\rho}}_i = \mathbf{0}. \tag{1.9.7}$$

These are only kinematical characteristics of the center of mass of the system. They do not depend on the forces acting on the particles.

We divide the forces acting on the system into two classes. The first, the class of *internal forces*, includes forces that arise between the particles of the system because of constraints or other effects. We denote a force acting on the ith particle from the jth particle by $\mathbf{F}_{ij}$ (noting, of course, that $\mathbf{F}_{ii} = \mathbf{0}$ for any i). Secondly, we have the *external forces*. We account for these by simply assuming that a resultant force $\mathbf{F}_i$ acts on the ith particle.

By Newton's third law, the internal forces should be balanced in the sense that

$$\mathbf{F}_{ij} = -\mathbf{F}_{ji}. \tag{1.9.8}$$

As is common in classical mechanics, we *assume* both of these forces share the same line of action connecting the particles. This is a restrictive assumption, but it makes sense in physics where in statics we meet only central forces between particles like gravitation or electrical attraction. Although the equality (1.9.8) makes sense, it is only an assumption; in mathematics we would call it an axiom with far-reaching consequences. Its validity is not so evident, say, for dynamical processes.

By this assumption, the sum of all internal forces is zero:

$$\sum_{i,j=1}^{n} \mathbf{F}_{ij} = \mathbf{0}. \tag{1.9.9}$$

This is often called d'Alembert's principle.[2] Although we obtained it from (1.9.8), it may be taken as an independent principle; in fact, it represents a weaker assumption than that in which the forces are assumed to be central and obey Newton's third law. We are about to see that (1.9.9) forms the background for the derivation of some principal conservation laws in mechanics.

Newton's second law for the ith particle is

$$m_i \ddot{\mathbf{r}}_i = \mathbf{F}_i + \sum_{j=1}^{n} \mathbf{F}_{ij}. \tag{1.9.10}$$

We have noted that the addition of forces acting on different particles is senseless, because the force acting on one particle cannot directly affect another one. However, this operation begins to make sense when we consider a system of particles, as it gives us some characteristics of the system as a whole. We have introduced the mass of the system and the position vector

---

[2] It is unrelated to the similarly named principle in § 1.8. d'Alembert's name appears throughout mechanics.

of the center of mass. Let us introduce, similarly, the following characteristics:

(1) the total linear momentum

$$\sum_{i=1}^{n} m_i \mathbf{v}_i \equiv \sum_{i=1}^{n} m_i \dot{\mathbf{r}}_i = M\dot{\mathbf{r}}_C; \qquad (1.9.11)$$

(2) the total angular momentum

$$\sum_{i=1}^{n} \mathbf{r}_i \times m_i \mathbf{v}_i \equiv \sum_{i=1}^{n} \mathbf{r}_i \times m_i \dot{\mathbf{r}}_i; \qquad (1.9.12)$$

(3) the resultant of external forces

$$\mathbf{F}_R = \sum_{i=1}^{n} \mathbf{F}_i; \qquad (1.9.13)$$

(4) the total moment of external forces (or total torque)

$$\sum_{i=1}^{n} \mathbf{r}_i \times \mathbf{F}_i. \qquad (1.9.14)$$

First we sum all the equations (1.9.10):

$$\sum_{i=1}^{n} m_i \ddot{\mathbf{r}}_i = \sum_{i=1}^{n} \left(\mathbf{F}_i + \sum_{j=1}^{n} \mathbf{F}_{ij} \right).$$

By (1.9.9) we have

$$\sum_{i=1}^{n} m_i \ddot{\mathbf{r}}_i = \sum_{i=1}^{n} \mathbf{F}_i.$$

Hence

$$\frac{d}{dt} \sum_{i=1}^{n} m_i \mathbf{v}_i = \mathbf{F}_R \qquad (1.9.15)$$

or

$$\frac{d}{dt}(M\mathbf{v}_C) = \mathbf{F}_R, \quad \mathbf{v}_C = \dot{\mathbf{r}}_C. \qquad (1.9.16)$$

So the motion of the center of mass of the system of particles depends only on the resultant external force acting on the system, and coincides with the motion of a particle of mass M under the same resultant force. From this we obtain a crucial result.

Conservation of linear momentum. *If $\mathbf{F}_R = \mathbf{0}$, then the linear momentum remains constant and the center of mass moves along a straight line with constant velocity.*

This follows from the fact that
$$\frac{d}{dt}(M\mathbf{v}_C) = \mathbf{0},$$
hence $M\mathbf{v}_C$ is a constant. Similar reasoning brings us to the conservation of total angular momentum. We begin with the elementary transformation
$$\frac{d}{dt}(\mathbf{r} \times m\mathbf{v}) = \mathbf{v} \times m\mathbf{v} + \mathbf{r} \times m\dot{\mathbf{v}} = \mathbf{r} \times m\ddot{\mathbf{r}}, \qquad \mathbf{v} = \dot{\mathbf{r}}. \qquad (1.9.17)$$
Next, cross the vector $\mathbf{r}_i$ into both sides of (1.9.10) and sum over i:
$$\sum_{i=1}^{n}(\mathbf{r}_i \times m_i\ddot{\mathbf{r}}_i) = \sum_{i=1}^{n}(\mathbf{r}_i \times \mathbf{F}_i) + \sum_{i=1}^{n}\mathbf{r}_i \times \left(\sum_{j=1}^{n}\mathbf{F}_{ij}\right).$$
By (1.9.17), the sum on the left is the time derivative of the total angular momentum:
$$\frac{d}{dt}\sum_{i=1}^{n}(\mathbf{r}_i \times m_i\mathbf{v}_i).$$
The first sum on the right is the total moment of the external forces. Consider the second sum on the right. Together with the term $\mathbf{r}_i \times \mathbf{F}_{ij}$ it contains a dual term $\mathbf{r}_j \times \mathbf{F}_{ji}$. By (1.9.8) their sum is
$$\mathbf{r}_i \times \mathbf{F}_{ij} + \mathbf{r}_j \times \mathbf{F}_{ji} = (\mathbf{r}_i - \mathbf{r}_j) \times \mathbf{F}_{ij} = \mathbf{0},$$
since $\mathbf{F}_{ij}$ is assumed to be parallel to the vector $\mathbf{r}_i - \mathbf{r}_j$ connecting the ith and jth particles. So the second sum on the right is zero and we have
$$\frac{d}{dt}\sum_{i=1}^{n}(\mathbf{r}_i \times m_i\mathbf{v}_i) = \sum_{i=1}^{n}(\mathbf{r}_i \times \mathbf{F}_i), \qquad (1.9.18)$$
which involves only the external forces. In particular we have the following.

Conservation of total angular momentum. *If the total moment of the external forces acting on a system of n particles is zero,*
$$\sum_{i=1}^{n}(\mathbf{r}_i \times \mathbf{F}_i) = \mathbf{0},$$
then the total angular momentum
$$\sum_{i=1}^{n}(\mathbf{r}_i \times m_i\mathbf{v}_i)$$

Exercise 1.9.2. *Supply the proof.*

Equation (1.9.18) describes the total angular momentum with respect to the origin of the inertial frame. Let us find the total angular momentum with respect to the center of mass of the system. We substitute

$$\mathbf{r}_i = \mathbf{r}_C + \boldsymbol{\rho}_i, \quad \mathbf{v}_i = \mathbf{v}_C + \dot{\boldsymbol{\rho}}_i, \quad \mathbf{v}_C = \dot{\mathbf{r}}_C,$$

into (1.9.18):

$$\frac{d}{dt}\sum_{i=1}^n [(\mathbf{r}_C + \boldsymbol{\rho}_i) \times m_i(\mathbf{v}_C + \dot{\boldsymbol{\rho}}_i)] = \sum_{i=1}^n [(\mathbf{r}_C + \boldsymbol{\rho}_i) \times \mathbf{F}_i].$$

This, after simple transformation, brings

$$\frac{d}{dt}\left\{\mathbf{r}_C \times \sum_{i=1}^n m_i \mathbf{v}_C + \mathbf{r}_C \times \sum_{i=1}^n m_i \dot{\boldsymbol{\rho}}_i + \sum_{i=1}^n m_i \boldsymbol{\rho}_i \times \mathbf{v}_C + \sum_{i=1}^n (\boldsymbol{\rho}_i \times m_i \dot{\boldsymbol{\rho}}_i)\right\}$$

$$= \mathbf{r}_C \times \sum_{i=1}^n \mathbf{F}_i + \sum_{i=1}^n (\boldsymbol{\rho}_i \times \mathbf{F}_i).$$

Using (1.9.5) and (1.9.6) we get

$$\left[\frac{d}{dt}(\mathbf{r}_C \times M\mathbf{v}_C) - \mathbf{r}_C \times \mathbf{F}_R\right] + \frac{d}{dt}\sum_{i=1}^n (\boldsymbol{\rho}_i \times m_i \dot{\boldsymbol{\rho}}_i) = \sum_{i=1}^n (\boldsymbol{\rho}_i \times \mathbf{F}_i).$$

The bracketed difference on the left is zero, which follows from (1.9.16) if we form the cross product with $\mathbf{r}_C$ and use (1.9.17). Thus

$$\frac{d}{dt}\sum_{i=1}^n (\boldsymbol{\rho}_i \times m_i \dot{\boldsymbol{\rho}}_i) = \sum_{i=1}^n (\boldsymbol{\rho}_i \times \mathbf{F}_i). \qquad (1.9.19)$$

Although (1.9.18) and (1.9.19) are similar in form, the latter shows no dependence on the motion of the center of mass. Hence the rotation of the system about the center of mass is independent of the motion of the center of mass. This holds for a system of particles and for a rigid body in particular.

Finally, we mention that by introducing inertial forces for each particle we can rewrite the equations of motion for the system in a form that coincides formally with the equations of equilibrium for the system. This set of equations constitutes, as in the case of one particle, d'Alembert's principle.

1.10 The Rigid Body

Let us reconsider the notion of a rigid body. In many books on theoretical mechanics, a rigid body is defined as a set of particles connected in such a way that their mutual distances are fixed. While this may seem acceptable, it fails to specify how the particles can react with one another (i.e., with something akin to (1.9.8)). The "definition" of a rigid body in mechanics should include not only the idea of constant shape, but the mechanism of force transmission between parts of the body.

The same textbooks often derive results for a rigid body consisting of a finite number of particles, then pass directly to the theory of a "continuous" body. It is supposed that this transition is done through an elementary limit passage. They first think of approximating the body as a finite collection of "particles"; each particle is actually an elemental volume over which the "mass density" is essentially constant. Assuming all mutual distances between particles are fixed during any motion, they calculate the total linear and angular momenta of the system. These are expressed in terms of finite summations, which are taken to become Riemann volume integrals during the subsequent limit passage. The listener is expected to accept the entire procedure without question, especially if he or she has a good calculus background. It turns out, however, that there is reason for concern: when we pass from a system containing finitely many objects to one containing infinitely many objects, we can encounter unexpected changes in qualitative behavior. In particular, we must question whether the properties of the internal forces should carry over. The interior state of a deformable solid cannot be described using only "central forces" (i.e., forces such as the $\mathbf{F}_{ij}$ that we used to describe a finite system of particles) or, indeed, using forces alone. We must employ a stress tensor: an entity that inherits some properties of force, but with other properties of its own.

The behavior of forces inside a rigid body cannot be derived straightforwardly from the corresponding picture for a finite system of particles. Rather, it must be formulated as a kind of axiom that reflects our primary interest in determining the integral characteristics of the motion. In particular, we can take the "simplest" case where the relations between parts of the body are thought to mimic those that would apply to the internal forces in a finite system. But again, this amounts essentially to formulating an axiom. We find that to describe the motion of a rigid body (which we consider as a system of n particles at fixed mutual separations while preserving our assumption on the internal forces $\mathbf{F}_{ij}$), it is necessary to invoke only (1.9.16) and (1.9.18), or equivalently

$$\frac{d}{dt}(M\mathbf{v}_C) = \mathbf{F}_R, \qquad \frac{d}{dt}\sum_{i=1}^{n}\boldsymbol{\rho}_i \times m_i\dot{\boldsymbol{\rho}}_i = \sum_{i=1}^{n}\boldsymbol{\rho}_i \times \mathbf{F}_i. \qquad (1.10.1)$$

Denoting

$$\mathbf{G}_C = \sum_{i=1}^{n}\boldsymbol{\rho}_i \times m_i\dot{\boldsymbol{\rho}}_i, \qquad \mathbf{M}_C = \sum_{i=1}^{n}\boldsymbol{\rho}_i \times \mathbf{F}_i,$$

where $\mathbf{G}_C$ is the angular (kinetic) momentum of the body with respect to the center of mass C and $\mathbf{M}_C$ the resultant angular moment of the external forces with respect to C, we can rewrite (1.10.1) as

$$\frac{d}{dt}(M\mathbf{v}_C) = \mathbf{F}_R, \qquad \frac{d}{dt}\mathbf{G}_C = \mathbf{M}_C. \qquad (1.10.2)$$

Using the kinematical analysis for a rigid body, which can be found in any textbook on classical mechanics, we can express $\mathbf{G}_C$ in terms of the angular velocity vector $\boldsymbol{\omega}$ for the body, which has three components in three-dimensional space. So in scalar form, we get six simultaneous differential equations in the six unknown components of $\mathbf{v}_C$ and $\boldsymbol{\omega}$. These components are uniquely defined by the equations and corresponding initial values, so we have a description of the motion of a rigid body. While the form of these equations seems simple, the study of rigid body motion occupies much space in a typical textbook. Our goal is to discuss only some main ideas.

We should add that here the use of a Cartesian frame for the description of relative motion of the body is not the best choice. And this leads us to the idea that to describe the motion of bodies, we must introduce other parameters such as the familiar Euler angles α, β, γ. In fact this is the first step toward Lagrangian mechanics. The latter is a consequence of Newtonian mechanics that makes use of generalized coordinates to describe the motion of objects. It simplifies the solution of many problems.

We may regard (1.10.1) as consequences of the dynamic equations for a system of particles. But normally (1.10.2) are applied to continuous rigid bodies that occupy volumes or that are idealized as surface or linear mass distributions. In such cases we may consider the terms of (1.10.1) as Riemann-sum-type approximations to corresponding volume, surface, or line integrals and obtain (1.10.2) using a limit passage. However, these equations must be regarded as new axioms for rigid body mechanics. Indeed, internal force terms such as $\mathbf{F}_{ij}$ are absent. This results from the assumptions on the nature of the internal forces. Inside a rigid body, however, it is strange to make any such assumptions; rather, it is preferable to begin with (1.10.2) as we will eventually use them to obtain relations for deformable bodies where such simple assumptions cannot be applied at all. So we will regard (1.10.2) as essentially axiomatic in nature.

Unique specification of the position of a rigid body requires six independent parameters. In contrast, the position of a particle in space can be specified by three coordinates or other parameters that define these coordinates uniquely. The position of a system consisting of some number of particles and rigid bodies can be uniquely defined through some minimal number of parameters. In this case we speak of the *configuration* of the system; the space in which these parameters take their values is known as the *configuration space*. The *minimal* number of *independent* parameters describing the system uniquely is called the *degree of freedom* of the system.

The degree of freedom is less than or equal to the total number of parameters that describes each item of the system. For example, if a particle can move only on some surface, then the description of its position requires just two parameters: coordinates of the point on the surface. If it can move only along some curve, a single parameter is required. The same holds for rigid bodies: if a point of a rigid body is fixed, we need just three parameters (e.g., the angles that define its position uniquely). Therefore the degree of freedom of a body with a fixed point is three.

The degree of freedom of a system of bodies in mechanics plays the same role for its configuration space as the dimension plays for a vector space: it shows the number of quantities we must know in order to determine some object in the space uniquely. The configuration space is not a vector space.

Exercise 1.10.1. *What is the degree of freedom of a moving "rigid" segment?*

1.11 Motion of a System of Particles; Comparison of Trajectories; Notion of Operator

A curve describing the motion of a particle in space is called a *trajectory*. If we mark, in configuration space, the points taken by a system of particles during its motion, we have what could be called the system trajectory. Since the ordinary differential equations of motion for particles are of second order, to define the motion uniquely we must provide two initial conditions for each component of the position vector for each particle. We have

$$m_i \ddot{\mathbf{r}}_i = \mathbf{F}_i + \sum_{j=1}^{n} \mathbf{F}_{ij} \qquad (i = 1, \ldots, n), \tag{1.11.1}$$

where the forces on the right can depend on time t and the position vectors of the particles (we continue to denote by $\mathbf{F}_{ij}$ the force exerted on the ith particle by the jth particle and take $\mathbf{F}_{ii} = \mathbf{0}$). We typically specify, for each particle, the position and velocity vectors at some initial time:

$$\mathbf{r}_i(0) = \mathbf{a}_i, \quad \dot{\mathbf{r}}_i(0) = \mathbf{b}_i \qquad (i = 1, \ldots, n), \tag{1.11.2}$$

where $\mathbf{a}_i, \mathbf{b}_i$ are given constants describing the initial position and velocity of the ith particle, respectively. The resulting problem is called a *Cauchy problem* or *initial value problem* in ordinary differential equation theory. The theory of Cauchy problems offers theorems covering existence and uniqueness of solution and providing for the continuous dependence of solutions on small changes in the initial conditions, masses, and external forces. There are problems to which these theorems do not apply, but they usually suffice for applications (note that the forces can depend on the $\mathbf{r}_i$ and their derivatives, so the equations can be complex).

In other kinds of problems, conditions are posed at various time instants. Such problems are called *boundary value problems*, because the conditions are commonly posed at two points often designated as "initial" and "final" points (although more than two points may be involved). Boundary value problems can have nonunique solutions, and their theory is not as clean as that of Cauchy problems. Nonetheless, a great deal of effort has been directed towards them and much is now known about both their theoretical and practical treatment.

Suppose we are dealing with a particle system whose motion problem has a unique solution under given initial or boundary conditions and under a set of forces. If these parameters change, so does the system trajectory or velocity. How should we measure this change on the time interval $[0, T]$? If we are interested only in the difference between the particle positions, we can measure the deviation between two trajectories $\mathbf{r} = \mathbf{r}_1(t)$ and $\mathbf{r} = \mathbf{r}_2(t)$ using a *uniform norm*. For a single particle we can use

$$\|\mathbf{r}_2 - \mathbf{r}_1\| = \max_{t \in [0,T]} \|\mathbf{r}_2(t) - \mathbf{r}_1(t)\|_{\mathbb{R}^3}, \tag{1.11.3}$$

where $\|\cdot\|_{\mathbb{R}^3}$ is a norm on $\mathbb{R}^3$. We require continuity of the vector functions $\mathbf{r}_1(t)$ and $\mathbf{r}_2(t)$ on the finite segment $[0, T]$, which is guaranteed by general theorems covering many motion problems for systems of particles having unique solutions. The rectilinear motion of a particle with given initial and final times a and b brings us to the uniform norm on the *space of continuous functions* on the segment $[a,b]$. The resulting normed space is denoted by $C(a,b)$. The norm of a function $f = f(t)$ in this space is given by

$$\|f\|_{C(a,b)} = \max_{t \in [a,b]} |f(t)|. \tag{1.11.4}$$

Later, when we consider a deformable body occupying a volume V, it will be convenient to introduce the uniform norm for functions $f = f(\mathbf{x})$ continuous on V. We will take V to be *compact* (i.e., closed and bounded) and use

$$\|f\|_{C(V)} = \max_{\mathbf{x} \in V} |f(\mathbf{x})| \tag{1.11.5}$$

for the norm.

Exercise 1.11.1. *Verify that* (1.11.5) *satisfies the norm axioms.*

If we must evaluate differences in velocities as well as positions, then for continuously differentiable vector functions $\mathbf{r} = \mathbf{r}_1(t)$ and $\mathbf{r} = \mathbf{r}_2(t)$ we can introduce another norm involving first derivatives. For the one-dimensional case, a norm on the space of functions continuously differentiable on $[a,b]$, denoted by $C^{(1)}(a,b)$, can be introduced in various ways. These include

$$\|f\|_{C^{(1)}(a,b)} = \max_{t \in [a,b]} |f(t)| + \max_{t \in [a,b]} |f'(t)| \tag{1.11.6}$$

and

$$\|f\|_{C^{(1)}(a,b)} = \max \left\{ \max_{t \in [a,b]} |f(t)|, \max_{t \in [a,b]} |f'(t)| \right\}, \tag{1.11.7}$$

which are equivalent in the sense of convergence of sequences of differentiable functions. (We will formalize this notion of equivalent norms in Definition 1.12.1.) They can be extended to the space of functions having continuous derivatives up to order m on a compact set $V \subset \mathbb{R}^n$. We have

$$\|f\|_{C^{(m)}(V)} = \max_{\mathbf{x} \in V} |f(\mathbf{x})| + \sum_{|\alpha| \leq m} \max_{\mathbf{x} \in V} |D^\alpha f(\mathbf{x})|, \tag{1.11.8}$$

where the *multi-index notation* D^α is understood as follows:

$$D^\alpha f = \frac{\partial^{|\alpha|} f}{\partial x_1^{\alpha_1} \cdots \partial x_n^{\alpha_n}}, \quad \alpha = (\alpha_1, \ldots, \alpha_n), \quad |\alpha| = \alpha_1 + \cdots + \alpha_n.$$

Such norms are used when we must characterize a solution beyond simply a deviation under a change in parameters of the problem. The importance of smoothness follows from the fact that the smoother a solution (i.e., the larger the number m in $C^{(m)}(V)$), the better the convergence of approximation schemes used to seek the solution numerically. Note that the norms are written out for dimensionless variables; otherwise, we would append coefficients to account for differences in units between the terms.

Exercise 1.11.2. *Verify that (1.11.8) satisfies the norm axioms.*

Let us return to the initial value problem (1.11.1)–(1.11.2) and suppose that the $\mathbf{F}_i$ depend only on t. Assume these vector functions can be defined independently. Then to a set of forces $\{\mathbf{F}_i\}$ given on $[0, T]$ there corresponds a unique set of trajectories $\{\mathbf{r}_i\}$ on $[0, T]$. This reminds us of the definition of a function, where to each point of some set called the domain of the function there corresponds a unique value of the function. In this case, however, a set of vector functions $\{\mathbf{F}_i\}$ stands in correspondence with another set of vector functions $\{\mathbf{r}_i\}$. We cannot say that to a value of $\{\mathbf{F}_i\}$ at an instant t there corresponds a value $\{\mathbf{r}_i\}$ at the same instant t, since $\{\mathbf{r}_i\}$ at t depends on all the values taken by $\{\mathbf{F}_i\}$ on $[0, t]$. This dependence cannot be described in terms of an ordinary function. We shall call the correspondence an *operator* (or sometimes *mapping*, or *map*) if it is uniquely defined.

To really define a function, we must specify not only a rule of correspondence between two sets, but the sets as well. The same holds for an operator. An operator A has a domain $D(A)$ and a range $R(A)$. In this book $D(A)$ will be a subset of a normed space X, while $R(A)$ will lie in a normed space Y (which may coincide with X). The correspondence itself, taking each point $x \in D(A)$ into a uniquely defined point $y \in R(A)$, will be denoted

$$y = A(x). \qquad (1.11.9)$$

We say that A *acts from* X *to* Y. If $Y = X$, we say that A *acts in* X.

A mapping A acting from a normed space X to the scalars $\mathbb{R}$ or $\mathbb{C}$ is called a *functional*. The branch of mathematics known as *functional analysis* had its origins in the study of such mappings.

In linear algebra we treat operators represented by matrices. An $n \times n$ matrix M applied to a vector $\mathbf{x} \in \mathbb{R}^n$ yields another vector $\mathbf{y} \in \mathbb{R}^n$. The reader must be aware that a matrix is not an operator but only a component representative of the operator corresponding to some basis in $\mathbb{R}^n$. The matrix elements play the same role as the components of a vector in a fixed basis: when we change the basis (and we can do this independently in the domain $\mathbb{R}^n$ and in the range $\mathbb{R}^n$), the matrix elements also change. They must do so in such a way that for any **x** given in some basis, application of this representative matrix yields **y** given in another basis — but this **y** must not depend on the choice of bases. Clearly, transformations of representative matrices must obey certain rules, but not necessarily those for transformations of vectors. The operators corresponding to such matrices are also known as tensors of the second rank. In the theory of tensors, vectors constitute tensors of the first rank and scalars constitute tensors of zero rank. Although we can consider the correspondence M as a function acting from $\mathbb{R}^n$ to $\mathbb{R}^n$, we prefer to call it a matrix operator. We know that a matrix operator is a linear transformation whose degree of continuity — that is, how a change in **x** affects the change in **y** — is measured by its norm $\|M\|$ (defined below). All such notions as continuity, linearity, and norm can be extended to the general case. They are mostly simple restatements of concepts from calculus or linear algebra.

So let us consider an operator A from a normed space X to a normed space Y. First we extend the ordinary notion of continuity at a point. Recall that a real-valued function f of a real variable x is said to be continuous at a point x_0 of its domain if to every positive number ε there corresponds a positive number δ (which may depend on ε) such that $|f(x) - f(x_0)| < \varepsilon$ whenever $|x - x_0| < \delta$.

Definition 1.11.1. An operator A acting from X to Y is *continuous* at $x_0 \in X$ if to every $\varepsilon > 0$ there corresponds $\delta = \delta(\varepsilon)$ such that
$$\|A(x) - A(x_0)\| < \varepsilon \text{ whenever } \|x - x_0\| < \delta.$$

The definition for an ordinary function was extended by using the norm in place of the absolute value operation. To emphasize that the spaces X and Y may have different norms, we could have written
$$\|A(x) - A(x_0)\|_Y < \varepsilon \text{ whenever } \|x - x_0\|_X < \delta$$
instead of the above. However, we shall follow the usual practice and attach subscripts to norm symbols only when the spaces involved may not be clear from the context.

Definition 1.11.2. Let $\{x_n\}$ be a sequence in X. We say that $\{x_n\}$ *converges* to x_0 and write
$$x_0 = \lim_{n \to \infty} x_n$$
if for any $\varepsilon > 0$ there exists $N = N(\varepsilon)$ such that
$$\|x_n - x_0\| < \varepsilon \quad \text{whenever } n > N.$$

We may also write $x_n \to x_0$ as $n \to \infty$.

Note that we have introduced the limit passage in a normed space using only the operations of classical analysis; indeed, only operations with numbers — the values of the norm — are involved.

Having the notion of sequence convergence, we may introduce

Definition 1.11.3. An operator A acting from X to Y is *sequentially continuous* at $x_0 \in X$ if $A(x_n) \to A(x_0)$ as $n \to \infty$ for any sequence $\{x_n\}$ such that $x_n \to x_0$ as $n \to \infty$.

Lemma 1.11.1. *The two types of continuity are equivalent.*

Proof. First let us show that continuity implies sequential continuity. Let A be continuous at x_0 and take any convergent sequence $\{x_n\}$ with $x_n \to x_0$. Let $\varepsilon > 0$ be given. By continuity there exists δ such that $\|A(x_n) - A(x_0)\| < \varepsilon$ whenever $\|x_n - x_0\| < \delta$. But, by convergence of $\{x_n\}$, there exists N such that this last inequality holds whenever $n > N$. Therefore, to any $\varepsilon > 0$ there corresponds N such that $\|A(x_n) - A(x_0)\| < \varepsilon$ whenever $n > N$. So A is sequentially continuous at x_0.

Conversely, let us show that sequential continuity implies continuity. Suppose A is not continuous at x_0. Then there exists $\varepsilon = \varepsilon_0$ with the property that for every positive integer n there is a point x_n inside the ball $\|x - x_0\| < 1/n$ such that $\|A(x_n) - A(x_0)\| \geq \varepsilon_0$. The sequence $\{x_n\}$ thus constructed is convergent to x_0, but it is false that $A(x_n) \to A(x_0)$. Therefore A is not sequentially continuous at x_0. $\square$

In view of the lemma, we will refer to sequential continuity as simply "continuity" and use whichever formulation is convenient.

Definition 1.11.4. We say that A is *linear* if

$$A(\alpha x + \beta y) = \alpha A(x) + \beta A(y) \qquad (1.11.10)$$

for any two scalars α and β and any two elements $x, y \in X$.

For such an operator we often write Ax instead of $A(x)$. One useful observation is

Lemma 1.11.2. *If a linear operator A is continuous at $x = 0$, it is continuous on the entire space X.*

This follows immediately from the relation

$$A(x) - A(x_0) = A(x - x_0),$$

since we can think of $x - x_0$ as a vector y that becomes arbitrarily small when we make x arbitrarily close to x_0.

Why is the notion of continuity seldom stressed for matrix operators in linear algebra? In fact the issue is trivial: any $n \times n$ matrix A having only finite elements, which represents an operator A in some basis, corresponds to a continuous operator.

Now we introduce

Definition 1.11.5. An operator A from a normed space X to a normed space Y is *bounded* if there is a positive constant c such that

$$\|Ax\| \leq c \|x\| \text{ for all } x \in X. \tag{1.11.11}$$

If (1.11.11) holds, then A is continuous at $x = 0$ by Definition 1.11.1 with $\delta = \varepsilon/c$ and hence is continuous on X.

For linear operators, we have

Theorem 1.11.1. *A continuous linear operator A from X to Y is bounded.*

Proof. A is continuous at $x = 0$. Take $\varepsilon = 1$; by definition there exists $\delta > 0$ such that $\|Ax\| \leq 1$ whenever $\|x\| < \delta$. For every nonzero $x \in X$, the norm of $x^* = \delta x/(2\|x\|)$ is

$$\|x^*\| = \|\delta x/(2\|x\|)\| = \delta/2 < \delta,$$

so $\|Ax^*\| \leq 1$. By linearity of A, this gives us

$$\|Ax\| \leq \frac{2}{\delta} \|x\|,$$

which is (1.11.11) with $c = 2/\delta$. □

Definition 1.11.6. The infimum of the set of constants c for which (1.11.11) holds is called the *norm* of A.

Alternatively the number $\|A\|$ is a norm if, for any $x \in X$, we have

$$\|Ax\| \leq \|A\| \|x\|, \tag{1.11.12}$$

and, for any $\varepsilon > 0$, we can find x_ε such that

$$\|Ax_\varepsilon\| > (\|A\| - \varepsilon) \|x_\varepsilon\|. \tag{1.11.13}$$

Clearly, when we use a term like operator "norm" we should prove that it really has all norm properties and is suitably defined on some linear space.

Here the space is the set of all continuous linear operators, with operations of addition and scalar multiplication patterned after those for matrices:

$$(A+B)x = Ax + Bx, \qquad (\alpha A)x = \alpha(Ax).$$

The reader should prove that $\|A\|$ satisfies N1–N3. The set of all continuous linear operators from a normed space X to a normed space Y is denoted $L(X,Y)$. If $Y = X$, the notation is $L(X)$.

Exercise 1.11.3. *Prove that the norm of an operator A can be defined as*

$$\|A\| = \sup_{\|x\|=1} \|Ax\| \quad or \quad \|A\| = \sup_{\|x\|\leq 1} \|Ax\|. \tag{1.11.14}$$

Although other norms can be placed on $L(X,Y)$, we will typically use the above norm which is related to the norms on X and Y.

1.12 Matrix Operators and Matrix Equations

Many continuum mechanics problems cannot be solved analytically. Even when analytic solution is possible, engineers often prefer approximate numerical simulations that yield instructive pictures of system behavior. For example, the motion of a particle system can be studied by applying approximate methods to the Cauchy problem for the corresponding ordinary differential equations; the problem is thereby reduced to a discrete one, to be integrated in finite time steps using finite difference approximations of the derivatives. For a Cauchy problem this is done successively beginning with the initial point, whereas for a boundary value problem we must satisfy the conditions at the final point of the system trajectory. The latter leads to a system of equations which, in the case of a general particle system with forces of attraction, etc., are transcendental as a rule. Treatment is seldom straightforward.

In continuum mechanics many engineering problems, like those of the theory of elasticity, are described by linear equations. Available solution methods include the finite element method, the boundary element method, the finite difference method, etc. All lead to systems of simultaneous linear equations that can be written as

$$a_{ij}x_j = f_i \tag{1.12.1}$$

or in matrix form as

$$A\mathbf{x} = \mathbf{f}. \tag{1.12.2}$$

In this way, a matrix A approximates the original operator of the boundary value problem. Formally, (1.12.2) looks like an equation for vector quantities $\mathbf{f}$ and $\mathbf{x}$ in a space $\mathbb{R}^n$ of higher dimension. But $\mathbf{x}$ and $\mathbf{f}$ are not real vectors, since their various components refer to different points of the body and we cannot work with them (for example, transforming between coordinate systems) as we do with vectors if we wish to properly preserve their physical meanings. Still, as soon as we write down the matrix equation it begins to live a life of its own and we can treat it using customary methods. For a time, we can virtually forget how the matrix equation originated — until, of course, we reach the point where we must interpret the final results in light of the original model.

Let us calculate a few matrix norms. Suppose, in analogy with (1.5.4), we define a norm on $\mathbb{R}^n$ using

$$\|\mathbf{x}\| = \max_{1 \leq i \leq n} |x_i| \qquad (1.12.3)$$

for $\mathbf{x} = (x_1, \ldots, x_n)$. Note the norm is written for a fixed basis of $\mathbb{R}^n$ that may be non-orthogonal. Writing $A = (a_{ij})$, which represents the operator A so that the ith component of the image $\mathbf{y} = A\mathbf{x}$ in the same basis is $\sum_{j=1}^{n} a_{ij} x_j$, we have

$$\begin{aligned}\|A\mathbf{x}\| &= \max_{1 \leq i \leq n} \left| \sum_{j=1}^{n} a_{ij} x_j \right| \\ &\leq \max_{1 \leq i \leq n} \sum_{j=1}^{n} |a_{ij} x_j| \\ &\leq \left(\max_{1 \leq i \leq n} \sum_{j=1}^{n} |a_{ij}| \right) \max_{1 \leq j \leq n} |x_j|,\end{aligned}$$

hence

$$\|A\mathbf{x}\| \leq \mu \|\mathbf{x}\| \qquad (1.12.4)$$

where

$$\mu = \max_{1 \leq i \leq n} \sum_{j=1}^{n} |a_{ij}|.$$

In linear algebra, μ is called the "max row sum" of the matrix A. Note that $\mu \geq \|A\|$ by definition of $\|A\|$. But we can go further and show that $\mu = \|A\|$. To see that equality holds in (1.12.4) for some nonzero $\mathbf{x}$, let k

be the value of i for which the max row sum occurs:

$$\mu = \sum_{j=1}^{n} |a_{kj}|.$$

Then choose the components of $\mathbf{x}$ according to the rule

$$x_i = \begin{cases} +1, & a_{ki} \geq 0, \\ -1, & a_{ki} < 0. \end{cases}$$

For this $\mathbf{x}$, we have $\|\mathbf{x}\| = 1$ and $\|A\mathbf{x}\| = \mu$, so equality holds.

Exercise 1.12.1. *Show that the quantity $\|\mathbf{x}\|$ specified in (1.12.3) satisfies norm axioms N1–N3. Repeat for*

$$\|\mathbf{x}\|_p = \left(\sum_{i=1}^{n} |x_i|^p \right)^{1/p} \tag{1.12.5}$$

where $1 \leq p < \infty$. The latter norm, of course, induces a metric given by $d_p(\mathbf{x}, \mathbf{y}) = \|\mathbf{x} - \mathbf{y}\|_p$. A special case of this metric was encountered in (1.3.2).

For the norm (1.12.5), we get

$$\|A\mathbf{x}\|_p = \left(\sum_{i=1}^{n} \left| \sum_{j=1}^{n} a_{ij} x_j \right|^p \right)^{1/p} \leq \left[\sum_{i=1}^{n} \left(\sum_{j=1}^{n} |a_{ij} x_j| \right)^p \right]^{1/p}.$$

Next we apply Hölder's inequality for sums. If $p > 1$, $q > 1$, and

$$\frac{1}{p} + \frac{1}{q} = 1,$$

then for any two sets of real numbers $a_1, \ldots, a_m$ and $b_1, \ldots, b_m$, we have

$$\sum_{i=1}^{m} |a_i b_i| \leq \left(\sum_{i=1}^{m} |a_i|^p \right)^{1/p} \left(\sum_{i=1}^{m} |b_i|^q \right)^{1/q}. \tag{1.12.6}$$

Hence

$$\left(\sum_{j=1}^{n} |a_{ij} x_j| \right)^p \leq \left(\sum_{j=1}^{n} |a_{ij}|^q \right)^{p/q} \left(\sum_{j=1}^{n} |x_j|^p \right)$$

where $q = p/(p-1)$, so

$$\|A\mathbf{x}\|_p \leq \left[\sum_{i=1}^{n} \left(\sum_{j=1}^{n} |a_{ij}|^q \right)^{p/q} \right]^{1/p} \|\mathbf{x}\|_p.$$

By this and the well-known conditions for equality in Hölder's inequality, we conclude that

$$\|A\| = \left[\sum_{i=1}^{n}\left(\sum_{j=1}^{n}|a_{ij}|^q\right)^{p/q}\right]^{1/p}.$$

If $p = 2$, then $q = 2$ and the formulas above reduce to

$$\|\mathbf{x}\|_2 = \left(\sum_{i=1}^{n}|x_i|^2\right)^{1/2}, \quad \|A\| = \left(\sum_{i=1}^{n}\sum_{j=1}^{n}|a_{ij}|^2\right)^{1/2}.$$

The norm $\|\mathbf{x}\|_2$ is induced by an inner product: viz.,

$$(\mathbf{x}, \mathbf{y}) = \sum_{i=1}^{n} x_i y_i.$$

In contrast, each of the norms $\|\mathbf{x}\|_p$ for $p \neq 2$ *cannot* be induced by an inner product.

It is worth emphasizing that the operator norm depends on the underlying norm imposed on $\mathbb{R}^n$.

Exercise 1.12.2. *Consider a matrix operator A acting between different normed spaces. Both spaces consist of n-tuples as above, but have respective norms given by*

$$\|\mathbf{x}\|_p = \left(\sum_{i=1}^{n}|x_i|^p\right)^{1/p}, \quad \|\mathbf{y}\|_r = \left(\sum_{i=1}^{n}|y_i|^r\right)^{1/r}.$$

Assuming $\mathbf{y} = A\mathbf{x}$*, estimate* $\|A\|$*.*

We mentioned equivalent norms for the spaces of differentiable functions and $\mathbb{R}^n$. Let us introduce a strict definition.

Definition 1.12.1. Two norms $\|\cdot\|_1$ and $\|\cdot\|_2$, imposed on the same set of vectors X, are *equivalent* if there exist positive constants c and C such that the inequality

$$c\|x\|_1 \leq \|x\|_2 \leq C\|x\|_1 \qquad (1.12.7)$$

holds for all $x \in X$.

Clearly, when a sequence $\{x_n\}$ converges to x_0 in one norm then it also converges to x_0 in any equivalent norm.

Exercise 1.12.3. *Show that any two norms in $\mathbb{R}^n$ are equivalent.*

Using some basis of a finite dimensional space, we can introduce a one-to-one correspondence between it and $\mathbb{R}^n$ or $\mathbb{C}^n$ that preserves algebraic operations and the norms. From this (and the equivalence of all norms on $\mathbb{R}^n$ or $\mathbb{C}^n$) it follows that on any finite-dimensional space all norms are equivalent. This is false for infinite-dimensional spaces. Moreover, *the equivalence of all norms on a linear space means the space is finite-dimensional.*

As we have said, numerical approaches to the linear problems of continuum mechanics effectively reduce these problems to matrix equations. Nonlinear problems also reduce, as a rule, to the solution of linear matrix equations that arise as intermediate problems. So an ability to solve matrix equations is essential. The relevant methods fall under the heading of numerical linear algebra and lie outside the scope of the present book. The reader can see any textbook on numerical analysis for a full discussion.

1.13 Complete Spaces

We expect a numerical approach to yield an approximation to a true solution. This approximation could be good or bad, however, and the best one can hope for is a reliable estimate of the error. Often we are merely assured that a method can, in principle, yield a sequence of approximations convergent to the true solution. Even so, it is evident that the numerical implementation of such a method may not converge to the true solution: roundoff error alone can prevent this. It can destroy a solution to a simultaneous system of equations when the dimension reaches a certain size.

Practitioners commonly judge the convergence of an approximation sequence by comparing successive terms of the sequence. When the difference seems "small enough" for the purpose at hand, computation is halted. So the analyst simply watches the successive differences between approximations and waits until some stopping criterion has been satisfied. For problems involving matrix equations, these differences are typically gauged using one of the norms on $\mathbb{R}^n$; either absolute errors or relative errors (obtained by dividing the difference by the norm of one of the solutions) can be used. If the calculations were perfect, without roundoff or truncation error, the analyst could continue the process indefinitely. The best he or she could hope for, however, is to observe the pattern typical of what we call a "Cauchy sequence" in calculus. This leads us to reframe the concept in the more general metric space setting.

Definition 1.13.1. Let $\{x_n\}$ be a sequence of points in a metric space (S, d). We say that $\{x_n\}$ is a *Cauchy sequence* if to each $\varepsilon > 0$, there corresponds a number $N = N(\varepsilon)$ such that $d(x_n, x_m) < \varepsilon$ whenever $m > N$ and $n > N$.

This definition practically coincides with the usual one given in calculus. We know that in $\mathbb{R}$ or $\mathbb{R}^n$ the concepts of "Cauchy sequence" and "convergent sequence" are essentially equivalent. Is this true in a general metric space? Let us explicitly generalize the idea of convergence.

Definition 1.13.2. Let $\{x_n\}$ be a sequence of points in a metric space (S, d). We say that $\{x_n\}$ is *convergent* if there is a point $x \in S$ having the property that, to each $\varepsilon > 0$, there corresponds a number $N = N(\varepsilon)$ such that $d(x_n, x) < \varepsilon$ whenever $n > N$. In this case the point x is called the *limit* of $\{x_n\}$, and we write

$$\lim_{n \to \infty} x_n = x$$

or $x_n \to x$ as $n \to \infty$.

Observe that $x_n \to x$ if and only if $d(x_n, x) \to 0$, by definition of the ordinary limit in $\mathbb{R}$. The reader should also be aware that we sometimes write

$$\lim_{m,n \to \infty} d(x_n, x_m) = 0, \quad \text{or} \quad d(x_n, x_m) \to 0 \text{ as } m, n \to \infty,$$

if $\{x_n\}$ is a Cauchy sequence.

Again, in the Euclidean space $\mathbb{R}^n$ every Cauchy sequence is convergent and vice versa. It is clear that even in a general metric space, every convergent sequence is a Cauchy sequence; we formulate this as

Exercise 1.13.1. *Suppose $x_n \to x$ as $n \to \infty$ in a general metric space (S, d). Show that $\{x_n\}$ is a Cauchy sequence in (S, d).*

What about the converse: are Cauchy sequences always convergent? We consider a couple of examples. Recall that $C(a, b)$ stands for the space of continuous functions defined on the closed interval $[a, b]$ with the "max norm"

$$\|f\| = \max_{t \in [a,b]} |f(t)|. \tag{1.13.1}$$

Here a sequence of functions $f_n = f_n(t)$ is a Cauchy sequence if

$$\max_{t \in [a,b]} |f_n(t) - f_m(t)| \to 0 \quad \text{as } m, n \to \infty.$$

For any $t \in [a,b]$, the sequence $\{f_n(t)\}$ is a numerical Cauchy sequence. Hence it has a unique limit that we denote as $f(t)$. This $f(t)$ is a function on $[a,b]$. Clearly $\{f_n(t)\}$ converges to $f(t)$ in the sense that

$$\sup_{t\in[a,b]} |f_n(t) - f(t)| \to 0 \quad \text{as } n \to \infty.$$

But we do not know whether $f(t)$ is continuous on $[a,b]$. We now refer to a theorem from classical analysis:

Theorem 1.13.1 (Weierstrass). *Suppose a sequence of continuous functions $\{f_n(t)\}$, defined on a closed and bounded interval $[a,b]$, is uniformly convergent to a limit function $f(t)$. Then $f(t)$ is also continuous on $[a,b]$.*

Because uniform convergence is precisely equivalent to convergence in the max metric of $C(a,b)$, every Cauchy sequence taken from $C(a,b)$ is convergent to an element of the space.

But now consider the linear space of functions continuous on $[-1,1]$ with the "L^1-norm"

$$\|f\|_1 = \int_{-1}^{1} |f(t)|\, dt. \tag{1.13.2}$$

Exercise 1.13.2. *Show that (1.13.2) satisfies N1–N3. The reason for the "L_1" designation and the subscript "1" on the norm symbol will become clear in § 1.15.*

In this new normed space we consider a sequence $\{f_n(t)\}$ given by

$$f_n(t) = \begin{cases} 0, & -1 \le t < 0, \\ nt, & 0 \le t \le 1/n, \\ 1, & 1/n < t \le 1, \end{cases} \quad (n=1,2,3,\ldots).$$

If $m > n$, then

$$d(f_m, f_n) = \int_0^{1/m} |mt - nt|\, dt + \int_{1/m}^{1/n} |1 - nt|\, dt$$

$$= \frac{1}{2}\left(\frac{1}{n} - \frac{1}{m}\right) \to 0 \quad \text{as } m,n \to \infty,$$

so $\{f_n(t)\}$ is a Cauchy sequence. But $f_n(t) \to U(t)$, where $U(t)$ is the Heaviside unit step function defined by

$$U(t) = \begin{cases} 1, & t \ge 0, \\ 0, & t < 0. \end{cases}$$

Indeed,
$$d(f_n, U) = \int_0^{1/n} |nt - 1|\, dt = \frac{1}{2n} \to 0 \quad \text{as } n \to \infty.$$
However, the limit function $U(t)$ is *not* continuous on $(-1, 1)$. We have established

Lemma 1.13.1. *Let S be the space of functions continuous on $[-1, 1]$ with the L^1-norm. There is a Cauchy sequence in S whose limit lies outside S.*

To establish a theorem we must provide a full proof. On the other hand, a proposition can be invalidated through just one counterexample. We have shown that, in general, not all Cauchy sequences converge. What shall we do then? It makes sense to select the class of spaces where the desired equivalence does hold.

Definition 1.13.3. A metric space S is *complete* if every Cauchy sequence taken from S converges to a limit in S. If S is not complete, it is *incomplete*. A normed space that is complete in its natural metric is a *Banach space*. An inner product space complete in its natural norm (i.e., complete in the metric induced by that norm) is a *Hilbert space*.

We pause to note that the idea of metric space completeness was known well before Stefan Banach (1892–1945). However, Banach was the first to perceive the true usefulness of complete normed spaces. Banach was educated as an engineer — he even published a book on classical mechanics. This background helped him understand the importance of the spaces that now bear his name. Unlike Banach, Hilbert was a pure mathematician. For many years he was considered the best in the world. His ideas set the stage for much of 20th century mathematics.

Whenever we encounter a new space, we should verify whether it is complete. Our mere introduction of the "completeness" notion does not mean such verification will be easy. We cannot, for example, immediately generalize our conclusion regarding $C(a, b)$ to the space $C(\Omega)$ where Ω is a compact subset of $\mathbb{R}^n$ — at least not without being aware that Weierstrass's theorem generalizes appropriately to the multivariable case. The same holds for a generalization to the space of continuous functions on Ω with metric

$$\|f - g\|_1 = \int_\Omega |f(\mathbf{x}) - g(\mathbf{x})|\, dx_1 \cdots dx_n. \tag{1.13.3}$$

In this case, however, incompleteness can indeed be shown.

We should make another point. We examined two metric spaces above, both constructed using the same base set S (i.e., the set of functions continuous on $[a,b]$). The metrics were different, however, and this allowed us to find a sequence $\{f_n(t)\}$ that happens to be a Cauchy sequence in one space but not in the other. In this sense there is a lack of equivalence between the norms on these two spaces.

It is evident that equivalent norms provide the same convergence properties. Recalling that all norms on $\mathbb{R}^n$ are equivalent, we could ask what makes our present examples non-equivalent. The answer is that, in contrast to $\mathbb{R}^n$, these spaces are infinite-dimensional.

Exercise 1.13.3. *Prove that the norms (1.13.1) and (1.13.2) on the set of continuous functions are not equivalent. (Hint: construct a sequence of elements that all have unit norm under (1.13.1), but whose norms under (1.13.2) tend to zero.)*

Of course, one might suggest that we simply avoid the L^1 norm. But norms of integral type are important. Is there a better way to circumvent the difficulties associated with the incompleteness of such spaces? It turns out that there is a powerful theorem which will permit us to "extend" an incomplete space to a resulting complete space, the latter containing (at least essentially — we shall clarify this below) the elements of the original incomplete space. This construction is not unfamiliar, since we tacitly make use of it when dealing with the real number system. We take up the full details in the next section.

1.14 Completion Theorem

Although irrational numbers such as π and $\sqrt{2}$ are truly numbers, we do not specify their values merely by giving them symbols. We can, however, approximate them to any desired accuracy. Indeed we can find a Cauchy sequence whose limit is irrational, but the best we can do to state the actual limit is to assign it a name (such as "π" or "$\sqrt{2}$"). So, from this viewpoint, an irrational number is *defined* through an approximation sequence. But the choice of sequence is obviously non-unique; many different sequences can define the same irrational number. One way around this difficulty is to introduce *equivalent Cauchy sequences*. In this approach, any two Cauchy sequences approaching the same limit are said to be equivalent; we can collect all these sequences into equivalence classes and identify each irrational number with one of the classes. Each rational number can be identified with an equivalence class as well. This idea lies at the base of the metric space completion theorem.

Before stating the theorem we introduce some terminology. Some of these concepts have been mentioned above, but we pause to formalize them.

Definition 1.14.1. Two sequences $\{x_n\}, \{y_n\}$ in a metric space (S, d) are said to be *equivalent* if $d(x_n, y_n) \to 0$ as $n \to \infty$. Given any Cauchy sequence $\{x_n\}$ in S, we can gather into an equivalence class X all Cauchy sequences in S that are equivalent to $\{x_n\}$. We then refer to any Cauchy sequence from X as a *representative* of X. Note that to any $x \in S$ there corresponds a *stationary* equivalence class containing the "stationary" Cauchy sequence $x, x, x, \ldots$.

We think in terms of metric spaces primarily in those situations (e.g., the study of convergence) where the distance between elements is crucial. If a one-to-one *distance preserving* correspondence exists between metric spaces, we can work with the elements of either space. This is true even if the spaces have elements of distinctly different natures — we can work with the elements of one space and understand that, since distances are of main concern, all statements we make relative to that space also hold for the corresponding elements of the other space. With this in mind we state

Definition 1.14.2. A mapping from one metric space to another is an *isometry* if it preserves distances; that is, f is an isometry from a metric space (S_1, d_1) to a metric space (S_2, d_2) if

$$d_2(f(x), f(y)) = d_1(x, y) \qquad (1.14.1)$$

for every pair of points x, y taken from S_1. If an isometry is also a one-to-one correspondence, it is a *one-to-one isometry* and the two metric spaces involved are *isometric*.

We know that a given real number can be approximated to any desired accuracy by a rational number. In short, "the rationals are dense in the reals." The notion of denseness can be extended to more general sets.

Definition 1.14.3. Let X and Y be two subsets of a metric space (S, d). We say that *X is dense in Y* if for any point $y \in Y$ and any $\varepsilon > 0$, we can find a point $x \in X$ such that $d(x, y) < \varepsilon$.

Now we can talk about metric space completion. The main result is

Theorem 1.14.1. *Let S be a metric space. There is a one-to-one isometry between S and a set $\tilde{S}$ which is dense in a complete metric space S^*.*

This is the *completion theorem*. It guarantees that any metric space can be completed. Of course, if S is already complete then there is nothing to prove, so the result is interesting only when S is incomplete. The proof is rather long and we subdivide it into digestible portions.

We begin by introducing the elements of S^*, using the idea of approximating irrational numbers with classes of equivalent Cauchy sequences of rational numbers. Given any particular Cauchy sequence $\{x_n\}$ in the original space S, we form the equivalence class X mentioned in Definition 1.14.1. We now view X as a single element of the space S^* and construct the remaining elements similarly. So S^* consists of equivalence classes of Cauchy sequences taken from the original space S.

Definition 1.14.4. We call S^* the *completion* of S.

Of course, we will have to define a suitable metric on S^* and then *show* that S^* is complete in this metric.

Now we define $\tilde{S}$. Corresponding to any $x \in S$, there is a stationary Cauchy sequence $x, x, x, \ldots$. This sequence would, by our procedure above, generate an equivalence class to be included in the completion space S^*. The subset of S^* consisting of only the stationary equivalence classes is denoted by $\tilde{S}$. There is clearly a one-to-one correspondence between S and $\tilde{S}$, and we will ultimately show that $\tilde{S}$ is dense in S^*. We begin by introducing a suitable metric on S^* (and $\tilde{S}$).

Lemma 1.14.1. *Let $X, Y \in S^*$. The function $D(X, Y)$ given by*

$$D(X, Y) = \lim_{n \to \infty} d(x_n, y_n), \qquad (1.14.2)$$

where $\{x_n\}$ and $\{y_n\}$ are arbitrary representatives of X and Y, respectively, is a metric on S^*.

Proof. We first show that the proposed metric is well-defined; i.e., that the indicated limit exists and is independent of the choice of representative sequences. The triangle inequality for the metric d (on S) allows us to write

$$d(x_n, y_n) \leq d(x_n, x_m) + d(x_m, y_m) + d(y_m, y_n)$$

so that

$$d(x_n, y_n) - d(x_m, y_m) \leq d(x_n, x_m) + d(y_m, y_n).$$

Interchanging m and n and using the symmetry of the metric, we obtain
$$-[d(x_n, y_n) - d(x_m, y_m)] \leq d(x_n, x_m) + d(y_m, y_n).$$
Therefore
$$|d(x_n, y_n) - d(x_m, y_m)| \leq d(x_n, x_m) + d(y_n, y_m). \tag{1.14.3}$$
Because $\{x_n\}$ and $\{y_n\}$ are Cauchy sequences, we know that $d(x_n, x_m) \to 0$ and $d(y_n, y_m) \to 0$ as $m, n \to \infty$. It follows from (1.14.3) that $\{d(x_n, y_n)\}$ is a Cauchy sequence of real numbers. So the limit in (1.14.2) exists by completeness of $\mathbb{R}$. To see that it does not depend on the choice of representatives, we take any two other representatives $\{x'_n\}$ and $\{y'_n\}$ from X and Y, respectively, and use the inequality
$$|d(x_n, y_n) - d(x'_n, y'_n)| \leq d(x_n, x'_n) + d(y_n, y'_n)$$
to get
$$|d(x_n, y_n) - d(x'_n, y'_n)| \to 0 \quad \text{as } n \to \infty.$$
This shows that
$$\lim_{n \to \infty} d(x'_n, y'_n) = \lim_{n \to \infty} d(x_n, y_n).$$
So $D(X, Y)$ is well-defined. Does it really satisfy the axioms of a metric? First, the inequality $D(X, Y) \geq 0$ follows from passage to the limit as $n \to \infty$ in the corresponding inequality $d(x_n, y_n) \geq 0$ that is satisfied by d for each n. If $X = Y$ then we certainly have $D(X, Y) = 0$ (since we can choose the same representative sequence from both X and Y). Conversely, the statement $D(X, Y) = 0$ implies that any two representatives $\{x_n\}$ and $\{y_n\}$ give $\lim_{n \to \infty} d(x_n, y_n) = 0$. By Definition 1.14.1 these representatives are equivalent and we conclude that $X = Y$. So metric axiom M1 is satisfied. Satisfaction of M2 follows from the definition of D and the symmetry of d:
$$D(X, Y) = \lim_{n \to \infty} d(x_n, y_n) = \lim_{n \to \infty} d(y_n, x_n) = D(Y, X).$$
Finally, the triangle inequality
$$D(X, Y) \leq D(X, Z) + D(Z, Y)$$
follows from passage to the limit as $n \to \infty$ in the inequality
$$d(x_n, y_n) \leq d(x_n, z_n) + d(z_n, y_n).$$
So D is a suitable metric on S^*. Since $\tilde{S}$ is a subset of S^*, we can also employ D as the metric on $\tilde{S}$. $\square$

Lemma 1.14.2. *The space* (S^*, D) *is complete.*

Proof. We will show that an arbitrary Cauchy sequence $\{X^i\}$ in S^* is convergent. From each X^i we take a representative $\{x_j^{(i)}\}$ and, from this, an element x_{k_i} such that $k_i > k_{i-1}$ and $d(x_{k_i}, x_j^{(i)}) < 1/i$ for all $j > k_i$. To see that $\{x_{k_i}\}$ is a Cauchy sequence in S, we denote by X_{k_i} the equivalence class containing the stationary sequence $(x_{k_i}, x_{k_i}, \ldots)$ and write

$$\begin{aligned} d(x_{k_i}, x_{k_j}) &= D(X_{k_i}, X_{k_j}) \\ &\leq D(X_{k_i}, X^{k_i}) + D(X^{k_i}, X^{k_j}) + D(X^{k_j}, X_{k_j}) \\ &\leq \frac{1}{i} + D(X^{k_i}, X^{k_j}) + \frac{1}{j} \\ &\to 0 \quad \text{as } k_i, k_j \to \infty. \end{aligned}$$

Denoting by X the class determined by $\{x_{k_i}\}$, we have

$$\begin{aligned} D(X^{k_i}, X) &\leq D(X^{k_i}, X_{k_i}) + D(X_{k_i}, X) \\ &\leq \frac{1}{i} + D(X_{k_i}, X) \\ &= \frac{1}{i} + \lim_{j \to \infty} d(x_{k_i}, x_{k_j}) \\ &\to 0 \quad \text{as } i \to \infty. \end{aligned}$$

So $X^{k_i} \to X$ in the metric of S^* and as $\{X^i\}$ is a Cauchy sequence then $X^i \to X$ as $i \to \infty$. □

Lemma 1.14.3. *The set* $\tilde{S}$ *is dense in the set* S^*, *relative to the metric* D.

Proof. Let $X \in S^*$ be given. We select a representative $\{x_n\}$ from X, and for each n denote by X_n the stationary equivalence class containing $(x_n, x_n, \ldots)$. Because

$$D(X_n, X) = \lim_{m \to \infty} d(x_n, x_m) \to 0 \quad \text{as } n \to \infty,$$

we can approximate X as closely as desired by the elements X_n taken from $\tilde{S}$. □

Lemma 1.14.4. *The spaces* (S, d) *and* $(\tilde{S}, D)$ *are isometric.*

Proof. The one-to-one correspondence between S and $\tilde{S}$ is defined by pairing with any $x \in S$ the element $X \in \tilde{S}$ that contains $(x, x, x, \ldots)$. Given

any $x, y \in S$, we can take their images X and Y under the correspondence and write
$$D(X, Y) = \lim_{n \to \infty} d(x, y) = d(x, y)$$
to see that it preserves distances. □

Theorem 1.14.1 follows from Lemmas 1.14.1–1.14.4. Since it is formulated for a general metric space, it holds for all particular cases. If a space has additional properties, these typically transfer to the completion space as well.

The most important property of this kind is linearity. Suppose S is a linear metric space with the additional operations of summation $x + y$ and multiplication λx by a (real or complex) number. It is clear that the same operations can be introduced in S^* for the elements X, Y and, moreover, that the above correspondence between S and $\tilde{S}$ preserves these algebraic operations. Hence the completion is also a vector space.

Particularly important are the normed spaces. Each is a vector space and, in addition, has a natural metric $d(x, y) = \|x - y\|$. Here we can also apply Theorem 1.14.1. Let us consider this case. Everything stated for metric spaces continues to hold, of course, but there are additional observations as well. The metric (1.14.2) now takes the form
$$D(X, Y) = \lim_{n \to \infty} d(x_n, y_n) = \lim_{n \to \infty} \|x_n - y_n\|.$$
This raises the question whether $D(X, Y)$ can be considered as the norm of the element $X - Y$: we denote this by $\|X - Y\|_*$. The reader should verify that $\|X\|_*$ satisfies N1–N3. Hence S^*, resulting from application of Theorem 1.14.1, is a Banach space.

Similar reasoning holds for an inner product space: Theorem 1.14.1 yields a Hilbert space S^* whose inner product between X and Y is defined by a limit passage in the inner product over representative Cauchy sequences for these elements. The result is independent of the choice of representatives and satisfies I1–I3.

In what follows, we will complete various normed and inner product spaces. Let us formulate appropriate versions of the theorem.

Theorem 1.14.2. *Let S be a normed space. There is a one-to-one isometry between S and a set $\tilde{S}$ which is dense in a Banach space S^*. Algebraic operations between elements are preserved under this correspondence. The norm in S^* is given by*
$$\|X\|_* = \lim_{n \to \infty} \|x_n\|, \tag{1.14.4}$$

where $\{x_n\}$ is any representative Cauchy sequence taken from the class X.

Theorem 1.14.3. *Let S be an inner product space. There is a one-to-one isometry between S and a set $\tilde{S}$ which is dense in a Hilbert space S^*. Algebraic operations between elements are preserved under this correspondence. The inner product in S^* is given by*

$$(X, Y)_* = \lim_{n \to \infty} (x_n, y_n), \qquad (1.14.5)$$

where $\{x_n\}$ and $\{y_n\}$ are any representative Cauchy sequences taken from the classes X and Y, respectively.

Exercise 1.14.1. *Denote by P the set of all polynomials with real coefficients on the closed segment $[0,1]$. Observe that P is a linear space. Supply it with the norm of $C(0,1)$, that is, $\|p\| = \max_{x \in [0,1]} |p(x)|$. Clearly the resulting normed space is not complete. Describe the space that results from the completion theorem in this case.*

Let us sketch a solution. Clearly, by the completion theorem, the reader will get a complete space consisting of all the classes of equivalent Cauchy sequences of polynomials. However, by the Weierstrass theorem, any continuous function in $C(0,1)$ can be uniformly approximated by a polynomial to within any desired accuracy. This means that P is dense in $C(0,1)$, which is a complete space. The latter means that any Cauchy sequence in $C(0,1)$, including sequences of elements of P, has a continuous function as a limit. It is easy to show that this limit does not depend on the choice of a representative sequence from a class of the completion space. So here any class of the completion space can be identified with a continuous function from $C(0,1)$. We can regard the completion space obtained by the theorem as another form of representation of the space $C(0,1)$.

1.15 Lebesgue Integration and the L^p Spaces

Suppose Ω is a closed and bounded (i.e., compact), Jordan measurable subset of $\mathbb{R}^n$. Let S be the collection of all functions $f(\mathbf{x})$ continuous on Ω and thus absolutely integrable over Ω in the Riemann sense:

$$\int_\Omega |f(\mathbf{x})| \, d\Omega < \infty. \qquad (1.15.1)$$

Since the integral expression above is a valid norm on S, we can consider the normed space $(S, \|\cdot\|_1)$ where

$$\|f\|_1 = \int_\Omega |f(\mathbf{x})|\, d\Omega. \tag{1.15.2}$$

Lemma 1.13.1 states that $(S, \|\cdot\|_1)$ is incomplete when Ω is a one-dimensional interval $[a,b]$. The same holds for $\Omega \subset \mathbb{R}^n$ with any finite n. We can apply Theorem 1.14.2 and extend the operation of integration to the elements of the resulting Banach space. The integral we obtain is called the *Lebesgue integral*, after H. Lebesgue who introduced it from another standpoint. Let us denote the elements of the completion by $F(\mathbf{x})$ and introduce

Definition 1.15.1. $L^1(\Omega)$ is the Banach space formed by completing the normed space $(S, \|\cdot\|_1)$. The norm on $L^1(\Omega)$ is given by

$$\|F\|_1 = \int_\Omega |F(\mathbf{x})|\, d\Omega. \tag{1.15.3}$$

For brevity, we often write $L(\Omega)$ instead of $L^1(\Omega)$.

The sense of integration on the right side of (1.15.3) is given by Theorem 1.14.2; it is therefore

$$\int_\Omega |F(\mathbf{x})|\, d\Omega = \lim_{n\to\infty} \int_\Omega |f_n(\mathbf{x})|\, d\Omega, \tag{1.15.4}$$

where $\{f_n(\mathbf{x})\}$ is a representative sequence (i.e., of continuous functions from S) taken from the class of equivalent Cauchy sequences $F(\mathbf{x})$. We will call the value

$$\int_\Omega |F(\mathbf{x})|\, d\Omega$$

the Lebesgue integral of $|F(\mathbf{x})|$.

We pause to observe that an element of $L(\Omega)$ is not an ordinary function. It is an *equivalence class* of Cauchy sequences of continuous functions. But we do employ function notation as indicated above. This is convenient if one maintains the correct interpretation of the symbolism. For example, the sum $F(\mathbf{x}) + G(\mathbf{x})$ of two elements is understood to be the equivalence class determined by a representative Cauchy sequence $\{f_n(\mathbf{x}) + g_n(\mathbf{x})\}$, where $\{f_n(\mathbf{x})\}$ and $\{g_n(\mathbf{x})\}$ determine $F(\mathbf{x})$ and $G(\mathbf{x})$, respectively.

We have introduced the integral for $|F(\mathbf{x})|$. Let us introduce the value of the integral

$$\int_\Omega F(\mathbf{x})\,d\Omega$$

for elements $F(\mathbf{x}) \in L(\Omega)$ themselves. The result will also be equivalent to the Lebesgue integral of a function. Taking a representative $\{f_n(\mathbf{x})\}$ from $F(\mathbf{x})$, we use the modulus inequality

$$\left|\int_\Omega f(\mathbf{x})\,d\Omega\right| \leq \int_\Omega |f(\mathbf{x})|\,d\Omega \qquad (1.15.5)$$

to show that the numerical sequence $\{\int_\Omega f_n(\mathbf{x})\,d\Omega\}$ is a Cauchy sequence:

$$\left|\int_\Omega f_n(\mathbf{x})\,d\Omega - \int_\Omega f_m(\mathbf{x})\,d\Omega\right| = \left|\int_\Omega [f_n(\mathbf{x}) - f_m(\mathbf{x})]\,d\Omega\right|$$
$$\leq \int_\Omega |f_n(\mathbf{x}) - f_m(\mathbf{x})|\,d\Omega$$
$$= \|f_n - f_m\|_1$$
$$\to 0 \quad \text{as } m, n \to \infty.$$

Definition 1.15.2. The quantity

$$\int_\Omega F(\mathbf{x})\,d\Omega \equiv \lim_{n \to \infty} \int_\Omega f_n(\mathbf{x})\,d\Omega \qquad (1.15.6)$$

is uniquely determined by $F(\mathbf{x})$ and is called the *Lebesgue integral* of $F(\mathbf{x})$ over Ω.

Again, this is equivalent to the Lebesgue integral as presented in real function theory.

Remark 1.15.1. The classical theory of Lebesgue integration begins with the notion of Lebesgue measurability of a set in $\mathbb{R}^n$. This differs from Jordan measurability in the way that the elementary domains, used for defining whether a domain is measurable, contain a countable set of elementary parallelepipeds (inscribed and circumscribed) unlike the Jordan case where only finite sets of these are used. The classical theory introduces the notion of Lebesgue integral, but the "functions" involved are not simple functions (just as they are not in our approach); rather, a function here is a collection of all those that are equal *almost everywhere* (i.e., except on a set of Lebesgue measure zero). Sets of Lebesgue measure zero can be complicated, and hence the corresponding "functions" for which

Lebesgue integration is introduced can differ substantially from ordinary functions. For example, the set of all rational numbers on the segment $[0, 1]$ has Lebesgue measure zero. Thus the functions participating in the classical theory of Lebesgue integration are also some classes of equivalent functions and offer no advantages over those used in our approach. When we say that the integrals are equivalent we mean that the resulting spaces can be placed in one-to-one correspondence in such a way that the integrals for the corresponding elements are equal. The correspondence is also isometric and preserves linear operations over the elements. Moreover, in the case of a continuous function for which the Riemann integral exists, both approaches to the Lebesgue integral yield a value equal to this Riemann integral. □

Note the following.

(1) If we take an element of $L(\Omega)$ that contains a stationary sequence of elements of S, this integral is equal to the ordinary Riemann integral of the function of the stationary sequence. Thus it extends the notion of Riemann integral.
(2) In $L(\Omega)$ all the functions (elements) are absolutely integrable.

$L(\Omega)$ belongs to a class of Banach spaces that are denoted by $L^p(\Omega)$, $p \geq 1$. In particular, $L(\Omega) \equiv L^1(\Omega)$. For a fixed $p \geq 1$, the space $L^p(\Omega)$ is the completion of S with respect to the metric induced by the norm

$$\|f\|_p = \left(\int_\Omega |f(\mathbf{x})|^p \, d\Omega \right)^{1/p}. \tag{1.15.7}$$

The norm of an equivalence class $F(\mathbf{x}) \in L^p(\Omega)$ is given by

$$\|F\|_p = \left(\int_\Omega |F(\mathbf{x})|^p \, d\Omega \right)^{1/p}. \tag{1.15.8}$$

Here integration is now understood in the Lebesgue sense:

$$\int_\Omega |F(\mathbf{x})|^p \, d\Omega = \lim_{n \to \infty} \int_\Omega |f_n(\mathbf{x})|^p \, d\Omega, \tag{1.15.9}$$

where $\{f_n(\mathbf{x})\}$ is any representative of $F(\mathbf{x})$.

Exercise 1.15.1. *Show that this integral is well-defined; i.e., that the limit on the right exists and is unique for any given $F(\mathbf{x}) \in L^p(\Omega)$. Hence the norm (1.15.8) is well-defined.*

We now state some important facts regarding the $L^p(\Omega)$ spaces. First, for compact Ω these spaces are nested in the sense that

$$L^p(\Omega) \subseteq L^r(\Omega) \quad \text{for} \quad 1 \leq r \leq p. \tag{1.15.10}$$

Second, a sufficient condition for existence of the integral

$$\int_\Omega F(\mathbf{x}) G(\mathbf{x})\, d\Omega$$

is that

$$F(\mathbf{x}) \in L^p(\Omega) \text{ and } G(\mathbf{x}) \in L^q(\Omega)$$

for p and q such that

$$\frac{1}{p} + \frac{1}{q} = 1 \text{ and } p > 1.$$

In this case Hölder's inequality for integrals

$$\left| \int_\Omega F(\mathbf{x}) G(\mathbf{x})\, d\Omega \right| \leq \left(\int_\Omega |F(\mathbf{x})|^p\, d\Omega \right)^{1/p} \left(\int_\Omega |G(\mathbf{x})|^q\, d\Omega \right)^{1/q} \tag{1.15.11}$$

holds, with equality if and only if $F(\mathbf{x}) = \lambda G(\mathbf{x})$ for some number λ.

Exercise 1.15.2. *Use* (1.15.11) *to prove* (1.15.10); *i.e., there exists a constant* $C_{p,r}$ *not dependent on* $F(\mathbf{x})$ *such that*

$$\left(\int_\Omega |F(\mathbf{x})|^r\, d\Omega \right)^{1/r} \leq C_{p,r} \left(\int_\Omega |F(\mathbf{x})|^p\, d\Omega \right)^{1/p}$$

when $1 \leq r \leq p$.

The space $L^2(\Omega)$ deserves mention since it is a Hilbert space. (The spaces $L^p(\Omega)$ for $p \neq 2$ are Banach spaces but their norms cannot be induced by suitable inner products.) If we begin with a base set S of complex functions, Theorem 1.14.3 yields a complex Hilbert space with inner product

$$(F, G) = \lim_{n \to \infty} \int_\Omega f_n(\mathbf{x}) \overline{g_n(\mathbf{x})}\, d\Omega = \int_\Omega F(\mathbf{x}) \overline{G(\mathbf{x})}\, d\Omega. \tag{1.15.12}$$

Complex conjugation can be omitted to obtain the correct expression for the inner product on the real version of $L^2(\Omega)$.

Fredholm's operator in $L^p(\Omega)$

Let us apply Hölder's inequality to find the norm of Fredholm's operator. Equations of the general form

$$U(\mathbf{x}) + \int_\Omega K(\mathbf{x},\mathbf{y})U(\mathbf{y})\,d\Omega = G(\mathbf{x}), \qquad (1.15.13)$$

where $K(\mathbf{x},\mathbf{y})$ and $G(\mathbf{x})$ are given and $U(\mathbf{y})$ is the unknown sought, occur commonly in mathematical physics (and continuum mechanics in particular). They are known as *Fredholm integral equations of the second kind*. We can write (1.15.13) in the slightly more abstract form

$$U(\mathbf{x}) + AU(\mathbf{x}) = G(\mathbf{x}) \qquad (1.15.14)$$

where A is the linear integral operator given by

$$AU(\mathbf{x}) = \int_\Omega K(\mathbf{x},\mathbf{y})U(\mathbf{y})\,d\Omega. \qquad (1.15.15)$$

This is Fredholm's integral operator. Depending on the application, A can be considered as acting in various spaces of functions. It may or may not be continuous, depending on the properties of its *kernel* $K(\mathbf{x},\mathbf{y})$. Let us obtain a condition on $K(\mathbf{x},\mathbf{y})$ sufficient to ensure that A is bounded when it acts in the space $L^p(\Omega)$ (i.e., maps elements $U \in L^p(\Omega)$ into images $AU \in L^p(\Omega)$). We have

$$\|AU\|_p = \left(\int_\Omega \left|\int_\Omega K(\mathbf{x},\mathbf{y})U(\mathbf{y})\,d\Omega\right|^p d\Omega\right)^{1/p}.$$

But by Hölder's inequality we can write

$$\left|\int_\Omega K(\mathbf{x},\mathbf{y})U(\mathbf{y})\,d\Omega\right|^p \leq \left(\int_\Omega |K(\mathbf{x},\mathbf{y})|^q\,d\Omega\right)^{p/q} \left(\int_\Omega |U(\mathbf{y})|^p\,d\Omega\right)$$

where $q = p/(p-1)$, so

$$\|AU\|_p \leq \left[\int_\Omega \left(\int_\Omega |K(\mathbf{x},\mathbf{y})|^q\,d\Omega\right)^{p/q} d\Omega\right]^{1/p} \|U\|_p.$$

The needed condition for continuity of A is

$$\left[\int_\Omega \left(\int_\Omega |K(\mathbf{x},\mathbf{y})|^q\,d\Omega\right)^{p/q} d\Omega\right]^{1/p} < \infty.$$

In fact, it can be shown that the quantity on the left equals $\|A\|$.

The idea of Lebesgue integration can be extended to the case of non-compact domains Ω. The above-stated fact about $\|A\|$ holds even if Ω is not compact, since Hölder's inequality continues to hold in that case. The reader should note the similarity in form between the corresponding norms of the Fredholm operator and the matrix operator in § 1.12.

Vectorial versions of (1.15.13) are also important in mechanics.

1.16 Orthogonal Decomposition of Hilbert Space

The Hilbert space $L^2(\Omega)$ plays an important role in modern theoretical investigations of partial differential equations. In this book we will encounter other Hilbert spaces that relate to the energy functionals of the various objects of continuum mechanics.

A Hilbert space possesses an important functional, the inner product, whose existence we have not used so far. However, many properties that hold in $\mathbb{R}^3$ — for example, those relating to projections, component representations of vectors, etc. — can be extended to general Hilbert spaces. Here we consider the decomposition of a Hilbert space into a sum of mutually orthogonal subspaces. In $\mathbb{R}^3$ a proper subspace might be a set of vectors acting in a direction parallel to a line or a plane. Suppose U is a subspace of vectors whose line of action is parallel to a plane through the origin, and that V consists of vectors whose line of action is parallel to a line perpendicular to that plane. Then any $\mathbf{v} \in V$ has the property that $(\mathbf{v}, \mathbf{u}) = 0$ for all $\mathbf{u} \in U$. Two subspaces related in this way are said to be *mutually orthogonal*. In this case any $\mathbf{x} \in \mathbb{R}^3$ can be written uniquely as a sum

$$\mathbf{x} = \mathbf{u} + \mathbf{v}, \qquad \mathbf{u} \in U, \ \mathbf{v} \in V. \tag{1.16.1}$$

We say that $\mathbb{R}^3$ has been decomposed as a *direct sum* of the subspaces U and V, and write

$$\mathbb{R}^3 = U \dotplus V. \tag{1.16.2}$$

These ideas extend to a general Hilbert space as follows.

Definition 1.16.1. Let V be a subspace of a Hilbert space H. A vector $x \in H$ is *orthogonal to V* if $(x, v) = 0$ for every $v \in V$. Let U be another subspace of H. We say that U and V are *mutually orthogonal subspaces*, and write $U \perp V$, if every $u \in U$ is orthogonal to V. Finally, if U and V have the property that any $x \in H$ can be expressed uniquely in the form

$x = u + v$ for some $u \in U$ and $v \in V$, we say that H has an *orthogonal decomposition* as the direct sum $H = U \dotplus V$.

The orthogonal projection of a vector $\mathbf{x} \in \mathbb{R}^3$ onto a subspace M of $\mathbb{R}^3$ has a few properties by which we can define the projection uniquely. One is that the difference between $\mathbf{x}$ and its projection $\mathbf{m}_0$, i.e., $\mathbf{x} - \mathbf{m}_0$, has the least length of all vectors $\mathbf{x} - \mathbf{m}$, where $\mathbf{m} \in M$. It turns out that the orthogonal projection of a vector on a subspace of a Hilbert space can be defined in the same way. Thus we begin with the question of minimization of the functional $\|x - m\|$ over $m \in M$, a closed subspace of H.

Exercise 1.16.1. *Verify that the inequality $ax^2 + bx + c \geq c$ holds for all real x if and only if $b = 0$ and $a \geq 0$. (This result will be needed in the proof of Theorem 1.16.1.)*

Theorem 1.16.1. *Suppose M is a closed subspace of a Hilbert space H and $x \in H$. There is a unique element $m_0 \in M$ such that $\|x - m\|$ is minimized when $m = m_0$. Furthermore, m_0 is the unique "minimizing vector" if and only if $x - m_0$ is orthogonal to M.*

Proof. We prove this when H is a real Hilbert space. The case where $x \in M$ is trivial (simply take $m_0 = x$), so we assume $x \notin M$. Define
$$\delta = \inf_{m \in M} \|x - m\|.$$
Now the distance between any two elements $m_i, m_j \in M$ can be expressed using
$$\|m_j - m_i\|^2 = \|(m_j - x) + (x - m_i)\|^2$$
where, by the parallelogram law, the right side satisfies
$$\|(m_j - x) + (x - m_i)\|^2 + \|(m_j - x) - (x - m_i)\|^2$$
$$= 2\|x - m_j\|^2 + 2\|x - m_i\|^2.$$
This means that
$$\|m_j - m_i\|^2 = 2\|x - m_j\|^2 + 2\|x - m_i\|^2 - 4\left\|x - \frac{m_i + m_j}{2}\right\|^2$$
$$\leq 2\|x - m_j\|^2 + 2\|x - m_i\|^2 - 4\delta^2. \qquad (1.16.3)$$
(Here we have used the fact that $(m_i + m_j)/2$ lies in M, since M is a subspace.) By definition of δ we can take a sequence $\{m_i\}$ in M such that $\|x - m_i\| \to \delta$. Such a sequence is a Cauchy sequence by (1.16.3).

Furthermore, because H is complete $\{m_i\}$ must converge and its limit m_0 must belong to M since M is closed. By continuity of the norm we have $\|x - m_0\| = \delta$.

Equation (1.16.3) can also be used to prove uniqueness of the minimizing vector. If $m_0, \bar{m}_0 \in M$ are any two minimizing vectors, we can set $m_i = m_0$ and $m_j = \bar{m}_0$ and obtain

$$\|\bar{m}_0 - m_0\|^2 \leq 2 \|x - \bar{m}_0\|^2 + 2 \|x - m_0\|^2 - 4\delta^2 \leq 2\delta^2 + 2\delta^2 - 4\delta^2 = 0,$$

hence $\bar{m}_0 = m_0$.

We finish the proof by showing that m_0 is the unique minimizing vector if and only if $x - m_0$ is orthogonal to M. Indeed, let $m \in M$. If m_0 is a minimizing vector then for any α we have

$$\|x - m_0\|^2 \leq \|x - m_0 - \alpha m\|^2 = \|x - m_0\|^2 - 2\alpha(x - m_0, m) + \alpha^2 \|m\|^2.$$

This is a quadratic inequality; by Exercise 1.16.1 it can hold for all α if and only if the coefficient of the first power of α, namely $(x - m_0, m)$, is zero. This means that any $m \in M$ is orthogonal to $x - m_0$, and thus $x - m_0$ is orthogonal to M.

Conversely, let $(x - m_0, m) = 0$ for any $m \in M$. Denote $m_1 = m - m_0$. Then

$$\|x - m\|^2 = \|x - m_0 - m_1\|^2 = \|x - m_0\|^2 - 2(x - m_0, m_1) + \|m_1\|^2.$$

As $(x - m_0, m_1) = 0$ we get

$$\|x - m\|^2 = \|x - m_0\|^2 + \|m_1\|^2$$

and so m_0 is the needed minimizer. $\square$

Definition 1.16.2. Given a subspace M of a Hilbert space H, the set of all $x \in H$ that are orthogonal to M is called the *orthogonal complement* of M and is denoted by $M^\perp$.

Exercise 1.16.2. *Let M be a closed subspace of H. Show that $M^\perp$ is a closed subspace of H. Hence M and $M^\perp$ are orthogonal subspaces of H.*

Now we can state the *orthogonal decomposition theorem*.

Theorem 1.16.2. *If M is a closed subspace of a Hilbert space H, then*

$$H = M \dotplus M^\perp. \tag{1.16.4}$$

Hence any $x \in H$ has a unique representation $x = m + n$, where $m \in M$ and $n \in M^\perp$.

Proof. Let $x \in H$. According to Theorem 1.16.1 there is a unique $m_0 \in M$ such that $x - m_0$ is orthogonal to M. Writing $x = m_0 + (x - m_0)$ we see that x is decomposed uniquely into a component in M and a component in $M^\perp$. □

Exercise 1.16.3. *Our proof was given for a real Hilbert space. Supply the proof for a complex Hilbert space.*

1.17 Work and Energy

It was not a simple task to devise a way of measuring the action performed by a person or machine. Eventually, the measure we now call *work* was introduced. It involves the notion that a force (or set of forces) acts on an object and thereby moves it. In the simplest case the force is constant and acts on a particle in the direction of its motion. Then the product of the force F and the distance s through which the force has shifted the particle is called its work:

$$\mathcal{W} = Fs. \qquad (1.17.1)$$

When the force is variable and so depends on the length parameter s as some function $F = F(s)$, it is reasonable to introduce work as an integral:

$$\mathcal{W} = \int_A^B F(s)\, ds, \qquad (1.17.2)$$

where A, B are the initial and final points of the particle trajectory.

When the direction of the force does not coincide with the direction of motion of the particle, a natural generalization is to introduce the force vector $\mathbf{F}$ and represent a small piece of the trajectory as an elemental displacement $d\mathbf{r}$. Then the work can be written as a dot product:

$$\mathcal{W} = \int_A^B \mathbf{F}(s) \cdot d\mathbf{r}. \qquad (1.17.3)$$

In fact, nothing prevents us from introducing other measures of the action of a force, but this particular one is intimately related to the quantity we call *energy*.

The notion of energy occurs in all the physical sciences. It is applied in many situations, but has no strict definition. We say that energy is conserved after observing a wide variety of processes in Nature. In mechanics, however, the law of conservation of mechanical energy follows directly from mathematical transformations that yield energy integrals in various situations; these integrals, in turn, are found to be related to work as introduced in (1.17.3).

The notion of mechanical energy is extremely important. We use it not only as a measure of something conserved during the motions and deformations of bodies, but to characterize the differences between states of a body. Moreover, we will introduce normed spaces that employ energy-related expressions for norms and inner products.

Let us illustrate how the above mentioned work–energy relation arises in simple problems. Consider the motion of a particle, having mass m, under the action of a force $\mathbf{F}$. We shall see how the simplest form of the energy conservation law arises in mechanics. The equation of motion is

$$m\ddot{\mathbf{r}} = \mathbf{F}. \tag{1.17.4}$$

We represent $d\mathbf{r}$ along the trajectory as

$$d\mathbf{r} = \dot{\mathbf{r}}\,dt.$$

Dot multiplying (1.17.4) by this and integrating with respect to time over $[t_0, t_1]$, we get

$$\int_{t_0}^{t_1} m\ddot{\mathbf{r}} \cdot \dot{\mathbf{r}}\,dt = \int_{t_0}^{t_1} \mathbf{F} \cdot d\mathbf{r}.$$

On the right we have the work of the force $\mathbf{F}$ acting during the time interval $[t_0, t_1]$. On the left, the simple transformation

$$m\ddot{\mathbf{r}} \cdot \dot{\mathbf{r}} = \frac{d}{dt}\frac{m\dot{\mathbf{r}}^2}{2} \equiv \frac{d}{dt}\frac{m\mathbf{v}^2}{2}$$

yields, after integration,

$$\left.\frac{m\mathbf{v}^2}{2}\right|_{t=t_1} - \left.\frac{m\mathbf{v}^2}{2}\right|_{t=t_0} = \int_{t_0}^{t_1} \mathbf{F} \cdot d\mathbf{r}. \tag{1.17.5}$$

It follows that if $\mathbf{F}$ is zero or orthogonal to the trajectory at all times, the quantity $m\mathbf{v}^2/2$ stays constant. This is the *kinetic energy* of the particle. The last equation states that the change in kinetic energy during some time interval is equal to the work of the force during that same interval. This is one formulation of the law of energy conservation.

We know that total energy (i.e., the sum of potential and kinetic energy terms) is conserved. Let us consider a simple problem: the oscillations, along a straight line, of a particle attached to a spring. *Hooke's law* relates the extension x of the spring to the applied force F:

$$F = kx. \tag{1.17.6}$$

By Newton's third law the spring exerts a force $-kx$ on the particle, and the equation of motion of a particle of mass m is

$$m\ddot{x}(t) = F_0(t) - kx(t). \tag{1.17.7}$$

The active force $F_0(t)$ is assumed given. Before repeating the transformations done above, let us introduce the potential

$$\mathcal{V} = \frac{1}{2}kx^2 \tag{1.17.8}$$

corresponding to the elastic force $-kx$. The name "potential" indicates that its derivative with respect to x gives us the force expression:

$$\frac{d\mathcal{V}}{dx} = -(-kx). \tag{1.17.9}$$

The equation of motion of the particle attached to the spring can be rewritten in the form

$$m\ddot{x}(t) + \frac{d\mathcal{V}(x)}{dx} = F_0(t). \tag{1.17.10}$$

Now we multiply through by

$$dx(t) = \dot{x}(t)\,dt$$

and integrate along the trajectory over the time $[t_0, t_1]$. We get

$$\int_{t_0}^{t_1} m\ddot{x}(t)\dot{x}(t)\,dt + \int_{t_0}^{t_1} \frac{d\mathcal{V}(x)}{dx}\,dx = \int_{t_0}^{t_1} F_0(t)\,dx(t)$$

which brings us, after integration, to

$$\left[\frac{mv^2}{2} + \mathcal{V}(x)\right]\bigg|_{t=t_1} - \left[\frac{mv^2}{2} + \mathcal{V}(x)\right]\bigg|_{t=t_0} = \int_{t_0}^{t_1} F_0(t)\,dx(t) \tag{1.17.11}$$

where

$$v(t) = \dot{x}(t).$$

The expression on the right-hand side of (1.17.11) is the work of the active force $F_0(t)$ during $[t_0, t_1]$. On the left we see the particle's kinetic energy $\mathcal{K} = mv^2/2$. We see the sum $\mathcal{K} + \mathcal{V}$ evaluated at the final and initial points of the time period under consideration. Calling $\mathcal{V}$ the potential energy and $\mathcal{K} + \mathcal{V}$ the total energy, we come to a well-known statement of energy conservation: the change in total energy of the "particle-spring" system during $[t_0, t_1]$ is equal to the work performed by the external force during that same period. Of course, the conservation of total energy holds when the work is zero, e.g., when $F_0 = 0$; in this case $\mathcal{K} + \mathcal{V}$ stays constant over time.

We called $\mathcal{V}$ the potential energy. It is clearly associated with the spring and not the particle. Of course, we are free to assign names in any desired way, but should have some justification. We have said that the energy relates to the work done by forces. So why is $\mathcal{V} = kx^2/2$ called potential energy? Consider the work done by the external force while stretching the spring by an amount x. At the final position, the extension of the spring is x, and the force to maintain this extension would be $F = kx$. A naive application of the "work equals force times distance" idea would yield a value of $kx \cdot x = kx^2$ for the work done. But $\mathcal{V}$ contains an additional factor of $1/2$. Why? The answer is that we cannot apply the force kx at once: at first we need only a small force, near zero, to produce a bit of extension. If we apply the force kx right away we will produce motion, but we suppose that there is no motion. So our external force should increase from zero to kx in such a way that at every moment we have a state of equilibrium.[3] Then the total work of the external force is

$$\mathcal{W} = \int_0^x k\xi \, d\xi = \frac{kx^2}{2}. \qquad (1.17.12)$$

This coincides with the value of $\mathcal{V}$ as introduced above.

The reader is aware of elementary physics problems in which a particle moves vertically through the Earth's gravitational field; in such cases we also consider the total energy of the particle to be the sum of its kinetic and potential energies, and the potential energy term is analogous to that found above for the mass-spring system. We also introduce a gravitational potential

$$\mathcal{V}(z) = mgy, \qquad (1.17.13)$$

where y is height above the Earth's surface. The total energy in this way

---

[3]Here we ran into a typical snag that is common in statics and thermodynamics: we essentially treat a moving system as though it were in true static equilibrium at any instant of the motion. Formally, this can be done in two ways: (1) we can consider certain masses that are involved to be zero (as we have done with the mass of the spring), or (2) we can assume extremely slow motion and consider all inertial forces to be zero (although it is not altogether clear that we can do this when we observe finite changes at the conclusion of the motion).

is

$$\mathcal{E} = \frac{mv^2}{2} + \mathcal{V}(y). \qquad (1.17.14)$$

So the two problems exhibit the structure for total energy, which is conserved during motion. In this sense they are analogous. Note that in both cases the total energy is related not only to the particle but to "the sources" of external force (i.e., the spring and gravitational field, respectively). In many problems we can regard the forces as "external" in nature and thereby introduce a potential-type function. Then $\mathcal{V}$ plays the role of the potential energy of the system under consideration, but in fact it relates to the energy of some external objects that "emanate" those forces somehow. This notion of the potential of external forces is extremely useful in Lagrangian mechanics.

1.18 Virtual Work Principle

For a system of n particles in equilibrium, the resultant force acting on the ith particle is zero:

$$\mathbf{F}_i = \mathbf{0} \qquad (i = 1, \ldots, n). \qquad (1.18.1)$$

If the motions of the particles are unconstrained, we can denote by $\delta \mathbf{r}_i$ the (arbitrary) permissible motion of the ith particle and write all the equilibrium equations as the single equation

$$\sum_{i=1}^{n} \mathbf{F}_i \cdot \delta \mathbf{r}_i = 0. \qquad (1.18.2)$$

This holds for all possible $\delta \mathbf{r}_i$, and from it we can recover (1.18.1) since we can appoint each $\delta \mathbf{r}_i$ independently. Equation (1.18.2) expresses the *virtual work principle* (VWP) for the equilibrium of a system of independent particles. We see that its terms express the work of the forces $\mathbf{F}_i$ over the displacements given by the vectors $\delta \mathbf{r}_i$. This is called *virtual work*, a name we shall soon explain.

The transformation from (1.18.1) to (1.18.2) offers no real advantages in this case. But the situation is different when constraints on the motion are present.

We have mentioned the constraints under which a system of particles becomes a rigid body. In mechanics there are also constraints of other types: supports, conditions of impenetrability, etc. We shall touch upon some of the problems in which friction is negligible.

Let us consider the equilibrium of a particle under the influence of a constraint that does not involve friction. The constraint itself is defined by an equation. For example, a particle may be constrained to move without friction on some surface expressed in Cartesian coordinates by

$$F(x, y, z) = 0. \tag{1.18.3}$$

In vectorial form this looks like $F(\mathbf{r}) = 0$. The absence of friction means the reaction force $\mathbf{R}$ of the surface on the moving particle is always directed along the surface normal $\mathbf{n}$. When the constrained particle is in equilibrium, the resultant force acting on it is zero. Let us call the remaining forces *active*, and denote their resultant by $\mathbf{F}$. We have

$$\mathbf{F} + \mathbf{R} = \mathbf{0}. \tag{1.18.4}$$

The reaction $\mathbf{R}$ lies along $\mathbf{n}$ and participates in the force balance along this direction, but has no component tangent to the surface; hence the projection of $\mathbf{F}$ on the local tangent plane must be zero. The equation of the tangent plane at a point $\mathbf{r}_0 = (x_0, y_0, z_0)$ on the surface is

$$\frac{\partial F(\mathbf{r}_0)}{\partial x}(x - x_0) + \frac{\partial F(\mathbf{r}_0)}{\partial y}(y - y_0) + \frac{\partial F(\mathbf{r}_0)}{\partial z}(z - z_0) = 0.$$

In vector form this is $\nabla F(\mathbf{r}_0) \cdot (\mathbf{r} - \mathbf{r}_0) = 0$. Let us denote $\mathbf{r} - \mathbf{r}_0$ by $\delta \mathbf{r}$ so that

$$\nabla F(\mathbf{r}_0) \cdot \delta \mathbf{r} = 0. \tag{1.18.5}$$

We call a vector $\delta \mathbf{r}$ satisfying (1.18.5) a *virtual displacement* of a particle at the point $\mathbf{r}_0$. In general the vector $\mathbf{r}_0 + \delta \mathbf{r}$ does not define a point on the surface, hence the displacement $\delta \mathbf{r}$ of the particle does not belong to the set of actual displacements. Usually $\delta \mathbf{r}$ is considered as an infinitesimal displacement of the particle, in this case it belongs to the surface tangent but (with the same success) it could be finite and so in general does not belong to the surface. This explains the curious term "virtual displacement": it does not belong, in general, to the set of real displacements of the particle but is "proportional" to one of the real infinitesimal displacements. The set of all virtual displacements $\delta \mathbf{r}$ covers all directions tangent to the surface, so the condition that the projection of the active force $\mathbf{F}$ onto the tangent plane is zero can be written in the form

$$\mathbf{F} \cdot \delta \mathbf{r} = 0 \tag{1.18.6}$$

as **R** is orthogonal to $\delta \mathbf{r}$. In equilibrium of the particle on a surface, (1.18.6) must hold for all virtual displacements $\delta \mathbf{r}$. It is the VWP equation in this case. In form it coincides with the VWP equation for a free particle, but the set of virtual displacement vectors $\delta \mathbf{r}$ is now restricted; because of this restriction, we excluded from the equation the reaction force **R**, and this offers some practical advantages.

If the particle is constrained to move only along a curve without friction, then the reaction cannot have components parallel to the tangent at each point. Here, the set of virtual displacements $\delta \mathbf{r}$ is restricted to the set of vectors parallel to the tangent line at any point. Reasoning similar to the above brings us to the equation of equilibrium, which coincides with (1.18.6) for the surface constraint.

In classical mechanics one also considers unilateral constraints. In the case of a surface, for example, a particle may be able to move on the surface or away from one side but cannot penetrate through to the other side. Here the set of virtual displacements is obviously not restricted to vectors lying in the tangent plane. When friction is absent, there are various arguments (more of the nature of axioms, really) supporting the notion that the work of the reaction force must be nonnegative:

$$\mathbf{R} \cdot \delta \mathbf{r} \geq 0. \qquad (1.18.7)$$

Then (1.18.4) gives

$$\mathbf{F} \cdot \delta \mathbf{r} \leq 0, \qquad (1.18.8)$$

which is regarded as the most general form of the virtual work principle for a particle whose position is restricted by a unilateral constraint. When treating a system of independent particles, we can write out the VWP equation (or inequality) for each particle and then add. The resulting equation (or inequality), by the independence of the virtual displacements for each particle, is equivalent to the complete set of equilibrium equations for the system. It does not contain the constraint reactions, however, so solution of the equations is simplified. Thus, for the case of a system of particles that can move only along certain curves or surfaces without friction (such systems are called *holonomic*) so that the constraints are expressed with equalities of the type $f(\mathbf{r}_1, \ldots, \mathbf{r}_n, t) = 0$ (such constraints are called *geometric* or *holonomic*, as opposed to the *kinematic constraints* whose equations include velocities of points) the virtual work principle is expressed by

$$\sum_{i=1}^{n} \mathbf{F}_i \cdot \delta \mathbf{r}_i = 0. \tag{1.18.9}$$

Of course, in problem solving there is no need to try all possible virtual displacements in this equation. For each i, it is enough to take only the vectors that constitute a basis in the corresponding space of virtual displacements.

If there are unilateral constraints without friction, the virtual work principle is given by

$$\sum_{i=1}^{n} \mathbf{F}_i \cdot \delta \mathbf{r}_i \leq 0. \tag{1.18.10}$$

It turns out that the virtual work principle can be used not only in cases involving independent particles, but with rigid bodies or systems of such bodies under the actions of forces. It is only necessary to observe that the virtual motions of different points of a rigid body are not independent. When we do this, we can obtain the equilibrium conditions for a rigid body. The reader can consult any textbook on classical mechanics for further details. In the same way, we can consider the virtual work principle for dynamic problems when "inertial" forces are present. So the virtual work principle applies in dynamics as well (cf., equation (1.19.2)).

For a mechanical system without friction, it seemingly does not matter whether we use the virtual work principle or Newton's laws to study particle motion. However, the virtual work principle has a broader range of application in classical mechanics. In general, the virtual work principle is not a direct consequence of Newton's laws, although in many cases it is possible to demonstrate their equivalence as was done above. Experience shows that the virtual work principle can be taken as the base formulation for the laws of equilibrium (and, with use of d'Alembert's principle, for the laws of motion) of particles and rigid bodies with constraints.

1.19 Lagrange's Equations of the Second Kind

Recall that the degree of freedom of a system of particles is the minimal number of independent parameters needed to uniquely specify the position of the system. This is often less than the formal number of Cartesian components associated with the position vectors of the particles. If a particle moves along a surface, it is sufficient to know two coordinates of the particle's position on the surface. If it moves along a line, knowledge of a single coordinate suffices. Finally, if particles make up a rigid body, then their mutual separation distances are constant, so fewer (six) position parameters are needed to describe the motion. A reduction in the number of needed parameters in comparison with the number of Cartesian components with which we may describe the system is normally due to constraints imposed on the system.

Let a system of r particles have n degrees of freedom so that its position is uniquely defined by the coordinate parameters $q_1, \ldots, q_n$. Because of possible ties between particles, we should suppose that the position vector for each particle depends on all the q_i, which are independent: that is,

$$\mathbf{r}_i = \mathbf{r}_i(q_1, \ldots, q_n, t). \tag{1.19.1}$$

Of course, we could use Newton's laws to describe the motion of a system of particles having a degree of freedom less than the total number of position vector components. But it is more reasonable to derive the minimal number of necessary equations while avoiding the equations of constraint, etc. This system of equations in the variables q_i is composed of *Lagrange's equations of the second kind*. Our derivation will proceed under relatively simple assumptions on the constraints. More complex cases are treated in fundamental textbooks on classical mechanics.

Note that for these equations the parameters $q_1, \ldots, q_n$ become functions of time t, and thus we can introduce corresponding velocities $\dot{q}_1, \ldots, \dot{q}_n$. In what follows we can consider $\dot{q}_i$ to be independent of $\dot{q}_j$ at any time t.

We begin by combining, for a system of r particles, the virtual work principle with d'Alembert's principle. Let us include inertial forces $-m_i \ddot{\mathbf{r}}_i$ in equation (1.18.2):

$$\sum_{i=1}^{r} (\mathbf{F}_i - m_i \ddot{\mathbf{r}}_i) \cdot \delta \mathbf{r}_i = 0. \tag{1.19.2}$$

Here $\delta \mathbf{r}_i$ is a virtual displacement for the ith particle. We assume the $\mathbf{F}_i$ depend on the positions and velocities of the particles. Thus, in terms of the q_i, they are

$$\mathbf{F}_i = \mathbf{F}_i(q_1, \ldots, q_n, \dot{q}_1, \ldots, \dot{q}_n, t). \tag{1.19.3}$$

We transform (1.19.2), with the virtual displacements represented as

$$\delta \mathbf{r}_k = \sum_{i=1}^{n} \frac{\partial \mathbf{r}_k}{\partial q_i} \delta q_i. \tag{1.19.4}$$

Note that the virtual displacements are taken at a fixed instant t. They are vectors proportional to infinitesimal vectors of admissible displacements at a fixed t, so we can find them by writing out the formal expression for the first differential while considering t as a fixed parameter. After writing everything in terms of the δq_i, we will select multipliers δq_i in (1.19.2). Using the mutual independence of the δq_i, we will then equate the coefficients of the δq_i to zero and obtain the needed equations.

The work of the active forces from (1.19.2) can be represented as

$$\sum_{i=1}^{r} \mathbf{F}_i \cdot \delta \mathbf{r}_i = \sum_{j=1}^{n} Q_j \, \delta q_j, \tag{1.19.5}$$

where Q_j is called the component of the generalized forces relating to the virtual "displacement" δq_j. By the above assumptions, Q_j depends on the $q_1, \ldots, q_n$, the $\dot{q}_1, \ldots, \dot{q}_n$, and t.

Now we transform the terms of (1.19.2) for the inertial forces. It turns out that these can be expressed in terms of the derivatives of the kinetic energy. We first use the relation (1.19.4):

$$\sum_{i=1}^{r} m_i \ddot{\mathbf{r}}_i \cdot \delta \mathbf{r}_i = \sum_{i=1}^{r} m_i \ddot{\mathbf{r}}_i \cdot \sum_{j=1}^{n} \frac{\partial \mathbf{r}_i}{\partial q_j} \delta q_j = \sum_{j=1}^{n} \delta q_j \left(\sum_{i=1}^{r} m_i \ddot{\mathbf{r}}_i \cdot \frac{\partial \mathbf{r}_i}{\partial q_j} \right). \tag{1.19.6}$$

Recall that the overdot denotes a total time derivative d/dt, which differs from $\partial/\partial t$. Note that

$$\frac{d}{dt}\left(\dot{\mathbf{r}}_i \cdot \frac{\partial \mathbf{r}_i}{\partial q_j} \right) = \ddot{\mathbf{r}}_i \cdot \frac{\partial \mathbf{r}_i}{\partial q_j} + \dot{\mathbf{r}}_i \cdot \frac{d}{dt}\left(\frac{\partial \mathbf{r}_i}{\partial q_j} \right),$$

and therefore

$$\ddot{\mathbf{r}}_i \cdot \frac{\partial \mathbf{r}_i}{\partial q_j} = \frac{d}{dt}\left(\dot{\mathbf{r}}_i \cdot \frac{\partial \mathbf{r}_i}{\partial q_j} \right) - \dot{\mathbf{r}}_i \cdot \frac{d}{dt}\left(\frac{\partial \mathbf{r}_i}{\partial q_j} \right). \tag{1.19.7}$$

Next we use two formulas, proved at the end of this section:

$$\frac{\partial \dot{\mathbf{r}}_i}{\partial \dot{q}_j} = \frac{\partial \mathbf{r}_i}{\partial q_j}, \tag{1.19.8}$$

$$\frac{\partial \dot{\mathbf{r}}_i}{\partial q_j} = \frac{d}{dt}\left(\frac{\partial \mathbf{r}_i}{\partial q_j} \right), \tag{1.19.9}$$

where q_i and $\dot{q}_j$ are considered as independent variables. Applying these to the right side of (1.19.7) we get

$$\ddot{\mathbf{r}}_i \cdot \frac{\partial \mathbf{r}_i}{\partial q_j} = \frac{d}{dt}\left(\dot{\mathbf{r}}_i \cdot \frac{\partial \dot{\mathbf{r}}_i}{\partial \dot{q}_j} \right) - \dot{\mathbf{r}}_i \cdot \frac{\partial \dot{\mathbf{r}}_i}{\partial q_j} = \frac{d}{dt}\left(\frac{1}{2} \frac{\partial \dot{\mathbf{r}}_i^2}{\partial \dot{q}_j} \right) - \frac{\partial}{\partial q_j}\left(\frac{\dot{\mathbf{r}}_i^2}{2} \right).$$

Substituting this into (1.19.6), we get

$$\sum_{i=1}^{r} m_i \ddot{\mathbf{r}}_i \cdot \delta \mathbf{r}_i = \sum_{j=1}^{n} \delta q_j \sum_{i=1}^{r} \left\{ \frac{d}{dt}\left[\frac{\partial}{\partial \dot{q}_i}\left(\frac{1}{2}m_i\dot{\mathbf{r}}_i^2\right)\right] - \frac{\partial}{\partial q_j}\left(\frac{1}{2}m_i\dot{\mathbf{r}}_i^2\right) \right\}$$

$$= \sum_{j=1}^{n} \delta q_j \left[\frac{d}{dt}\left(\frac{\partial \mathcal{E}}{\partial \dot{q}_j}\right) - \frac{\partial \mathcal{E}}{\partial q_j} \right]$$

where

$$\mathcal{E} = \sum_{i=1}^{r} \frac{1}{2} m_i \dot{\mathbf{r}}_i^2. \tag{1.19.10}$$

Combining this and (1.19.5) with (1.19.2), we derive

$$\sum_{j=1}^{n} \delta q_j \left[Q_j - \frac{d}{dt}\left(\frac{\partial \mathcal{E}}{\partial \dot{q}_j}\right) + \frac{\partial \mathcal{E}}{\partial q_j} \right] = 0.$$

Finally, using the independence of the δq_i, we obtain

$$\frac{d}{dt}\left(\frac{\partial \mathcal{E}}{\partial \dot{q}_j}\right) - \frac{\partial \mathcal{E}}{\partial q_j} = Q_j \qquad (j = 1, \ldots, n). \tag{1.19.11}$$

These are Lagrange's equations of the second kind. They constitute a system of n ordinary differential equations with respect to the unknown functions $q_j(t)$. In general they contain terms involving $q_j(t)$, $\dot{q}_j(t)$, and $\ddot{q}_j(t)$, so the system is of order $2n$.

Before finishing this section, we demonstrate how the assumption of potentiality for the generalized forces leads us to something resembling the Euler–Lagrange equations for the problem of minimum of a functional (considered in § 1.20). Potentiality of the set of Q_j means there is a potential function $\mathcal{V} = \mathcal{V}(q_1, \ldots, q_n)$ such that

$$Q_j = -\frac{\partial \mathcal{V}}{\partial q_j}. \tag{1.19.12}$$

In many cases $\mathcal{V}$ is called the potential energy; it is related to the energy of the "source" of the external forces Q_j. Substituting into (1.19.11), we get

$$\frac{d}{dt}\left(\frac{\partial \mathcal{E}}{\partial \dot{q}_j}\right) - \frac{\partial \mathcal{E}}{\partial q_j} = -\frac{\partial \mathcal{V}}{\partial q_j} \qquad (j = 1, \ldots, n).$$

By assumption, $\mathcal{V}$ does not depend on $\dot{q}_j$ so $\partial \mathcal{V}/\partial \dot{q}_j = 0$. Introducing a new function

$$\mathcal{L} = \mathcal{E} - \mathcal{V} \tag{1.19.13}$$

called the *kinetic potential* or the *Lagrangian*, we transform (1.19.11) to

$$\frac{d}{dt}\left(\frac{\partial \mathcal{L}}{\partial \dot{q}_j}\right) - \frac{\partial \mathcal{L}}{\partial q_j} = 0 \qquad (j = 1, \ldots, n). \tag{1.19.14}$$

These are Lagrange's equations of the second kind for the case of active forces having potential.

Although we derived the Lagrange equations under simplified conditions for the mechanical system, they can be extended to much less restrictive conditions. Moreover, they are used not only in classical mechanics: physicists use these and similar equations in a variety of areas.

Finally, let us derive (1.19.8) and (1.19.9). Writing

$$\frac{\partial \dot{\mathbf{r}}_i}{\partial \dot{q}_j} = \frac{\partial}{\partial \dot{q}_j}\left(\sum_{k=1}^n \frac{\partial \mathbf{r}_i}{\partial q_k}\dot{q}_k + \frac{\partial \mathbf{r}_i}{\partial t}\right)$$

and noting that $\mathbf{r}_i$ does not depend on $\dot{q}_j$ (and hence neither do $\partial \mathbf{r}_i/\partial q_j$ or $\partial \mathbf{r}_i/\partial t$), we immediately obtain (1.19.8). Relation (1.19.9) holds because we can interchange the order in mixed partial differentiation:

$$\frac{d}{dt}\left(\frac{\partial \mathbf{r}_i}{\partial q_j}\right) = \sum_{k=1}^n \frac{\partial}{\partial q_k}\left(\frac{\partial \mathbf{r}_i}{\partial q_j}\right)\dot{q}_k + \frac{\partial}{\partial t}\left(\frac{\partial \mathbf{r}_i}{\partial q_j}\right)$$

$$= \frac{\partial}{\partial q_j}\left(\sum_{k=1}^n \frac{\partial \mathbf{r}_i}{\partial q_k}\dot{q}_k + \frac{\partial \mathbf{r}_i}{\partial t}\right)$$

$$= \frac{\partial \dot{\mathbf{r}}_i}{\partial q_j}.$$

To present Lagrange's equations from a variational standpoint, as a consequence of Hamilton's principle, we will need some basic notions from the calculus of variations.

1.20 Problem of Minimum of a Functional

Lagrange's equations are important in mechanics. We wish to show that they can be derived from a variational principle that can serve as the starting point for mechanics instead of Newton's laws. We begin with some elementary facts from the calculus of variations.

As with a function, we can consider the problem of extremum of a functional. For definiteness let us take a functional $F = F(x)$ given on a Banach space X. The definitions of such concepts as a point of local maximum or minimum, global maximum or minimum, etc., can be extended from functions to functionals. For example, a point x is called a point of local minimum of $F(x)$ if there is a δ-neighborhood of x such that $F(x) \leq F(z)$ for all z in the δ-neighborhood, i.e., whenever $\|z - x\| < \delta$.

We will examine particular minimization problems for functionals in detail when considering the equilibrium problems for elastic models. Now we can find an equation that a local extreme point must satisfy (i.e., a *necessary condition* for its existence) if $F(x)$ is sufficiently smooth (differentiable). We shall not pause to define differentiability of a functional, but will proceed rather formally instead.

Suppose x_0 is a point of local minimum of $F(x)$. This means there is an $\varepsilon > 0$ having the property that if we take an element ty such that $\|ty\| \leq \varepsilon$, where t is a real number, then

$$F(x_0) \leq F(x_0 + ty). \tag{1.20.1}$$

Now let us fix y in addition to x_0. The functional $F = F(x_0 + ty)$ becomes a simple function of the real variable t which, according to (1.20.1), is minimized when $t = 0$. If it is differentiable, the necessary condition of minimum is simply

$$\left.\frac{dF(x_0 + ty)}{dt}\right|_{t=0} = 0. \tag{1.20.2}$$

This must hold for any $y \in X$. It is, in fact, a necessary condition for x to be a minimum or maximum point of the functional.

We should note that the expression on the left-hand side of (1.20.2) is called the *Gâteaux derivative* of the functional $F(x)$. If $F(x)$ is a function in n variables, i.e., if $x = (x_1, x_2, \ldots, x_n)$, this expression gives us the directional derivative in the direction $y = (y_1, y_2, \ldots, y_n)$. In particular, when $y = (1, 0, \ldots, 0)$ it yields the partial derivative $\partial F/\partial x_1$.

Now let us apply (1.20.2) to a simple functional that appears in any textbook on the calculus of variations:

$$F(y) = \int_a^b f(x, y, y')\, dx. \tag{1.20.3}$$

We shall consider $F(y)$ over a set of functions $y = y(x)$ satisfying the *Dirichlet boundary conditions*

$$y(a) = c_0, \qquad y(b) = c_1. \tag{1.20.4}$$

Let us employ the space $C^{(2)}(a, b)$. The problem of minimum is formulated as

Problem 1.20.1. *Minimize the functional*

$$\int_a^b f(x, y, y')\, dx$$

over the set of functions $y(x) \in C^{(2)}(a, b)$ *that satisfy the boundary conditions* $y(a) = c_0$ *and* $y(b) = c_1$.

Suppose $y(x)$ is a solution. Satisfaction of (1.20.4) by the sum $y + t\varphi$ for any value of the parameter t requires that the function $\varphi = \varphi(x)$ vanish at the endpoints $x = a, b$. According to the scheme discussed above, we set

$$\frac{d}{dt} \int_a^b f(x, y + t\varphi, y' + t\varphi')\, dx \bigg|_{t=0} = 0. \qquad (1.20.5)$$

We can pass the derivative operator d/dt through the integral sign if $f(x, y, y')$ is continuously differentiable in y and y' (these being regarded as independent variables). The result, after taking the total derivative, is that the equation

$$\int_a^b [f_y(x, y, y')\varphi + f_{y'}(x, y, y')\varphi']\, dx = 0 \qquad (1.20.6)$$

must hold for any smooth function φ vanishing at $x = a, b$. Here partial derivatives with respect to y and y' are denoted by the subscripts. The left side of (1.20.6) is called the *first variation* of the functional and is denoted by $\delta F(y, \varphi)$, while φ is called the *variation* of y and in mechanics books is denoted δy. From (1.20.6), which must hold for arbitrary admissible $\varphi(x)$, we can derive a differential equation for $y(x)$. We first integrate by parts to obtain

$$\int_a^b \left[f_y(x, y, y') - \frac{d}{dx} f_{y'}(x, y, y') \right] \varphi\, dx = 0, \qquad (1.20.7)$$

where the terms $f_{y'}\varphi \big|_a^b$ vanish because $\varphi(a) = \varphi(b) = 0$. Now we need the *Main Lemma* of the calculus of variations. We introduce

Definition 1.20.1. *By* $\mathcal{D}(0, l)$ *we mean the set of functions infinitely differentiable on* $(0, l)$ *and vanishing in some neighborhood of the endpoints* 0 *and* l *(this neighborhood can differ for different functions in the set).*

Lemma 1.20.1. *If* $G = G(x)$ *is continuous on* $[0, l]$ *and satisfies*

$$\int_0^l G(x)\varphi(x)\, dx = 0$$

for any $\varphi \in \mathcal{D}(0, l)$, *then* $G(x) = 0$ *on* $[0, l]$.

Proof. We suppose $G(x^*) \neq 0$ at some $x^* \in [0, l]$ and obtain a contradiction. By continuity there is a neighborhood $[x^* - \varepsilon, x^* + \varepsilon]$ of x^* throughout which the sign of $G(x)$ persists (either strictly positive or strictly negative). If we choose $\varphi(x)$ so that it, too, has a constant sign in this neighborhood and vanishes elsewhere, then the integral

$$\int_0^l G(x)\varphi(x)\,dx = \int_{x^*-\varepsilon}^{x^*+\varepsilon} G(x)\varphi(x)\,dx$$

must be nonzero since its integrand $G(x)\varphi(x)$ never changes sign. This is the desired contradiction; the proof will be complete if we can display a function $\varphi \in \mathcal{D}(0, l)$ satisfying the condition stated above. The "bell-shaped" function

$$\varphi(x) = \begin{cases} \exp\left(\dfrac{\varepsilon^2}{(x-x^*)^2 - \varepsilon^2}\right), & |x - x^*| < \varepsilon, \\ 0, & |x - x^*| \geq \varepsilon, \end{cases}$$

is one example. $\square$

Because the set of admissible functions φ in (1.20.7) includes $\mathcal{D}(0, l)$, we can apply Lemma 1.20.1 to (1.20.7) and obtain

$$f_y - \frac{d}{dx}f_{y'} = 0. \tag{1.20.8}$$

This *Euler equation* is analogous to Fermat's condition $f' = 0$ for an ordinary function. In general it is an ordinary differential equation of second order. Since the functional F will arise from a mechanical problem where we usually seek a unique solution, we understand why we specified two boundary conditions earlier. The derivative with respect to x in (1.20.8) is a total derivative: i.e., we have

$$f_y - f_{y'x} - f_{y'y}y' - f_{y'y'}y'' = 0. \tag{1.20.9}$$

In (1.20.8) we recognize Lagrange's equation (1.19.14) when $f = \mathcal{L}$ and $y = q_i$.

Subsequently, we will need to pose a minimum problem for (1.20.3) without specifying a boundary condition at an endpoint. In fact, in addition to (1.20.8), from (1.20.5) there follow two boundary conditions called *natural boundary conditions*. To see this we return to (1.20.6) and integrate by parts *without* imposing (1.20.4). We get

$$\int_a^b \left[f_y(x,y,y') - \frac{d}{dx}f_{y'}(x,y,y')\right]\varphi\,dx + f_{y'}(x,y(x),y'(x))\varphi(x)\Big|_{x=a}^{x=b} = 0. \tag{1.20.10}$$

If we temporarily limit our consideration to those smooth functions $\varphi(x)$ that *do* satisfy $\varphi(a) = \varphi(b) = 0$, then we have

$$\int_a^b \left[f_y(x,y,y') - \frac{d}{dx} f_{y'}(x,y,y') \right] \varphi \, dx = 0. \qquad (1.20.11)$$

By Lemma 1.20.1 we once again find that (1.20.8) must hold in (a,b). Clearly, now relation (1.20.11) holds for *any* continuous φ. Thus, returning to (1.20.10), we find that

$$f_{y'}(x, y(x), y'(x)) \varphi(x) \Big|_{x=a}^{x=b} = 0$$

for *any* smooth function $\varphi(x)$. The particular choices $\varphi(x) = x - b$ and $\varphi(x) = x - a$ yield, respectively,

$$f_{y'}|_{x=a} = 0, \qquad f_{y'}|_{x=b} = 0. \qquad (1.20.12)$$

These are the natural boundary conditions for (1.20.3).

So one can minimize F in the absence of a boundary condition on y at one of the endpoints a or b. At that endpoint the corresponding natural condition applies automatically. Note that, although the Euler equation is of second order, we *cannot* in general introduce two *initial* conditions for y; we cannot prescribe $y(a)$ and $y'(a)$, for example, because the natural condition at b would still apply. The result would be too many conditions to define a solution of a second order ordinary differential equation. Hence we always have two boundary conditions, one at each endpoint.

The Euler equation (1.20.8) appears in all the minimum problems for the functional (1.20.3), regardless of the boundary conditions chosen, if the problem is correctly posed. For more general functionals it changes form. Listed below are additional functionals of interest along with their corresponding Euler equations.

1. If $y(x)$ is replaced by a vector function $\mathbf{y}(x) = (y_1(x), y_2(x), \ldots, y_n(x))$, then a functional of the type

$$F(\mathbf{y}) = \int_a^b f(x, \mathbf{y}, \mathbf{y}') \, dx \qquad (1.20.13)$$

results. The Euler equation can be written in vector form as

$$\nabla_{\mathbf{y}} f - \frac{d}{dx} \nabla_{\mathbf{y}'} f = 0 \qquad (1.20.14)$$

or, alternatively, in scalar form as the system of equations

$$f_{y_i} - \frac{d}{dx} f_{y'_i} = 0 \qquad (i = 1, \ldots, n). \qquad (1.20.15)$$

These, in fact, follow easily from (1.20.8): if we fix all but the ith component of the minimizing function, we obtain a functional of exactly the same form as before (but with respect to the function y_i). Natural boundary conditions for this functional can be stated as

$$f_{y'_i}\big|_{x=a} = 0, \qquad f_{y'_i}\big|_{x=b} = 0 \qquad (1.20.16)$$

for $i = 1, \ldots, n$.

2. An extreme point of the functional

$$F_n(y) = \int_a^b f(x, y, y', \ldots, y^{(n)})\, dx \qquad (1.20.17)$$

will satisfy the *Euler–Lagrange equation*

$$f_y - \frac{d}{dx} f_{y'} + \frac{d^2}{dx^2} f_{y''} - \cdots + (-1)^n \frac{d^n}{dx^n} f_{y^{(n)}} = 0 \qquad (1.20.18)$$

with $2n$ natural boundary conditions. The method of obtaining this is similar to that for the simplest functional. Namely, supposing y to be a minimizer of the functional, we find that $F_n(y + t\varphi)$, being a function of the real parameter t when φ is sufficiently smooth, takes its minimum at $t = 0$. This leads to a result analogous to (1.20.6). Subsequent integration by parts yields a result analogous to (1.20.10) but having $2n$ boundary terms outside the integrals. This equation holds, in particular, for all $\varphi \in \mathcal{D}(0, l)$. Now all the boundary terms vanish, and as a consequence of Lemma 1.20.1 we get the above Euler–Lagrange equation. Next, considering all the smooth functions φ, we will derive the natural boundary conditions for the functional.

In a similar manner we can find the necessary conditions of minimum for a functional defined on functions in many variables.

3. In the two-dimensional case where

$$F(u) = \iint_S f(x, y, u(x, y), u_x(x, y), u_y(x, y))\, dx\, dy, \qquad (1.20.19)$$

the equation

$$f_u - \left(\frac{\partial f_{u_x}}{\partial x} + \frac{\partial f_{u_y}}{\partial y}\right) = 0 \qquad (1.20.20)$$

plays the role of the Euler equation. Here the subscripts u_x and u_y denote partial differentiation with respect to these quantities as independent variables. The operations $\partial/\partial x$ and $\partial/\partial y$, on the other hand, are *complete* partial derivatives where all the arguments of f (i.e., u, u_x, u_y) are regarded as functions of x and y and the chain rule is applied. Natural boundary conditions can be stated as

$$\left(f_{u_x} n_x + f_{u_y} n_y\right)\big|_\Gamma = 0, \qquad (1.20.21)$$

where Γ is the boundary of S.

Note that at each point of the boundary there is exactly one natural boundary condition; this was the case for the earlier simple problem for $f(y)$. When we use minimum energy principles to set up mechanics problems, the natural conditions represent, as a rule, force conditions on the boundary.

Many other functionals can be found in textbooks along with sufficient conditions for a function to be a minimum point. But we know enough about the calculus of variations for our immediate purposes.

Exercise 1.20.1. *Derive the natural boundary conditions for $F(\mathbf{y})$.*

Exercise 1.20.2. *Derive the component-form Euler equations for a functional containing derivative terms up to $\mathbf{y}^{(n)}$. How many boundary conditions should be given?*

Example

Consider a simple equilibrium problem for a bar (a structure that will be considered in more detail later) of length l stretched by both a distributed load $t(x)$ and forces F_0 and F_1 applied to its ends (Fig. 1.3). We first consider the case of a "free" bar under these forces, which means the bar is not clamped at any point. We will encounter similar equations, and even boundary conditions, in the equilibrium problem for a string.

The linear model assumes that during deformation the external forces do not change and are applied at the same points in space. This will be the case for all linear models in this book.

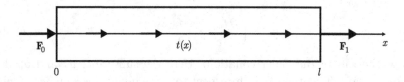

Fig. 1.3 A bar under axial loading.

When the bar is rigid and there are no geometrical constraints on its motion, the condition for equilibrium is that the resultant force must vanish:

$$\int_0^l t(x)\,dx + F_0 + F_1 = 0. \quad (1.20.22)$$

Let us see what happens when the bar is linearly elastic. We will frequently use the *minimum total energy principle*. For the present problem, this states that the equilibrium of the bar is reached at the point (i.e., the set of values of the displacement function $u(x)$) where the functional

$$\mathcal{E}_b(u) = \frac{1}{2}\int_0^l ES[u'(x)]^2\,dx - \int_0^l t(x)u(x)\,dx - F_0 u(0) - F_1 u(l)$$

$$(1.20.23)$$

takes its minimum on the set of all sufficiently smooth functions $u(x)$. In deriving the functional we use Hooke's law, which relates the tension σ in a cross section of the bar with the strain $\varepsilon = u'(x)$. So the force F in the cross section due to deformation is $F = ESu'(x)$, where E is *Young's modulus* and S is the cross-sectional area.

Note that $\mathcal{E}_b(u)$ contains the non-integrated terms $F_0 u(0) + F_1 u(l)$. This means we cannot simply use the above formulas, but must repeat the steps that led to the necessary conditions for a minimum. We should arrive at an equilibrium equation and a set of natural boundary conditions.

So we assume a state of equilibrium described by the displacement function $u(x)$. First we consider how the *self-balance condition* (1.20.22) appears in the equilibrium problem for the bar. We fix an arbitrary $u(x)$ and consider $\mathcal{E}_b(u)$ over the set of functions $u(x) + c$ where c is an arbitrary constant representing the strain energy of the bar. The first term of $\mathcal{E}_b(u)$, the strain energy for the displacement $u(x) + c$, does not depend on c:

$$\frac{1}{2}\int_0^l ES[(u(x)+c)']^2\,dx = \frac{1}{2}\int_0^l ESu'^2(x)\,dx.$$

Because c is arbitrary, we can get any large negative value for the quantity

$$\int_0^l t(x)[u(x)+c]\,dx + F_0[u(0)+c] + F_1[u(l)+c],$$

which is the work of external forces over $u(x) + c$. So the minimization problem makes sense only if the coefficient of c vanishes:

$$\int_0^l t(x)\,dx + F_0 + F_1 = 0.$$

Hence (1.20.22) is a necessary condition for the existence of a minimum of $\mathcal{E}_b(u)$. This equation for the elastic bar matches that for the rigid bar, confirming that the elastic model inherits the properties of the more elementary rigid model. We will see that not all elastic models (that of the membrane, for example) preserve all of the equilibrium conditions for the corresponding free rigid objects.

The requirement for external force balance can be explained in another way, if we apply the forces exactly as in Fig. 1.3; the resultant is along the x-axis. Clearly the body should move in the same direction. But we neglected the mass of the bar ($m = 0$). Consequently the bar should experience infinite acceleration, since its mass is zero but the resultant force is not. We could prevent this only by assuming the resultant force is zero. This happens for the equilibrium problems for all the free structures we will consider.

Thus we have found the force balance condition to be necessary for equilibrium. Assuming this, let us continue. According to the above theory, we implement the equation

$$\frac{d\mathcal{E}_b(u + \lambda\varphi)}{d\lambda}\bigg|_{\lambda=0} = 0$$

for all sufficiently smooth $\varphi(x)$. Appropriate calculations yield

$$\int_0^l ESu'(x)\varphi'(x)\,dx - \int_0^l t(x)\varphi(x)\,dx - F_0\varphi(0) - F_1\varphi(l) = 0.$$

Integrating by parts in the first integral, we obtain

$$-\int_0^l [(ESu'(x))' + t(x)]\,\varphi(x)\,dx$$
$$+ [ESu'(l) - F_1]\,\varphi(l) + [-ESu'(0) - F_0]\,\varphi(0) = 0. \qquad (1.20.24)$$

Since this holds for all smooth $\varphi(x)$, it holds in particular for those that satisfy $\varphi(0) = 0 = \varphi(l)$. For these we get

$$-\int_0^l [(ESu'(x))' + t(x)]\,\varphi(x)\,dx = 0.$$

By Lemma 1.20.1 then, we have the equilibrium equation

$$(ESu'(x))' + t(x) = 0. \qquad (1.20.25)$$

This is the Euler equation for the functional. Substituting it into (1.20.24) we obtain

$$[ESu'(l) - F_1]\,\varphi(l) + [-ESu'(0) - F_0]\,\varphi(0) = 0,$$

where φ is arbitrary. By this we get two natural boundary conditions:

$$ESu'(l) = F_1, \qquad ESu'(0) = -F_0. \tag{1.20.26}$$

These have a clear mechanical sense: the external forces at the endpoints equal the tension forces at those same points. The negative sign in the second condition stands in accord with the algebraic sign rule for the strength of materials. We derived it without reference to that rule (cf., § 3.2).

Again, we have derived exactly two natural boundary conditions. What if we fix the end at $x = 0$? We should repeat all the above, requiring $\varphi(x) = 0$ at $x = 0$. The derivation would yield the equilibrium (Euler) equation for $0 < x < l$ and the right-end natural condition $ESu'(l) = F_1$. There would still be precisely one condition for each endpoint, as expected, since the Euler equation is of second order.

Mechanical considerations are used to explain why an equilibrium solution for a free bar exists only for special choices of the external loads. In general, they can throw light on the physical origins of many conditions that seem to arise artificially in pure mathematics. Newton said that it is useful to solve problems, meaning the problems of real life.

Exercise 1.20.3. *Prove that if a solution of the Euler equation with natural boundary conditions exists, it minimizes the functional $\mathcal{E}_b(u)$.*

1.21 Hamilton's Principle

The pioneers of modern mechanics were sure that the universe was created in the most economical fashion and that all processes occur in an optimal manner. The existence of optimum principles was part of the ideology of Metaphysics and their manifestations were everywhere sought. While it is not our goal to discuss all the extremal principles of mechanics, we should touch on one closely related to the calculus of variations. This is *Hamilton's principle of stationary action*, which can be regarded as a simple consequence of the equations derived in § 1.20. On the other hand, it can also be regarded as the basis from which some portion of classical mechanics can be developed. It has many restrictive assumptions and the interested reader can pursue these in various textbooks.

This principle allows us to obtain the Lagrangian equations by seeking the stationary point of a functional called the *action*. The action $\mathcal{W}$ is given by

$$\mathcal{W} = \int_{t_1}^{t_2} \mathcal{L}\, dt, \tag{1.21.1}$$

where $\mathcal{L}$ is the Lagrangian of a system of particles; here we regard the integration variable t as time and its limits as arbitrary but temporarily fixed instants. It should be emphasized that $\mathcal{W}$ is required to have a *stationary value* (not necessarily a minimum), which means that its first variation must vanish and the Euler–Lagrange equations must hold. To exclude any additional conditions at the endpoints of the time interval, we consider only motions in which all particles in the system start and finish at the positions taken by the real particles in the actual motion. So the trajectories of admissible motions in the space of parameters q_j, given by the functions $q_1(t), \ldots, q_n(t)$ on the segment $[t_0, t_1]$, must start at $(q_1(a), \ldots, q_n(a))$ and finish at $(q_1(b), \ldots, q_n(b))$. We also take them to be sufficiently smooth as usual.

Let us compose the Euler–Lagrange equations for the problem of minimum of the functional

$$\int_{t_0}^{t_1} \mathcal{L}(q_1, \ldots, q_n, \dot{q}_1, \ldots, \dot{q}_n, t)\, dt$$

under the stated assumptions. By (1.20.15) we obtain

$$\frac{d}{dt}\left(\frac{\partial \mathcal{L}}{\partial \dot{q}_j}\right) - \frac{\partial \mathcal{L}}{\partial q_j} = 0 \quad (j = 1, \ldots, n),$$

which coincides with (1.19.14). We may now state

Hamilton's stationary action principle. *Among all trajectories that start and finish along with the real trajectory, the actual trajectory yields a stationary value for the action functional $\mathcal{W}$.*

For an ordinary function, the fact that the derivative vanishes at some point does not mean the function takes a minimum value there. A similar statement can be made for a variational problem. A real trajectory is not necessarily a point of maximum or minimum of the functional. Hamilton's principle shows that a real trajectory of the system is one of its extremals (i.e., satisfies (1.19.14)).

We conclude this section by mentioning the "variational principles of mechanics." The calculus of variations deals with the minimization and maximization of functionals. The derivation of necessary conditions of minimum leads to equations similar to (1.20.10), which contain admissible perturbations of the unknown functions. These functionals are linear with respect to the variations. The use of various versions of the Main Lemma yields differential equations, as a consequence.

In mechanics, certain equations can be obtained as Euler–Lagrange equations of functionals; however, the variational problems are only to find stationary points of a functional, not necessarily minima or maxima. Moreover, in mechanics there are integro-differential equations that resemble the equality of the first variation of a functional to zero but such that the expressions are not the first variation of any functional. Nonetheless, from such equations we can still use the Main Lemma to derive mechanically meaningful differential equations. They fall under the heading of the *variational principles of mechanics*. An example is the virtual work principle as applied to non-elastic bodies. Such "variational principles" are widely used in the generalized setup of boundary value problems, and in the construction of numerical solution methods.

1.22 Energy Conservation Revisited

The notion of energy plays a central role in science. In "physics" books whose contents are mainly mathematical, we find discussions of energy and its transformations — largely offered as illustrations of how mathematical tools can generate physically meaningful relations. In this book we take energy as a central quantity. We omit many important portions of classical mechanics and consider only those that relate to the contents of the book. Not surprisingly, the idea of energy turns out to be extremely fruitful in the mathematical analysis of mechanical problems. We therefore return to the main principle of physics: that of energy conservation. Let us examine, in general form, the equations that give rise to the notion of energy and to its all-important conservation law. The reader is asked to work the following preparatory exercise.

Exercise 1.22.1. *We say that a function $F = F(x_1, \ldots, x_n)$ is homogeneous of degree r if there is a constant r such that $F(cx_1, \ldots, cx_n) = c^r F(x_1, \ldots, x_n)$ whenever $c > 0$. Euler's theorem states that if F is differentiable and homogeneous of degree r, then*

$$\sum_{k=1}^{n} x_k \frac{\partial F}{\partial x_k} = rF.$$

Show that $g(x,y) = x^2 + 2xy + 3y^2$ is homogeneous of degree 2 and verify Euler's theorem for this function.

We would like to derive the law of energy conservation for a system of material particles having n degrees of freedom. Let us consider the relatively simple case of a system under stationary *holonomic constraints*; this means the constraint equations do not depend explicitly on time t. So we can express the position vector of a particle as a function of the variables $q_1, \ldots, q_n$:

$$\mathbf{r}_i = \mathbf{r}_i(q_1, \ldots, q_n) \qquad (i = 1, \ldots, n).$$

We recall that the kinetic energy $\mathcal{E}$ of a system of r particles having masses m_i and position vectors $\mathbf{r}_i$ is

$$\mathcal{E} = \sum_{i=1}^{r} \frac{1}{2} m_i \dot{\mathbf{r}}_i^2.$$

Substituting

$$\dot{\mathbf{r}}_i = \sum_{j=1}^{n} \frac{\partial \mathbf{r}_i}{\partial q_j} \dot{q}_j$$

into $\mathcal{E}$, we find that $\mathcal{E}$ is a quadratic form with respect to the variables $\dot{q}_j$ and having coefficients that depend only on the variables q_k. That is,

$$\mathcal{E} = \frac{1}{2} \sum_{i,j=1}^{n} a_{ij} \dot{q}_i \dot{q}_j$$

where $a_{ij} = a_{ij}(q_1, \ldots, q_n)$. The time derivative of $\mathcal{E}$ is

$$\frac{d\mathcal{E}}{dt} = \sum_{i=1}^{n} \left(\frac{\partial \mathcal{E}}{\partial \dot{q}_i} \ddot{q}_i + \frac{\partial \mathcal{E}}{\partial q_i} \dot{q}_i \right)$$

$$= \frac{d}{dt} \left(\sum_{i=1}^{n} \frac{\partial \mathcal{E}}{\partial \dot{q}_i} \dot{q}_i \right) - \sum_{i=1}^{n} \left(\frac{d}{dt} \frac{\partial \mathcal{E}}{\partial \dot{q}_i} - \frac{\partial \mathcal{E}}{\partial q_i} \right).$$

Since $\mathcal{E}$ is homogeneous of degree two with respect to the $\dot{q}_i$, Euler's theorem yields

$$\frac{\partial \mathcal{E}}{\partial \dot{q}_i} \dot{q}_i = 2\mathcal{E}.$$

Substituting this, and using the Lagrange equations (1.19.11) for the second term

$$\frac{d}{dt}\left(\frac{\partial \mathcal{E}}{\partial \dot{q}_j} \right) - \frac{\partial \mathcal{E}}{\partial q_j} = -\frac{\partial \mathcal{V}}{\partial q_j},$$

we get
$$\frac{d\mathcal{E}}{dt} = \frac{d(2\mathcal{E})}{dt} + \sum_{i=1}^{n} \frac{\partial \mathcal{V}}{\partial q_i} \dot{q}_i.$$

If $\mathcal{V}$ does not depend on t explicitly, then
$$\frac{d\mathcal{V}}{dt} = \sum_{i=1}^{n} \frac{\partial \mathcal{V}}{\partial q_i} \dot{q}_i$$

and hence
$$\frac{d}{dt}(\mathcal{E} + \mathcal{V}) = 0. \qquad (1.22.1)$$

So $\mathcal{E} + \mathcal{V}$ is time-independent, which is a statement of energy conservation. Again, however, we proceeded under certain assumptions: (1) $\mathcal{E}$ is a homogeneous quadratic form with respect to the $\dot{q}_i$, having coefficients that do not depend explicitly on time t; (2) the forces are potential, and the potential does not depend explicitly on t either. A system satisfying these assumptions is said to be *conservative*. Thus we have established that

> For a conservative system of particles, $\mathcal{E} + \mathcal{V}$ the sum of kinetic energy and the potential function is conserved over time.

The function $\mathcal{V}$ is often called the potential energy. This is reasonable in view of its role in the sum above. But there is a deeper reason for this name, which we will understand through consideration of the following problem.

In § 1.17 we mentioned energy conservation for a point mass m projected vertically upward through the gravitational field. The result is
$$\frac{mv^2}{2} + mgh = \frac{mv_0^2}{2} + mgh_0,$$

where v and h are the vertical velocity and height of the body, respectively, and v_0, h_0 are their initial values. Here $\mathcal{V} = mgy$; taking the $+y$-direction upwards, we see that $\partial \mathcal{V}/\partial y = -(-mg)$ where $-mg$ is the weight force acting on the body. We recall that mgh is sometimes called the potential energy of the particle. It is, however, related to the work of the gravitational force and the expression $mg(h - h_0)$ is the corresponding change in the kinetic energy of the Earth. Despite the fact that this results in a negligible effect on the Earth's motion, the problem really does involve two bodies in principle. Here the Earth is something like an infinite source of produced work whose state does not change because of the work done. So the term "potential energy" is a convenient way to talk about the change in kinetic energy of the other bodies acting on the system under consideration, without explicitly mentioning those bodies or their states.

Chapter 2

Simple Elastic Models

2.1 Introduction

Theoretical mechanics deals with particles and rigid bodies. Its role in continuum mechanics is similar to that of arithmetic in calculus: certain statements are accepted as axioms and various consequences are derived.

Unfortunately, we cannot directly transfer the laws of rigid bodies to deformable bodies. The forces that restrict the relative displacements between points of a deformable body (and which, together with the inertial forces in dynamics, must be in balance) are of a more complex nature. We will see that on a macroscopic level they must be described using a stress tensor. Despite this, we shall continue to refer to them as *internal forces*.

The equations describing a deformable body should reduce to those for a rigid body when deformation is negligible. Some models of continuum mechanics, like those of linear elasticity, satisfy this requirement completely. In other cases we encounter technical models of objects that hold only under additional restrictions.

We will present a few classical models of continuum mechanics, beginning with the oldest and simplest: the model for a spring.

2.2 Two Main Principles of Equilibrium and Motion for Bodies in Continuum Mechanics

When we wish to apply calculus to real bodies within the framework of continuum mechanics, we should neglect atomic structure. This is the first — and possibly roughest — assumption to be regarded as an axiom. It is formulated along with the solidification principle below. Since these are not absolute and cannot be derived logically from other simpler statements, we call them "central principles" of continuum mechanics.

Continuum principle. *Continuum mechanics considers the interior structure of a deformable object to be continuous. The distribution of any material characteristic can be described using sufficiently smooth functions.*[1]

Basic continuum mechanics consists of the problems of equilibrium and motion of deformable bodies. The second principle, that of *solidification*, adopts the equations of equilibrium and motion for a rigid body as the main tool for investigating the equilibrium and motion of any part of the body at each time instant.

We can split the set of all forces acting on any selected portion of a body into

(1) internal forces, by means of which the constituents of the portion react with each other, and
(2) external forces. These can be further subdivided into

   (a) contact reactions acting across the boundary of the body,
   (b) forces due to fields (e.g., gravitational) of other bodies, and
   (c) forces due to fields of distant portions of the same body.

The last contribution is negligible in many practical problems, and as a rule is not represented in the describing equations.

We will limit ourselves to internal forces of contact type. When we isolate a portion of a body, the reaction forces from the remainder will act only across the surface of contact between the two portions.

Solidification principle. *Let a deformable body be in equilibrium. Any portion of the body can be treated as a rigid body in equilibrium under the action of the set of all external forces and internal forces applied to the points of that portion by the remainder of the body.* This means that if we "cut away" the rest of the body, the deformed portion of interest must be in equilibrium as a fictitious rigid body under the action of all external forces acting on it, as well as the reaction forces produced by the "cutaway" part.

The equations of motion are derived from the equilibrium equations via d'Alembert's principle; i.e., by formal inclusion of the inertia forces.

---

[1](a) The notion of "sufficiently smooth function" is rather fluid, depending on the situation and the tools being employed. (b) In some continuum problems the material involved is assumed at the outset to be quite irregular (e.g., porous). But in later stages, an averaging technique is employed, and the problem becomes a typical problem of continuum mechanics (i.e., with a continuous representation of material characteristics).

We would like to mention two points. First, recall that the rigid body equations rest on the assumption that all internal forces are self-balanced and do not affect the motion of the body. In the theory of elasticity, central forces are insufficient to describe the internal interactions of particles; we will have to introduce a more complex object called a stress tensor. But the introduction of this latter object is based on the same main assumption: in the state of equilibrium of any finite portion of the body, the internal stresses (playing the role of internal forces for a rigid body) are self-balanced and do not participate in the equilibrium equations (although they affect the shape of the portion of material). Of course, nothing except its subsequent success in applications can justify this assumption; it therefore has the character of an axiom.

Second, we must remember that in deriving the equations of equilibrium or motion in continuum mechanics, we use the tools of calculus. These involve working with infinitesimally small portions of bodies and limit passages which, for real bodies having atomic structure, are invalid. Hence the models of continuum mechanics *cannot be precise*, regardless of whether they are linear or nonlinear. All equations of continuum mechanics are approximate; because of this, some finite (or discrete) models may be just as accurate as the continuous versions that mathematicians regard as precise.

2.3 Equilibrium of a Spring

We begin with a simple model of a spring under load. We show the structure of the relations used to construct the model, and then discuss the energy relations used to analyze the problem based on the model.

First we should introduce a **measure of internal forces**. In this problem there is only F_{elast}, the reaction of the spring.

Next we should obtain the **equation of equilibrium**. For a spring clamped at one end under an external force F applied to the free end, we have

$$F_{\text{elast}} = F. \tag{2.3.1}$$

Next we present a **measure of deformation** of the spring. We could use the *strain* $\varepsilon = \Delta l / l$, where Δl is the change in the length l of the spring. At present, however, it will suffice to use Δl itself and for convenience we rename this x.

The equation that relates the force and deformation characteristics is called a **constitutive law**. In our case it is Hooke's law

$$F_{\text{elast}} = kx, \qquad (2.3.2)$$

where k is the spring constant. Combining (2.3.1) and (2.3.2), we obtain the **equilibrium equation in terms of the displacement**:

$$kx = F. \qquad (2.3.3)$$

The result will seldom be so elementary. Whether simple or complex, however, the equilibrium equation is the starting point for those who seek solutions of any given problem. Our present interest is not so much in explicit solutions but in their existence and uniqueness (trivial issues in the present case) and in certain qualitative properties of the model and problem.

Let us consider the spring from an **energy** standpoint. We find the additional internal energy of the spring due to elongation, understanding that this comes from the work needed to extend the spring by an additional amount x. Over an infinitesimal extension ds, the extension force does work $ks\,ds$. The total work is

$$\int_0^x ks\,ds = \frac{1}{2}kx^2. \qquad (2.3.4)$$

An equal amount of **strain energy** is stored by the spring during deformation. An extension x caused by a constant force F produces an amount of work Fx. The difference

$$\mathcal{E} = \frac{1}{2}kx^2 - Fx \qquad (2.3.5)$$

is called the **total energy** of the spring. Since

$$F = -\frac{d}{dx}(-Fx),$$

we conclude by definition of potential that $-Fx$ is the potential of a constant force F. We will find that $-Fx$ is a potential energy similar to that of a particle in the gravitational field. The total energy will play a central role in each of our problems. In the present case it can be found by integrating the difference $ks - F$ over s between 0 and x.

Remark 2.3.1. We have introduced two types of energy without worrying too much. But the careful reader may raise a concern. First, how can we obtain the additional extension ds, and hence the additional work $ks\,ds$, if the force $F = ks$ corresponds exactly to the equilibrium state? We could answer as follows. Let us increase the applied force by a very small quantity δF. This will extend the spring length by $\delta s = \delta F/k$. The work for the increased force is between $F\delta s = ks\,\delta s$ and $(F+\delta F)\,\delta s$. In terms of infinitesimals, $dF = k\,ds$, we get $ks\,ds$ for the infinitesimal increase of work just as above. The same trick is used in other situations in continuum mechanics when deriving expressions for the differentials of the density of the work and strain energy.

However, there is still a question of how to change the spring length without changing the kinetic energy. This way of loading — when dynamic effects are absent and the deformation is produced in such way that at any moment the body-system is in equilibrium — is common in continuum mechanics and thermodynamics. It would certainly be worth carrying out a more careful analysis regarding this strange assumption of the existence of changeable motion and deformation where dynamic effects can be fully neglected! □

Let us consider the formula for $\mathcal{E}$. Evidently the point of minimum of $\mathcal{E}$ gives us (2.3.3). This represents a general principle of the theory of elasticity, the **minimum total energy principle**; it allows us to derive equilibrium equations for much more complex elastic objects.

Next we do a trivial transformation that leads to nontrivial consequences. We multiply both sides of the equilibrium equation (2.3.3) by an arbitrary variable δx:

$$-kx\,\delta x + F\,\delta x = 0. \qquad (2.3.6)$$

(We typically refer to δx as the *variation* of x, but it is an ordinary variable and could just as well be denoted by z, say. It can be subject to kinematical constraints, although that is not the case here.) If we say that (2.3.6) has a solution x if it holds for all virtual displacements δx that we will regard as additional spring elongations, then it becomes equivalent to (2.3.3). But (2.3.6) is a statement of the **virtual work principle** because $F\,\delta x$ is the work of F over δx and $-kx\,\delta x$ can be regarded as the work of internal forces over that same virtual displacement. Hence, the virtual work principle could be used as a starting point for investigating the elastic model of the spring.

It is worth noting that we can formally obtain the virtual work principle from the principle of minimum total energy. The differential of $\mathcal{E}$, which must be zero at the point of minimum, is

$$d\mathcal{E} = kx\,dx - F\,dx = 0.$$

This yields (2.3.6) up to the notation $dx = \delta x$.

Although the virtual work principle appears to be a consequence of the minimum total energy principle, it is actually more general and can be regarded as independent. For example the equation of the virtual work principle for a spring, which is

$$-F_{\text{elast}}\delta x + F\delta x = 0, \qquad (2.3.7)$$

holds even if the spring is inelastic. In this case the internal energy cannot be defined with formulas depending only on x; the same holds for viscoelastic materials, for which F_{elast} depends on the history of deformation.

We have presented the structure of the main relations that underpin the mathematical study of equilibrium problems in elasticity. In the next section we will show how to use the VWP equation or the minimum total energy principle to introduce a generalized setup of an equilibrium problem for an elastic model and to investigate existence and uniqueness of generalized solutions. Since the model of a spring is differential, we need a more complex model to demonstrate that.

Now we wish to consider the dynamical problem for a massless spring. Suppose a point mass m is attached to the free end. Let us include dynamical effects in the spring model. We apply both F and the elastic reaction to the mass. By **d'Alembert principle** we should add the inertial force $-md^2x/dt^2$ to the equilibrium equation for the mass m; hence the equation of motion for the mass is

$$m\frac{d^2x}{dt^2} + kx - F = 0. \qquad (2.3.8)$$

This holds at all times during the motion.

Next, we wish to obtain the energy relations for this spring-mass model when F is constant. Let us multiply by $dx = (dx/dt)\,dt$ and integrate over an arbitrary time interval $[t_0, t_1]$:

$$\int_{t_0}^{t_1} m\frac{d^2x}{dt^2}\frac{dx}{dt}\,dt + \int_{t_0}^{t_1} kx\frac{dx}{dt}\,dt - \int_{t_0}^{t_1} F\frac{dx}{dt}\,dt = 0.$$

Using

$$m\frac{d^2x}{dt^2}\frac{dx}{dt} = \frac{m}{2}\frac{d}{dt}\left(\frac{dx}{dt}\right)^2, \qquad kx\frac{dx}{dt} = \frac{k}{2}\frac{d}{dt}x^2,$$

and integrating, we get

$$\left.\frac{m}{2}\left(\frac{dx}{dt}\right)^2\right|_{t_0}^{t_1} + \left.\frac{kx^2}{2}\right|_{t_0}^{t_1} - \left.Fx\right|_{t_0}^{t_1} = 0.$$

We can rewrite this as
$$\left[\frac{m}{2}\left(\frac{dx}{dt}\right)^2 + \frac{kx^2}{2} - Fx\right]\bigg|_{t=t_1} = \left[\frac{m}{2}\left(\frac{dx}{dt}\right)^2 + \frac{kx^2}{2} - Fx\right]\bigg|_{t=t_0}.$$
Because t_0 and t_1 are arbitrary, the quantity
$$\frac{m}{2}\left(\frac{dx}{dt}\right)^2 + \frac{kx^2}{2} - Fx$$
stays constant during the motion. The first term is the kinetic energy of m, while the remaining terms constitute the total (i.e., strain plus potential) energy of the spring-force system. So we have derived the law of energy conservation for the spring-mass-force system (valid only for constant F) and introduced the total energy that is related to the static problem. The situation will be similar for each elastic model we examine.

In the statics problems treated in this book, the total energy will be the sum of the strain energy and the potential energy $\mathcal{V}$. In this book $-\mathcal{V}$ will be the work of all the external forces over the field of displacements. The principle of minimum total energy will play a central role.

2.4 Equilibrium of a String

Any concrete model in continuum mechanics rests on the main principles and additional assumptions that reflect real properties of the object under consideration. Often the assumptions contradict more exact theories — or even each other — but the result is always an approximation having a certain degree of accuracy under given circumstances. When the circumstances change, the model must often change.

The vibrating string equation has been known for centuries. Violin designers do not use it in their work, since the sound of a violin depends on more than just the motion of its strings. Nevertheless, the equation provides a good model for string deflections, and we shall find it useful in our study of continuum mechanics.

First we examine the main assumptions behind the derivation of the string equation. We shall eventually examine the setup of a corresponding boundary value problem and show how functional analytic tools are used. The problems of equilibrium and oscillation of a string allow us to consider the main principles of continuum mechanics in a relatively easy setting.

Let us derive the simplest equilibrium equation for an initially straight string, with fixed ends, under load. Classically, the string is assumed to be tightly stretched in advance. The additional assumptions of the classical theory are as follows.

(1) The string is represented as a curve in the plane (Fig. 2.1). Its change in shape under a normally-directed load $f = f(x)$ is described by a function $w = w(x)$ giving the deflection from the initially horizontal state.
(2) The angle of inclination of the string under load, represented by the derivative $w_x(x)$, is sufficiently small: squared and higher degrees of $w_x(x)$ are negligible in comparison with linear terms. (Note that w_x is dimensionless, so comparison of the degrees makes sense even if we consider variables having units.)
(3) The string is absolutely flexible: at any cross section, the reaction force of one part of the string on another — the tension force T_0 — is tangent to the curve $w = w(x)$ at the point x representing the string after deformation.
(4) The tension T_0 remains unchanged under any normal load distribution.

By Assumption (2), the length of a piece $[a, b]$ of the string after deformation is

$$l = \int_a^b \sqrt{1 + w_x^2}\, dx. \tag{2.4.1}$$

Because w_x is small,

$$b - a \le l \le (b - a)\left(1 + \frac{1}{2} \max_{x \in [a,b]} w_x^2\right). \tag{2.4.2}$$

So if we take the length of any part $[a, b]$ to be unchanged during deformation — i.e., if we assume that $l = b - a$ — then the error we introduce is of the second order of smallness with respect to $\max |w_x|$. But we assumed we can neglect such terms. A consequence of this is Assumption (4): if we accept Hooke's law for additional elongation of the string, the constancy of the length of each piece implies the constancy of the tension force T_0. So why do we refer to (4) as an assumption? The answer is that the more exact theory of a stretched string is essentially nonlinear, and in order to reach such conclusions we would have to study a nonlinear problem. Such an undertaking is not our goal here. So we consider the string to be such that the length of any of its parts is unchangeable; hence our only measure of deformation is the inclination angle, and the theory is simplified a priori.

Regarding the string as a continuous curve, we have utilized the continuum principle of § 2.2. We will use the solidification principle to obtain the

Simple Elastic Models

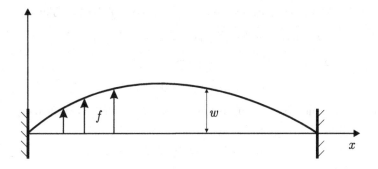

Fig. 2.1 Deflection of a loaded string.

equilibrium equations. Consider a small portion of the string $[x, x + \Delta x]$, with reactions and external forces as shown in Fig. 2.2. We denote the density of the distributed vertical load by $f(x)$.

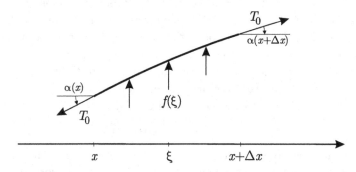

Fig. 2.2 Derivation of the string equation.

For the planar equilibrium problem we need three equations: two for the force components and one for the moment. We begin by projecting forces onto the vertical axis. Figure 2.2 shows that the tension reactions from the right and left are, respectively, $T_0 \sin \alpha(x + \Delta x)$ and $-T_0 \sin \alpha(x)$, where $\alpha(x)$ is the angle at x between the initial x-direction of the string and its tangent for the loaded state. We take the external load to be continuously distributed so that $f(x)$ is continuous. That load should be calculated as the integral

$$\int_x^{x+\Delta x} f(\xi)\, d\xi,$$

so we can use the mean value theorem and obtain $f(\xi_1)\Delta x$. Thus

$$T_0 \sin \alpha(x + \Delta x) - T_0 \sin \alpha(x) + f(\xi_1)\Delta x = 0.$$

Dividing by Δx and letting $\Delta x \to 0$, we get

$$T_0 \alpha'(x) \cos \alpha(x) + f(x) = 0.$$

From this and the above assumptions, it follows that the classical equation of equilibrium for a string is

$$T_0 w''(x) + f(x) = 0. \tag{2.4.3}$$

In particular, this follows from the equivalence of infinitesimally small quantities when suppressing the terms of a higher order of smallness with respect to $w'(x)$:

$$\sin \alpha(x) \sim \alpha(x) \sim \tan \alpha(x) = w'(x), \quad \cos \alpha(x) \sim 1.$$

Equation (2.4.3) can be found in textbooks on mathematical physics and applied mathematics.[2] At this point the typical textbook stops. We, however, are interested in all aspects of the derivation, including subtleties like validity of the hypotheses, potential contradictions, etc. So we pose a reasonable question. There should be two other equilibrium equations for the same portion of the string. Where are they? Are they trivial, or are there substantial issues that require examination?

Let us project the forces onto the x-direction:

$$T_0 \cos \alpha(x + \Delta x) - T_0 \cos \alpha(x) = 0.$$

Textbooks reasonably ignore this equation; because $\alpha(x)$ is small, it seems to involve only quantities of second order in α. But let us proceed further. Dividing by Δx and letting $\Delta x \to 0$, we have

$$[T_0 \sin \alpha(x)]\, \alpha'(x) = 0.$$

It follows that $\alpha(x)$ must be constant. What has gone wrong? The answer lies in our assumptions. We have now derived an equation in which the quantities are of a higher order of smallness, but such quantities are suppressed by the two main assumptions for the string: that of "inextensibility" and the constancy of T_0. Assuming $T_0 = T_0(x)$ depends on x, we get another equilibrium equation:

$$T_0(x + \Delta x) \cos \alpha(x + \Delta x) - T_0(x) \cos \alpha(x) = 0,$$

---

[2] In textbooks they normally derive dynamical equations at once, which can be obtained from this one via d'Alembert's principle. Here we could replace $f(x)$ by $f - \rho w_{tt}$ and get the wave equation.

leading to

$$[T_0(x)\cos\alpha(x)]' = 0.$$

From this we can find $T_0(x)$. If we invoke Hooke's law or a nonlinear constitutive relation for this quantity, we will find that points of the string can move horizontally as well as vertically, and that T_0 can be expressed in terms of the horizontal (longitudinal) displacement $u(x)$.

We have found two equations governing w and u, but have not considered the moment equation for equilibrium. Is it trivial? Actually no; the reader should derive it and observe that it also involves quantities of a higher order of smallness. So the solidification principle yields a third nontrivial equation for the same two variables w, u. What is the source of this contradiction? It is the assumption that the string is absolutely flexible. A real string offers some relatively small bending resistance and also has thickness; consideration of this would give us a third unknown variable in our set of three equations. It can also appear in the other equations, depending on how it is introduced, so all three equations may have to be rewritten. Some readers may recognize our procedure as the reverse of that used to derive the equations for a beam in the strength of materials, which will be considered in Chapter 3. In that theory we examine the linear problem of beam bending. However, with our present approach it is possible to get a nonlinear boundary value problem for the ordinary differential equations that describe bending of a prestretched beam.

We often hear the term "elastic" applied to a string modeled this way. For many students, this term implies the use of Hooke's law. But we are not using this law here. In fact, the term "elastic" is fairly difficult to define precisely. It is somewhat naive to say that it describes solids for which (at least for simple problems like extension of a rod) the deformation does not depend on the history of how the object was loaded. This description also applies to the membrane idea, which itself applies to certain liquid films (e.g., soap films). The classification of various materials is actually a rather delicate problem.

For a string, however, we really can derive the equations from nonlinear elasticity. We assume some constant initial stretching. Then we linearize the equations with respect to the additional deformation of the string due to a normal load, which is assumed small. Clearly we should place some assumptions on the order of various terms such that the dependence between the strains and stresses disappears from the equilibrium equation. So the ordinary theory of the string results from linearization of the more general equations of nonlinear elasticity. In fact, the theory of linear elasticity is sometimes referred to as "linearized elasticity". This is because in large part we do not know the initial state of the body before loading; it can be strained. We can find deformations that result only from the action of forces. In this sense, the situation is worse than that with the string; we know nothing (or almost nothing) about the initial state of the body and must find the additional deformation.

An additional comment. In deriving the equilibrium equation we tacitly assumed the reactions are contact forces only. All other models in this book are treated similarly. This is the usual assumption in continuum mechanics, although other approaches do exist.

Further digging into unstated hypotheses would uncover other interesting issues. One is that the point to which the load $f(x)$ was initially related has moved. This means that the force should be changed somehow to reflect not only this fact, but also a change in density, since the length of the elementary part of the string also changes. However, all these amount to a change of the second order of smallness.

We see that even with this simple mechanical object, we could pursue numerous extensions of the model. In reality we must stop somewhere. For practical purposes a crude theory often works as well as a "more accurate" one, since something important may be missing from the latter theory as well. Because the classical string model given by (2.4.3) suffices for many purposes, we shall continue to discuss this equation in the context of how to formulate a good boundary value problem for equilibrium of the string.

2.5 Equilibrium Boundary Value Problems for a String

We have derived a simple linear equation governing the equilibrium of a string:

$$T_0 w''(x) + f(x) = 0. \qquad (2.5.1)$$

This is a familiar second-order differential equation. It is easily integrated, and the general solution involves two free constants which, in the textbooks, are fixed by two initial conditions. But our task is to consider a real object rather than an abstract equation. So let us return to the initial problem after a few general comments.

It is worth noting that we will deal with a model of the string and not the string itself. The model was based on observations of real strings; however, after a model has been formulated, it will often take on a life of its own. We have imposed some restrictive assumptions during the derivation above — requiring, for example, the inclination angle to be "infinitesimally small" everywhere. What does this mean practically? After all, the angles we obtain from a numerical solution are always finite. Should we restrict them? If so, how? Would a restriction of 0.01 or 0.0000001 be enough? From the infinitesimal viewpoint, both of these numbers are "large". There are sticky issues here, but it is common in mathematics to simply forget about the original assumption and proceed according to the usual approaches of one's field. An engineer may avoid certain loads that he knows will cause the string to leave its elastic regime or even snap. A mathematician may choose to study the equation under conditions that violate all assumptions made in deriving it. The results of such a study can be useful for engineers, revealing qualitative properties and bounds of applicability of the model. Through direct empirical activity, of course, engineers also add much to our understanding of a model's range of applicability.

The last general point: any model is approximate. Although designed using ideal assumptions that are exact, its parameters can be found only approximately; this means that even exact solutions found on the basis of the model will be approximate. However, what is their accuracy? We could consider the models of physics as perturbations of ideal models (such models do not exist, of course, but there are certainly more accurate models that we can regard as "exact"), so our question concerns the dependence of ideal solutions on those perturbations and the difference between an ideal solution and our approximate one. As one of the first steps in studying the accuracy of our solution with respect to real objects, we can pose a problem of the dependence of solutions on parameters of the object, and thus of the model.

Many mathematics books examine the dependence of functions, or problem solutions, on parameters. They emphasize results pertaining to discontinuity, irregularities like branching solutions, etc. This is because continuity is relatively simple, while discontinuities can arise in various ways. But the number of discontinuity-related results seems disproportionate to real life: familiar objects seldom fall apart because of tiny perturbations in their shapes or material properties. We are fortunate that things are this way, and that small errors during production of objects seldom lead to catastrophe (by which we mean the irregular behavior of an object when, for small reasons, we see large deviations in behavior). So, in large part, engineering models display continuity of the output with respect to the input parameters; in other words, their solutions depend continuously on the parameters of the object. However, irregularities in the dependence of solutions on parameters for real objects are of great interest: some do lead to catastrophic failures. Moreover, we sometimes exploit them in the design of objects. So the relevant theorems are important.

Let us return to our simple problem. Our derivation assumed that the string carries a significant tension T_0. Hence we must fix the ends of the string in the horizontal direction. Since we should also restrict them in the vertical direction, we are led to consider a boundary value problem.

Let the string correspond to the interval $(0,l)$. Assume one end is fixed vertically:

$$w(0) = a. \qquad (2.5.2)$$

A similar condition $w(l) = b$ can be placed on the other end. These define a solution uniquely. Of course, when $b \neq a$ we find that $|w'|$ is at least $|b-a|/l$ at some points, hence engineers should watch carefully for the failure of assumptions. If we take into account the change in tension due to this clamping (which can be done by recalculating T_0 through the use of Hooke's or other constitutive laws) then only values in w' additional to $(b-a)/l$ play a role in the smallness assumptions. So the same equation can be used when $(b-a)/l$ is not small.

We can also "clamp" an end in such a way that free vertical motion is possible, say by a frictionless button that moves inside a channel (Fig. 2.3). We must express this as a boundary condition for the string equation. Let us consider the condition for equilibrium of the button. Here the only active force is the tension T_0; this has a horizontal component balanced by the reaction of the channel walls, and a vertical component $-T_0 w'$ (approximately, since the sine of the angle of inclination of the string is approximately equal to w', and in this problem we must use this as an exact relation). So

$$T_0 w' = 0 \quad \text{at } x = l, \qquad (2.5.3)$$

and thus

$$w' = 0 \quad \text{at } x = l. \qquad (2.5.4)$$

Geometrically, this end of the string is perpendicular to the clamping line AB.

We can suppose an external vertical force P acts on the button. This yields another approximate condition of equilibrium for the vertical com-

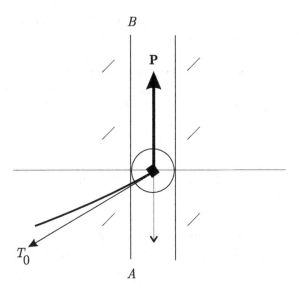

Fig. 2.3 Use of a button sliding in a channel to "fix" a string end without impeding its vertical deflection.

ponents of the forces at the point $x = l$:

$$T_0 w' = P \quad \text{at } x = l. \tag{2.5.5}$$

Finally, suppose the last mode of end clamping is realized at both ends. In mathematical physics these boundary conditions would be called *Neumann conditions*; in mechanics, we say the problem is posed "without geometrical constraints on the boundary" (or, in the theory of elasticity, as a problem "with force conditions on the boundary"). Vertical motion is not geometrically restricted. Since, in equilibrium, the string should satisfy the equilibrium equations for a rigid body, the forces acting on the string should be *self-balanced*; otherwise the string will move. So the following question is significant. In this version of the problem we neglect the mass of the string; the body (string) moves under the action of forces while having no mass. What happens? It is clear that the acceleration will be infinite if the external forces are not self-balanced, but we have not involved acceleration at all. How is this potentially bad behavior of the body reflected in the equilibrium problem? The answer is that if the external forces are not self-balanced, the boundary value problem cannot have a solution. *This holds for any equilibrium problem for an object that can move freely in some direction or rotate about an axis: the corresponding external loads must be self-balanced with respect to all possible free motions;* otherwise the statics problem will not have a solution.

For equilibrium of a planar rigid body, classical mechanics specifies three equations for self-balance of the external forces: two for the force components — vertical and horizontal — and one for the moment. In this setup the horizontal components are balanced by boundary devices, so we have two conditions. For the vertical forces we have

$$\int_0^l f(x)\,dx + P_0 + P_l = 0,$$

where P_0 and P_l are forces applied at the respective endpoints $x = 0$ and $x = l$. The moment with respect to the left end seems to be

$$\int_0^l x f(x)\,dx + l P_l = 0$$

but also appears to be unnecessary: in this model, the string resists any rotation as a rigid body.

We will consider Neumann problems for other objects. The situation with the self-balance conditions is similar for the approximate theories describing beams, plates, and shells: depending on the theory, some self-balance relations may be lost because the theoretical assumptions introduce additional properties into the model (this will happen in the membrane theory). In elasticity theory, however, the self-balance conditions of classical mechanics are fulfilled "exactly".

A final remark for this section: we could consider other boundary conditions and thereby run into fairly difficult problems even for this simple differential equation. Some would be hard to pose correctly; others would lead to complex nonlinearities. For example, we could assume nonzero friction between the button and the clamping device; the theory of friction is difficult, and the law depends on the concrete structure of this device. We could introduce forces that depend on the string deflection, etc. So even with a simple object, the variety of possible boundary value problems is significant. It becomes truly daunting if we consider a nonlinear equation for the string.

2.6 Generalized Formulation of the Equilibrium Problem for a String

We reconsider the equilibrium problem for a string with fixed ends:

$$T_0 w''(x) + f(x) = 0, \qquad x \in (0, l), \tag{2.6.1}$$

$$w(0) = 0, \quad w(l) = 0. \tag{2.6.2}$$

Let us try to forget that we can easily solve this via double integration; we seek general methods appropriate for more complex problems. The simplicity of the present problem will give us a good look at these tools without obscuring them with too much detail.

Besides differential equations and boundary conditions, a mathematical setup for a boundary value problem must delineate a class in which we will seek solutions. Since the existence or uniqueness of a solution can turn on the choice of this class, it is of obvious importance.

Historically it was common to consider only those solutions to a boundary value problem that possess as many continuous derivatives as appear in the equations of the problem. We call such a setup *classical*.[3] For the present problem, a classical solution should be a function belonging to $C^{(2)}(0, l)$; that is, it should be continuous on $[0, l]$ together with its first two derivatives, in addition to satisfying (2.6.1) and (2.6.2). Of course, to have such a solution we must require that the load density f be continuous: $f \in C(0, l)$. But we can also assume f is discontinuous. The load can contain even lumped forces. Now let us forget that we can solve the problem using simple integration. We wish to change the setup in order to imbed non-smooth solutions. We do this to demonstrate how we will handle more complex problems where direct integration is impossible. How should we alter the setup to consider such problems? The methods of the calculus of variations prompt us to use integro-differential equations instead of pure differential equations. This approach is based on Lemma 1.20.1. If the equation

$$\int_0^l F(x)\varphi(x)\,dx = 0,$$

with respect to a function $F(x)$ continuous on $[0, l]$, holds for any $\varphi \in \mathcal{D}(0, l)$, the set of functions infinitely differentiable on $(0, l)$ and vanishing

---
[3] Of course, the theory considered solutions having singular points as well. But those points were regarded as exceptional, and the solutions containing them as supplementary to the smooth solutions of the truly classical problems.

in some neighborhood of each endpoint, then $F(x) = 0$ on $[0,l]$. An examination of the proof shows that Lemma 1.20.1 still holds if $F(x)$ is only piecewise continuous; then $F(x) = 0$ at each point of continuity of $F(x)$. This suggests a way in which we could reduce the requirements on the smoothness of a solution to (2.6.1)–(2.6.2).

Supposing w is a solution to the problem (2.6.1)–(2.6.2), let us multiply (2.6.1) by $\varphi \in \mathcal{D}(0,l)$ and integrate over $[0,l]$:

$$\int_0^l [T_0 w''(x)\varphi(x) + f(x)\varphi(x)]\, dx = 0. \tag{2.6.3}$$

When $T_0 w''(x) + f(x)$ is continuous on $[0,l]$, then (2.6.3), formulated as a result that must hold for all $\varphi \in \mathcal{D}(0,l)$, is equivalent to (2.6.1). This follows from Lemma 1.20.1. Thus we can consider this new integrodifferential equation, which contains the unknown function $w(x)$, given load $f(x)$, and an arbitrary function $\varphi(x)$, as the basis for formulating the boundary value problem. It uniquely defines a smooth $w(x)$ when $f(x)$ is smooth, as for smooth $f(x)$ it is equivalent to the equation $T_0 w''(x) + f(x) = 0$. We can go further and consider the integral equation when $f(x)$ is not continuous but merely integrable. We expect $w''(x)$ to be integrable as well, and can try to reduce the smoothness requirements on the solution by saying that the integrodifferential equation defines a generalized solution. This new setup seemingly will not allow us to include lumped forces, which are common, so we will proceed.

Integrating by parts in the first term of (2.6.3) we get

$$\int_0^l [-T_0 w'(x)\varphi'(x) + f(x)\varphi(x)]\, dx = 0, \tag{2.6.4}$$

since $\varphi(0) = 0 = \varphi(l)$. Each term in (2.6.4) has a clear meaning. First,

$$\int_0^l f(x)\varphi(x)\, dx \tag{2.6.5}$$

is the work of external forces over the virtual displacement φ. Because the term

$$-\int_0^l T_0 w'(x)\varphi'(x)\, dx \tag{2.6.6}$$

arose from the expression

$$\int_0^l T_0 w''(x)\varphi(x)\, dx, \tag{2.6.7}$$

it is called the *work of internal forces* over the same virtual displacement φ. Indeed, the integral (2.6.7) shares the form of (2.6.5) with the internal force $T_0 w''(x)$ replacing the external loading function $f(x)$.

We can include lumped forces F_i at points $x = c_i$ by rewriting (2.6.4) as

$$\int_0^l [-T_0 w'(x)\varphi'(x) + f(x)\varphi(x)]\, dx + \sum_i F_i \varphi(c_i) = 0. \qquad (2.6.8)$$

We stop at this stage of generalization. A pure mathematician might continue on to the *theory of distributions*: a subject that finds applications to the theory of differential equations. However, in the class of merely integrable functions we would lose not only uniqueness of solution but the possibility of formulating boundary conditions as w would not be continuous. As we will see, this setup can be done directly if we start with the virtual work principle for the string. This is another reason we do not try to generalize further.

In what follows we will introduce *generalized solutions* to various problems of mechanics. It is, of course, desirable for a generalization to have mechanical roots as well. Our definitions will be based on the minimum total energy principle or the virtual work principle. When introducing a new setup, we should clearly understand the features of such a generalization.

A generalized solution represents an extension of a classical solution in the sense that the former can be much less smooth; however, when a classical solution of the problem exists, the generalized one must coincide with it. We typically adhere to the following guidelines when introducing generalized solutions.

(1) A smooth classical solution of a problem should be a generalized solution as well.
(2) The notion of generalized solution is related to mechanical principles and truly extends the problem setup to regimes of parameters, loads, etc., for which classical solutions do not exist.
(3) The generalized setup is well-posed; this means that a generalized solution exists for all mechanically meaningful external data, and is unique for those problems expected to have this property.

The third item holds for linear problems and clearly fails for nonlinear ones where nonuniqueness is expected.

So, according to our mechanical viewpoint on the problem, we should stop (at least) at the level of generalization given by (2.6.4).

The approach we have demonstrated is typical of pure mathematics. A problem is transformed: when the new form appears useful, another theory begins to develop. In the present case, however, we can arrive at the same equation (2.6.4) or its extension (2.6.8) via mechanical considerations. So we shall consider the principle of minimum total energy for the string.

2.7 Virtual Work Principle for a String

Let us obtain the same equilibrium equation from a mechanical viewpoint. We begin with the total energy functional for the "string–load" system:

$$\mathcal{E}(w) = \int_0^l \frac{T_0}{2} w_x^2(x)\, dx - \int_0^l f(x) w(x)\, dx. \quad (2.7.1)$$

The first term is the strain energy possessed by the string because of its transverse deflection. The second accounts for the work of external forces. To see the parallels with the elementary problems we studied earlier, we could regard this term as a kind of potential of the load, or as the potential energy of the "string–load" system. However, we will not need this viewpoint in what follows.

Suppose the string ends are fixed:

$$w(0) = w_0, \qquad w(l) = w_1. \quad (2.7.2)$$

Let us seek a point of minimum of $\mathcal{E}(w)$ and see how it relates to a solution of the equilibrium problem for the string. Denoting[4] by $\delta w(x)$ an admissible variation of w, which is a twice-differentiable function vanishing at the endpoints, we get the expression for the first variation of $\mathcal{E}(w)$. At a minimum point it should vanish:

$$\delta\mathcal{E}(w, \delta w) = \int_0^l T_0 w_x \, \delta w_x \, dx - \int_0^l f \, \delta w \, dx = 0. \quad (2.7.3)$$

Because this matches (2.6.4) for $\delta w = \varphi$, we obtain implicit confirmation that the total energy can attain a minimum value on the solution of the equilibrium problem. We will confirm this later (as a consequence of Theorem 2.13.2).

---

[4] It is traditional in mechanics to denote such a function by two symbols such as δw, although it is an ordinary function. We sometimes encounter situations in which the symbol δ is interpreted as meaning that the deviation δw, being an admissible (virtual) displacement, is small or even infinitesimal. That is why the notation is used in the mechanical literature. But the equations depend on δw linearly, hence remain valid for non-small δw as well.

We now state

Virtual work principle for a string. *In the state of equilibrium of the string, the sum of the work of internal and external forces over any admissible field of displacements δw is equal to zero.*

We see that the equilibrium equation (2.7.3), which we restate as

$$-\int_0^l T_0 w_x \, \delta w_x \, dx + \int_0^l f \, \delta w \, dx = 0, \qquad (2.7.4)$$

expresses the virtual work principle for a string with fixed ends.

Such formulations are typical in elasticity; people gloss over "small" details such as smoothness of the displacement field. It is taken to be "sufficiently smooth". But for us the issue of smoothness is crucial. We will use the virtual work principle as a basis for defining a generalized (energy) solution to the string problem, but the smoothness issue will be treated carefully.

The equilibrium problems of continuum mechanics satisfy a similar virtual work principle.[5] The reader may wonder why we have borrowed a term from classical mechanics, even though the form of this statement differs markedly from the form of the virtual work principle used there. It turns out that the difference is not so great. In classical mechanics, we see only the work of forces which, in our present terms, could be called external forces. Our "internal" forces are absent in classical mechanics because it employs a model of a rigid body in which the work of the internal forces that keep the particles of the body mutually stationary is always zero. In continuum mechanics, because of deformation, work due to small disturbances in the mutual positions of the points of a body appears, and this brings new terms into the expression for virtual work. In this case, the work of internal forces is obtained by calculating the first variation of the strain energy functional. When we come to the stress and strain tensors, we will introduce it from another viewpoint, deriving it as the work of stresses inside the body over the field of infinitesimally small virtual displacements. This approach works even if one cannot derive an expression for the strain energy functional.

We will use the above principle and similar formulations for generalized setups of corresponding problems. For additional rigor, we will need material from functional analysis; we must specify the classes in which we seek

---

[5] We have "derived" the virtual work principle from the minimum energy principle. The former is more general, however, and can be applied in problems for which a total energy functional does not exist.

solutions and introduce tools for proving solvability and uniqueness. This will brings us to the idea of energy spaces which, in large part, turn out to be subspaces of Sobolev spaces. The latter are spaces in which the norm is of integral type, involving not only the functions themselves, as in the space L^p, but also their derivatives. The main tool for proving existence–uniqueness will be Riesz's representation theorem for a linear continuous functional in a Hilbert space. These topics will be presented below, but let us further discuss the virtual work principle for a string.

We know that a solution to the boundary value problem for a string with fixed ends satisfies the virtual work principle stated above. Let us seek the conditions of minimum for the energy functional $\mathcal{E}(w)$ with no endpoint constraints. Of course, we get the same equation (2.7.4) of the virtual work principle from which, by the general theory of the calculus of variations, follows the equilibrium equation (2.6.1) and the natural boundary conditions

$$T_0 w_x(0) = 0, \quad T_0 w_x(l) = 0.$$

(Later we derive these for a more general case and therefore do not pause to treat them here.) These coincide with the equilibrium conditions for "free" ends of the string (equation (2.5.3)). This suggests that the natural boundary conditions for the string are related to the equilibrium condition. We verify this as follows.

We extend the formulation of the virtual work principle for the string, including the action of some forces at the string ends. Clearly these forces, P_0 and P_1 say, are external to the string, and their work values over a virtual displacement field $\delta w(x)$ are $P_0 \, \delta w(0)$ and $P_1 \, \delta w(l)$. We know we must supplement the total energy functional with new terms. This leads us to the following equation, which we can write directly using the above word formulation of the virtual work principle: in equilibrium, the sum of the work of internal and external forces over any virtual displacement field of the string is equal to zero:

$$-\int_0^l T_0 w_x \, \delta w_x \, dx + \int_0^l f \, \delta w \, dx + P_0 \, \delta w(0) + P_1 \, \delta w(l) = 0. \qquad (2.7.5)$$

The presence of terms outside the integrals distinguishes this from the problems we considered in the calculus of variations. However, the method of obtaining the necessary conditions that follow from the equation is practically the same. We integrate by parts on the left, and rearrange:

$$-\int_0^l T_0 w_{xx}\, \delta w_x\, dx - \int_0^l f\, \delta w\, dx + T_0 w_x(l)\, \delta w(l) - T_0 w_x(0)\, \delta w(0)$$
$$= P_0\, \delta w(0) + P_1\, \delta w(l).$$

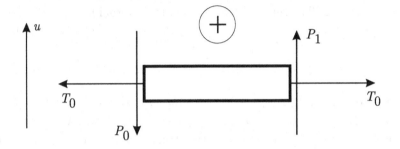

Fig. 2.4 Sign convention for reaction forces.

First we consider only those virtual displacements for which $\delta w(0) = 0$ and $\delta w(l) = 0$. For these, the problem does not differ from that for the string with fixed ends; hence the differential equation of equilibrium (2.6.1) holds. This also means that the sum of the integral terms equals zero independently of the values of δw at the ends. Consequently,

$$T_0 w_x(l)\, \delta w(l) - T_0 w_x(0)\, \delta w(0) = P_0\, \delta w(0) + P_1\, \delta w(l).$$

Taking $\delta w = x$ and $\delta w = l - x$, respectively, we get two end conditions:

$$w_x(l) = P_1/T_0, \quad w_x(0) = -P_0/T_0. \qquad (2.7.6)$$

The first coincides with the Neumann condition (2.5.5). The negative sign is a simple consequence of Newton's third law. When we use a surface to "cut away" a portion of a spatial object for study, we must replace the action of any removed portions by suitable force reactions. But a cut can be viewed from either side, and by Newton's third law the mutual reaction forces should be equal and opposite. We will see a similar sign pattern for the contact reactions in all our elastic models. So let us pause to develop a convenient formal rule for this. We must introduce a sign convention for the reaction forces. We temporarily imagine the string to have nonzero thickness. Figure 2.4 shows positive directions for forces acting at cross sections of a small portion, depending on the side of the cut. Contact forces due the string inclination are regarded as positive when they act as shown in Figure 2.4. The situation is the same with the forces P_i acting at the ends of the string; i.e., the positive directions for P_0 and P_1 at the ends are opposite. We will introduce a similar sign convention for stresses in the theories of beam bending and elasticity.

So we have obtained the natural boundary conditions (2.7.6) for a free string. These turned out to be the equilibrium conditions for the endpoints of free clamping. The self-balance equation for external forces now follows from the fact that (2.7.5) must hold, in particular, for $\delta w = 1$. So we have

$$\int_0^l f\,dx + P_0 + P_1 = 0,$$

which is the self-balance equation stated earlier. We could consider more general string problems using the virtual work principle. We could, say, place elastic supports at points along the string. These would be considered as part of a combined structure ("string + supports") and a term accounting for the work of the support reactions included in the work of "internal" forces. For example, if a spring acts on the string at point c with reaction $kw(c)$ (i.e., proportional to $w(c)$), then the virtual work principle reads

$$\int_0^l T_0 w_x\,\delta w_x\,dx + kw(c)\,\delta w(c) = \int_0^l f\,\delta w\,dx + P_0\,\delta w(0) + P_1\,\delta w(l).$$

To derive the dynamical equations we also apply d'Alembert's principle; we add the inertia forces to study motion in the presence of, say, air resistance or point masses attached to the string.

However, to consider the mathematical consequences of the virtual work principle we require additional machinery as noted above.

2.8 Riesz Representation Theorem

We first demonstrate that a continuous linear functional defined on $\mathbb{R}^n$ can be represented as a scalar product in $\mathbb{R}^n$. We know that any vector $\mathbf{x} \in \mathbb{R}^n$ can be expressed as

$$\mathbf{x} = \sum_{i=1}^n x_i \mathbf{e}_i, \qquad (2.8.1)$$

where the $\mathbf{e}_i$ are the vectors of the standard basis. This allows us to see that a given linear functional $F(\mathbf{x})$ (i.e., a linear function of the n variables

x_i with no additive constant term) can be written in the form

$$F(\mathbf{x}) = \sum_{i=1}^{n} x_i a_i \tag{2.8.2}$$

with scalars a_i that do not depend on $\mathbf{x}$. Indeed, with $a_i \equiv F(\mathbf{e}_i)$ we have

$$F(\mathbf{x}) = F\left(\sum_{i=1}^{n} x_i \mathbf{e}_i\right) = \sum_{i=1}^{n} x_i F(\mathbf{e}_i) = \sum_{i=1}^{n} x_i a_i.$$

Therefore we can write (2.8.2) as

$$F(\mathbf{x}) = \mathbf{x} \cdot \mathbf{a} \tag{2.8.3}$$

where the vector $\mathbf{a}$ is uniquely determined by $F(\mathbf{x})$. This is the aforementioned representation of $F(\mathbf{x})$ as an inner product. Note that the equation

$$\mathbf{x} \cdot \mathbf{a} = a_1 x_1 + \cdots + a_n x_n = 0$$

specifies a hyperplane through the origin in $\mathbb{R}^n$. The orientation of this hyperplane is described by its normal vector $\mathbf{a}$ with components a_i.

These ideas can be extended to a general Hilbert space H. Of course, we cannot use the technique outlined above to determine the coefficients in an infinite-dimensional setting. However, two other facts come to our aid: For a linear continuous functional F in H: (1) the *kernel* of F, which is the set of all x for which $F(x) = 0$, is a hyperplane (subspace) of the Hilbert space as was the case in $\mathbb{R}^n$ above, and (2) the orthogonal complement of the kernel in H is one-dimensional and contains a unique vector f such that F is uniquely defined through the inner product: $F(x) = (x, f)$. This was proved by F. Riesz. The *Riesz representation theorem* will play a significant role in our presentation; it will allow us to analyze linear problems of continuum mechanics by recasting them as operator equations.

Theorem 2.8.1 (Riesz representation). *Let $F(x)$ be a continuous linear functional given on a Hilbert space H. There is a unique element f in H such that*

$$F(x) = (x, f) \quad \text{for every } x \in H. \tag{2.8.4}$$

Furthermore, this element satisfies $\|f\| = \|F\|$.

Proof. Denote by M the kernel of $F(x)$: i.e., the set of all x for which $F(x) = 0$. It follows from the linearity of $F(x)$ that M is a subspace of H.

Furthermore, M is closed; if $\{m_k\} \subset M$ is convergent in H to a limit m^*, then $m^* \in M$ because

$$F(m^*) = F\left(\lim_{k\to\infty} m_k\right) = \lim_{k\to\infty} F(m_k) = 0$$

by continuity of F. Applying Theorem 1.16.2 we obtain the unique decomposition

$$H = M \dot{+} M^\perp.$$

In this case the orthogonal complement $M^\perp$ is one-dimensional. To prove this, take any two elements n_1 and n_2 from $M^\perp$ and examine the linear combination

$$n_3 = F(n_1)n_2 - F(n_2)n_1. \tag{2.8.5}$$

We know this element belongs to $M^\perp$ because $M^\perp$ is a subspace. But it also belongs to M because $F(n_3) = F(n_1)F(n_2) - F(n_2)F(n_1) = 0$. Therefore it must be the zero vector, and (2.8.5) shows that n_2 must be a scalar multiple of n_1. Because $n_2 \in M^\perp$ is arbitrary, $M^\perp$ is one-dimensional.

Now let n_0 be a normalized element of the one-dimensional subspace $M^\perp$. We can represent any $x \in H$ as

$$x = m + \alpha n_0$$

for some element m from M and some scalar α. Taking the inner product of both sides with n_0 we find that $\alpha = (x, n_0)$, hence

$$F(x) = F(m) + \alpha F(n_0) = (x, n_0)F(n_0) = (x, \overline{F(n_0)}n_0).$$

This is the desired representation (2.8.4) with $f = \overline{F(n_0)}n_0$. Uniqueness is proved by supposing the existence of two "representers" f_1 and f_2 so that

$$F(x) = (x, f_1) = (x, f_2) \quad \text{holds for all } x.$$

This yields $(x, f_1 - f_2) = 0$ for all x, and we can set $x = f_1 - f_2$ to deduce that $f_1 = f_2$.

Finally, we must establish $\|F\| = \|f\|$. We have

$$\|f\|^2 = (f, f) = F(f) \le \|F\| \|f\|$$

where the inequality holds by definition of the norm of a functional. Therefore $\|f\| \le \|F\|$. On the other hand

$$\|F\| = \sup_{\|x\|\ne 0} \frac{|F(x)|}{\|x\|} = \sup_{\|x\|\ne 0} \frac{|(x, f)|}{\|x\|} \le \sup_{\|x\|\ne 0} \frac{\|x\| \|f\|}{\|x\|} = \|f\|$$

by the Schwarz inequality. Since $\|f\| \le \|F\|$ and $\|f\| \ge \|F\|$, we have $\|f\| = \|F\|$. □

2.9 Generalized Setup of the Dirichlet Problem for a String

Let us return to equation (2.7.4),

$$\int_0^l T_0 w_x \, \delta w_x \, dx = \int_0^l f \, \delta w \, dx, \qquad (2.9.1)$$

and for simplicity fix the ends of the string as follows:

$$w(0) = 0, \qquad w(l) = 0. \qquad (2.9.2)$$

We have seen that (2.9.1), when it holds for all admissible $\delta w(x)$, is equivalent to the differential equation of equilibrium (2.6.1). So we exploit (2.9.1) for the generalized setup of the equilibrium problem.

In (2.9.1), an admissible $\delta w(x)$ can have the same smoothness as $w(x)$ and satisfy the same boundary conditions (2.9.2):

$$\delta w(0) = 0, \qquad \delta w(l) = 0. \qquad (2.9.3)$$

Thus we enjoy a full symmetry between the properties of $\delta w(x)$ and $w(x)$. When posing a problem we should specify the class of functions in which we seek a solution. If we momentarily forget the equilibrium equation and consider only (2.9.1), then all we can expect of the solution are the properties that make sense for all the terms of (2.9.1); hence we can require only some integrability of w, w_x, δw, and δw_x. With the same thing in mind, we require integrability of f.

To formulate the generalized setup of the problem we construct the energy space starting with the subspace S_0 of functions in $C^{(2)}(0, l)$ satisfying (2.9.2). Let us exploit the symmetry of the left side of (2.9.1) with respect to w and δw. We introduce

$$(w, \delta w)_S = \int_0^l T_0 w_x \, \delta w_x \, dx. \qquad (2.9.4)$$

Considering the properties of this functional when $w, \delta w \in S_0$, we see that it can act as an inner product on S_0.

Exercise 2.9.1. *Show this in detail.*

It turns out that the resulting inner product space is incomplete. The situation is unchanged if we impose a lesser requirement on the elements of S_0, that they merely have continuous first derivatives on $[0, l]$. We shall introduce generalized solutions that, in general, lack two continuous derivatives but always have finite strain energy. The class in which we will seek generalized solutions is introduced as follows.

Definition 2.9.1. The completion of the set S_0 of functions with respect to the norm

$$\|w\|_S = (w,w)_S^{1/2} \tag{2.9.5}$$

is the *energy space for the Dirichlet string problem*. We denote it by E_{SD}.

Remark 2.9.1. We will denote all energy spaces using similar notation: the first subscript will refer to the object and the second to the type of boundary conditions. Currently, S stands for "string" and D for "Dirichlet boundary conditions". A second subscript N will mean "Neumann" (or "natural") conditions. □

The quantity $\|w\|_S^2$ is the double strain energy functional for the string and is clearly defined for each element of E_{SD}. This property is mechanically meaningful. But the E_{SD} norm involves w_x and not w. Hence we do not know the properties of w, and cannot determine conditions on f in order to ensure that the term $\int_0^l f\,\delta w\,dx$ makes sense for all $\delta w \in E_{SD}$.

So we would like to consider our equilibrium problem in E_{SD}, at each point of which the strain energy is defined. We will use the minimum energy principle (or, equivalently, the virtual work principle). Again, we know nothing about the properties of w. We must also know how to treat another term in the equation: the work of external loads. Certain results, called *imbedding theorems*, are useful here: they show us that the elements of one space can be identified with those of another and specify the properties of this identification. They permit us to formulate conditions under which the work of external loads makes sense and, in addition, demonstrate that the problem of minimization of the total energy functional has a unique solution in E_{SD}. Of primary interest is a particular case of the *Sobolev theorem* regarding the equilibrium of a membrane (Theorems 2.16.2–2.16.4). Now let us consider a simple imbedding theorem for the space E_{SD}.

2.10 First Theorems of Imbedding

To understand the imbedding concept we begin with a simple example. Consider the space $C(\Omega)$ of all continuous functions $f = f(\mathbf{x})$ on a compact domain $\Omega \subset \mathbb{R}^n$. Any $f \in C^{(1)}(\Omega)$ belongs to $C(\Omega)$. It might seem that little is gained by regarding such a function first as an element of $C^{(1)}(\Omega)$ and then as an element of $C(\Omega)$. We should be careful, however, not to think of this correspondence as the result of applying a simple "identity operator" — this would only be the case if the domain and range of the operator were the same space. In fact the norms on the two spaces differ: the norm of $C(\Omega)$ does not implicate the derivative of its argument. So the operator that pairs a given function in $C^{(1)}(\Omega)$ with the same function but regarded as an element of $C(\Omega)$ (including with respect to its norm!) cannot be called an identity operator. Instead it is called the *imbedding* of $C^{(1)}(\Omega)$ into $C(\Omega)$. Let us denote it by T and write the correspondence as

$$f = Tf. \qquad (2.10.1)$$

Again, both occurrences of the symbol f stand for the same function $f(\mathbf{x})$. But — and this is crucial — the function on the right is considered as an element of $C^{(1)}(\Omega)$, and on the left as an element of $C(\Omega)$.

It is clear that T is linear. What about its other properties? According to the definitions of the norms on the two spaces,

$$\|f\|_{C(\Omega)} = \max_{\mathbf{x} \in \Omega} |f(\mathbf{x})| \qquad (2.10.2)$$

and

$$\|f\|_{C^{(1)}(\Omega)} = \max_{\mathbf{x} \in \Omega} |f(\mathbf{x})| + \sum_{k=1}^{n} \max_{\mathbf{x} \in \Omega} \left|\frac{\partial f(\mathbf{x})}{\partial x_n}\right|, \qquad (2.10.3)$$

we have

$$\|f\|_{C(\Omega)} \leq \|f\|_{C^{(1)}(\Omega)}. \qquad (2.10.4)$$

In terms of T this is

$$\|Tf\|_{C(\Omega)} \leq \|f\|_{C^{(1)}(\Omega)}. \qquad (2.10.5)$$

So T is bounded and thus continuous. It is not continuously invertible, however, since there are continuous functions that do not have continuous derivatives on Ω.

The *imbedding theorems* specify the properties of imbedding operators. In the present case we can state the following theorem, the second part of which (on compactness) is a consequence of Exercise 3.25.1. The notions of compact set and compact operator are detailed in § 3.25.

Theorem 2.10.1. *Let Ω be compact. The imbedding operator from $C^{(1)}(\Omega)$ to $C(\Omega)$ is continuous and compact.*

The situation with Sobolev and energy space imbedding results is slightly different from the one above. We will continue to see a one-to-one correspondence–identification between the elements of two different spaces;

the image of an element under this correspondence, however, may not be that same element (indeed the natures of the elements of the two spaces may differ). We will continue to use the term "imbedding" when referring to theorems that provide additional properties of elements of a space when these elements can be regarded as belonging to another "wider" space.

Since many of the spaces we use are based on the completion theorem, it is instructive to consider

Exercise 2.10.1. *Suppose X is a Banach space. What we can say about its completion? What are the relations between the elements of X and those of its completion?*

The essential properties of the energy space elements are obtained from certain inequalities between the elements ("functions") or their derivatives and the energy norms. A standard trick for obtaining such inequalities is as follows. Recall that an energy space element is really an equivalence class of Cauchy sequences of smooth functions. We take one such smooth function, derive the needed inequality for it, and then extend the inequality to the actual energy space element (by exploiting properties of the norm of the space). So the first step requires only simple calculus, while the second requires justification of limit passages in the inequality. In § 1.15 we saw how such passages were justified: we established the existence of the limiting inequality and showed that it was independent of the choice of representative sequence. In the interest of brevity we omit such details from our subsequent development.

Let us consider the space E_{SD}. The expression for $\|u\|_S$ involves only the derivative of u, which is square integrable. To establish the properties of the elements of E_{SD}, we first establish some properties of the base functions, then extend these to all elements of the space. Now we prove a simple inequality for $u(x)$ from the base class S_0:

$$\max_{x \in [0,l]} |u(x)| \le l^{1/2} \left(\int_0^l [u'(x)]^2 \, dx \right)^{1/2}. \tag{2.10.6}$$

For this, we write out the integral representation

$$u(x) = \int_0^x u'(t) \, dt. \tag{2.10.7}$$

We now invoke the Schwarz inequality, which states that

$$\left(\int_a^b f(x) g(x) \, dx \right)^2 \le \left(\int_a^b f^2(x) \, dx \right) \left(\int_a^b g^2(x) \, dx \right). \tag{2.10.8}$$

We have

$$|u(x)| = \left| \int_0^x 1 \cdot u'(t)\, dt \right|$$
$$\leq \left(\int_0^x 1^2\, dt \right)^{1/2} \left(\int_0^x [u'(t)]^2\, dt \right)^{1/2}$$
$$\leq l^{1/2} \left(\int_0^l [u'(x)]^2\, dx \right)^{1/2}$$

as required. Now take an element $U(x) \in E_{SD}$. Inequality (2.10.6) holds for any function $u_n(x)$ in its representative sequence. We can rewrite this inequality for differences $u_n(x) - u_m(x)$:

$$\max_{x \in [0,l]} |u_n(x) - u_m(x)| \leq l^{1/2} \left(\int_0^l [u_n'(x) - u_m'(x)]^2\, dx \right)^{1/2}.$$

On the right we have, up to a constant multiplier, the norm of the difference $u_n(x) - u_m(x)$ in the space E_{SD}. This tends to zero as $m, n \to \infty$. On the left, we get the norm of this difference in the space $C(0, l)$. It follows that $\{u_n(x)\}$ is a Cauchy sequence in both E_{SD} and $C(0, l)$. By completeness of $C(0, l)$, there is a limiting continuous function $u(x)$ to which $\{u_n(x)\}$ converges in $C(0, l)$. If we take another representative of $U(x)$ and repeat the above reasoning, we arrive at the same limit function $u(x)$ (the reader should verify this). So we have a correspondence between an element $U \in E_{SD}$ and a function $u \in C(0, l)$.

What are the properties of this correspondence? If $U(x)$ contains a stationary sequence $\{u(x)\}$ where $u \in S_0$, then $u(x)$ is also the limiting element obtained in the above correspondence. The correspondence we have described — wherein each $U \in E_{SD}$ is paired with a function u — is an operator of the same nature as the operator T mentioned previously. This new operator, G say, takes any element of E_{SD} into $C(0, l)$ and, again, is such that any element $U(x)$ containing a stationary sequence composed of an element $u(x)$ of the base set S_0 is mapped to that same function $u(x)$. We call it the operator of imbedding of E_{SD} into $C(0, l)$. A consequence of (2.10.6) is that for $u = GU$ we get

$$\max_{x \in [0,l]} |u(x)| \leq c \|U\|_S$$

with $c = l^{1/2}/T_0$, so G is continuous. We will also find (Lemma 3.25.1) that it is compact.

Theorem 2.10.2. *The imbedding operator from E_{SD} into $C(0,l)$ is continuous and compact.*

Because of this limiting correspondence between the elements of E_{SD} and $C(0,l)$, we will use the name of the limiting function $u(x)$ for the corresponding element $U(x)$ from E_{SD} and write equations and other relations in terms of $u(x)$. So the first statement of Theorem 2.10.2 (about continuity of the operator) we shall express as

$$\|u\|_{C(0,l)} \leq m \, \|u\|_S \qquad (2.10.9)$$

with some constant m independent of $u \in E_{SD}$.

There are, however, situations in which we cannot blur the distinction between $u(x)$ and $U(x)$; then we must deal with the actual element $U(x)$ (carrying out appropriate limit passages when necessary). Similar comments hold for the other energy spaces we encounter subsequently.

We should add something about the nature of the derivatives of the elements $U \in E_{SD}$. Again, these elements are paired with continuous functions. Not all these functions have derivatives of the type to which we become accustomed in classical calculus. However, to each such continuous function $u(x)$ there corresponds a class of sequences $\{u'_n(x)\}$, equivalent in $L^2(0,l)$, each being a Cauchy sequence in $L^2(0,l)$. When the sequences have a continuous limit function, it is possible to prove that the latter is an ordinary derivative of $u(x)$. If such a continuous limit does not exist, we still have a correspondence between $u(x)$ and some element of $L^2(0,l)$ that contains the above sequences $\{u'_n(x)\}$. This element is called the *generalized derivative* of $u(x)$. In § 2.16 we will consider in more detail how to introduce function spaces containing generalized derivatives. The space E_{SD} is one of the simplest examples where we must do this. Moreover, it is one of the so-called Sobolev spaces.

2.11 Generalized Setup of the Dirichlet Problem for a String, Continued

Having introduced the notion of imbedding, we return to our previous problem.

Definition 2.11.1. By a *generalized solution of the Dirichlet problem of equilibrium of a string* is meant an element $w \in E_{SD}$ that satisfies the equation of the virtual work principle for the string:

$$\int_0^l T_0 w_x(x)\, \delta w_x(x)\, dx = \int_0^l f(x)\, \delta w(x)\, dx, \qquad (2.11.1)$$

for any $\delta w \in E_{SD}$.

Observe, on the left side of (2.11.1), the inner product between u and δw in the space E_{SD}. Note also the assumption that the smoothness properties of $w(x)$ and $\delta w(x)$ are the same. This has a meaning, since virtual displacements can include the class of real displacements and so it makes no sense to restrict the class of virtual displacements only to very smooth functions. Such symmetry in the properties of w and δw in the definition allows us to use Hilbert space tools.

Now consider the right side. Clearly, it is a linear functional with respect to δw in E_{SD}. Earlier we required that the load f be integrable: $f \in L(0, l)$. Thus

$$\left| \int_0^l f\, \delta w\, dx \right| \le \max_{x \in [0,l]} |\delta w(x)| \int_0^l |f(x)|\, dx,$$

and by (2.10.9) we have

$$\left| \int_0^l f\, \delta w\, dx \right| \le m_1 \|w\|_S \qquad (2.11.2)$$

with a constant m_1 depending on the norm of f in $L(0, l)$ only. Inequality (2.11.2) means that the integral

$$\int_0^l f\, \delta w\, dx$$

is a linear continuous functional in E_{SD} with respect to δw. Hence we can apply Theorem 2.8.1 in the Hilbert space E_{SD} and get

$$\int_0^l f\, \delta w\, dx = (w^*, \delta w)_S,$$

where $w^* \in E_{SD}$ is uniquely defined by f. Finally, we can substitute this representation into (2.11.1) and get an equivalent relation

$$(w, \delta w)_S = (w^*, \delta w)_S$$

valid for all $\delta w \in E_{SD}$. Rewriting it as

$$(w - w^*, \delta w)_S = 0$$

and using the arbitrariness of δw, we see that there exists a unique generalized solution w^* of the equilibrium problem for the string. Let us formulate this as

Theorem 2.11.1. *Let $f \in L(0,l)$. There is a unique generalized solution $w^* \in E_{SD}$ of the boundary value problem of equilibrium for a string under load f, in the sense of Definition 2.11.1.*

We could extend the set of admissible loads to include point forces acting on the string.

Exercise 2.11.1. *Repeat all the above steps for such loads, supplementing the work of external forces with the term $\sum_{i=1}^{m} F_i \delta w(c_i)$, where the force F_i acts at the point $x = c_i$ (cf., equation (2.6.8)).*

A final remark concerns the minimum energy principle for this problem. Using the above notation, we can rewrite the total energy functional in terms of the inner product as

$$\mathcal{E}(w) = \frac{1}{2} \int_0^l T_0 w_x^2 \, dx - \int_0^l fw \, dx = \frac{1}{2}(w,w)_S - (w^*,w)_S.$$

This can be easily transformed as

$$\mathcal{E}(w) = \frac{1}{2}(w - w^*, w - w^*)_S - \frac{1}{2}(w^*, w^*)_S.$$

It is clear that $\mathcal{E}(w)$ takes its minimum at w^*, which is a generalized solution of the problem under consideration. Thus we formulate

Theorem 2.11.2. *Under the conditions of Theorem 2.11.1, the total energy functional takes its minimum value at $w^* \in E_{SD}$, which is a generalized solution of the Dirichlet problem for the string.*

Next we consider the problem of equilibrium for the string with free ends.

2.12 Neumann Problem for the String

Now we include external concentrated forces acting at the string ends. In this case the virtual work principle takes the form

$$\int_0^l T_0 w_x \, \delta w_x \, dx = \int_0^l f \, \delta w \, dx + P_0 \, \delta w(0) + P_1 \, \delta w(l), \qquad (2.12.1)$$

and there are no restrictions on the endpoint values of $w(x)$ or $\delta w(x)$. In the equilibrium state $u = u(x)$, with given forces $f(x)$, P_0, and P_1, this equation must hold for any admissible δw. Let us adapt our presentation of the Dirichlet problem to this new problem.

Our base set S_1 will consist of all the functions that are twice continuously differentiable on $[0, l]$, with no additional restrictions. (Note that here we use only the set of functions and not the notation $C^{(2)}(0, l)$, which assumes a certain norm!) As before, we wish to employ the energy inner product

$$(w, \delta w)_S = \int_0^l T_0 w_x \, \delta w_x \, dx. \qquad (2.12.2)$$

A difficulty lurks here, however: the set of smooth functions satisfying $(w, w)_S = 0$ is not empty; it consists of functions such that $w_x = 0$ on $[0, l]$, hence of the constant functions $w = c$. These represent parallel translations of the string as a rigid body. When studying the equilibrium of the free string, we are interested only in its deformation, and can try to remove "rigid motions" from consideration.

The algebraic notion of factor space is useful in this regard. We can unite all constant functions into a class M_R, and let this class play the role of "zero" in a new factor space $S_{1F} = S_1/M_R$. Each element of the latter contains all functions whose differences are constant.

It is clear that the functional $(w, \delta w)_S$ is an inner product on S_{1F}, hence S_{1F} is a suitable base space for an energy space for the Neumann problem. Before proceeding, we should verify that our setup — defined by (2.12.1) — makes sense. When we take $\delta w = c$, for a solution $w(x)$ the VWP equation must hold. Since the left side is zero, so is the right side. When $\delta w = 1$, the self-balance condition

$$\int_0^l f(x) \, dx + P_0 + P_1 = 0 \qquad (2.12.3)$$

follows. Since the functional for the work of external forces is linear with respect to δw, we see that (2.12.1) holds for any $\delta w = c$, and thus makes sense in the space S_{1F}.

At this point the reader may ask why, for a free string, we do not encounter the self-balance condition for the moment of external forces. The answer is that because of tension T_0 the string resists any inclination as a whole body. This can easily be seen by watching the reactions at the ends of the inclined string. So the self-balanced moment of the external forces is not a necessary condition for the problem.

Now we can introduce the energy space for the problem.

Definition 2.12.1. The energy space E_{SN} for the Neumann problem is the completion of S_{1F} in the norm $\|\cdot\|_S$.

At this point we could continue to follow the steps taken above for the Dirichlet problem.[6] Instead we will simplify the analysis by introducing an equivalent problem. The idea is to select from each element of S_{1F} a representative function in such a way that the collection of those functions forms a linear space. Then we reformulate the problem in terms of representative functions and establish an existence–uniqueness theorem for this new problem. Finally, we use the one-to-one correspondence between problems to formulate a corresponding theorem for the original problem. As M_R is one-dimensional, the selection of representative functions will be done using a linear functional such that over the set of representative functions $\|\cdot\|_S$ is actually a norm. Since this procedure is common to all Neumann problems we will consider, we should explain the properties of such a functional. Each time we will show that the energy space, which we obtain as the completion of some set of smooth functions with respect to an energy norm, is a subspace of a Sobolev space (a topic to be considered later). So the functional should be taken as continuous in this space. Here, on the set of representative functions, the energy norm will be equivalent to the norm of $W^{1,2}(0,l)$. Hence the functional should be taken continuous in $W^{1,2}(0,l)$. This problem for a free string is typical. In more advanced theories like the theory of elasticity, the representative elements are selected using linear functionals of integral form. But now, because of the imbedding of $W^{1,2}(0,l)$ to $C(0,l)$ we simply fix the end $x=0$ of the string (note that the value $w(0)$ defines a linear functional acting on $W^{1,2}(0,l)$). Everything is clear mechanically: we preclude free movement of the string. Of course, the result is not a Neumann problem: the two problems stand in one-to-one correspondence, however, as do their respective energy spaces. Hence we can obtain the needed solution of the Neumann problem by studying the problem for the string with one fixed end.

So we consider the problem of equilibrium under the same load as for the Neumann problem. Because the left end is fixed, we have $w(0)=0$

---

[6]A slight difficulty lurks, however. We need an inequality along the lines of (2.10.6), but must account for the fact that elements u now are defined up to a constant. So we cannot simply use the space $C(0,l)$ and its norm in the inequality, but rather must use a factor space $C(0,l)/M_R$ with norm $\|f\| = \min_c \max_{x \in [0,l]} |f(x) - c|$. The reader is encouraged to prove a corresponding imbedding theorem and existence-uniqueness theorem.

and $\delta w(0) = 0$. Under this condition the virtual work principle for the Neumann problem changes to

$$\int_0^l T_0 w_x \, \delta w_x \, dx = \int_0^l f \, \delta w \, dx + P_1 \, \delta w(l). \qquad (2.12.4)$$

Note that the end force P_0 has disappeared from the equilibrium equation. In order to have (2.12.4) with $w(0) = 0 = \delta w(0)$ be equivalent to the previous setup of the Neumann problem when "rigid" motions are permitted, we should maintain the condition (2.12.3) that "restores" P_0.

Therefore, let us obtain an existence-uniqueness theorem for a string with clamped left end and free right end. At this point we can repeat our steps from the Dirichlet problem. This is done as follows. First we replace S_0 with the set S_{01} of twice continuously differentiable functions that vanish at $x = 0$, and impose the energy norm on this set just as we did for the Dirichlet problem. Then we introduce the energy space E_{SD1} as the completion of S_{01} in the energy norm. Definition 2.11.1 is rephrased as

Definition 2.12.2. By a *generalized solution of the new problem of equilibrium of a string with fixed end at $x = 0$* is meant an element $w \in E_{SD1}$ that satisfies the equation of the virtual work principle (2.12.4) for any $\delta w \in E_{SD1}$.

Everything else, including the imbedding theorem, follows with only small changes in notation.

Theorem 2.12.1. *Let $f \in L(0, l)$. There is a unique generalized solution $w^* \in E_{SD1}$ of the boundary value problem of equilibrium of a string under load f, P_1, in the sense of Definition (2.12.2).*

The one-to-one correspondence between the above problems is achieved when we suppose that the self-balance equation (2.12.3) holds. So we also formulate

Theorem 2.12.2. *Let $f \in L(0, l)$, and let the load be self-balanced as in (2.12.3). Then there is a unique generalized solution $w^* \in E_{SN}$ for the boundary value problem of equilibrium of a string under load f, P_0, P_1.*

The reader can formulate corresponding theorems on the principle of minimum total energy for these problems.

Note, finally, that for Theorem 2.12.1 we need not impose (2.12.3). But it turns out that in order to formulate the existence theorem for a free string, we must impose the self-balance condition as we did in Theorem 2.12.2.

Exercise 2.12.1. *Establish the existence-uniqueness theorem for the equilibrium problem of a string with free ends that is fixed at (a) one intermediate point c, (b) n intermediate points. Suppose the load includes a finite number of external point forces. Is it necessary to add self-balance conditions for the external forces? Explain using mechanical and mathematical reasoning.*

2.13 The Generalized Solution of Linear Mechanical Problems and the Principle of Minimum Total Energy

In § 2.12 we saw that existence-uniqueness results could be established for very different equilibrium problems in parallel fashion. This also holds for more complex problems of linear elasticity. In this section we formulate an abstract theorem to which we shall reduce all the linear problems of this book.

Let us consider the structure of the equations describing the two boundary value problems for the string. The goal is to recognize those properties necessary to prove existence and uniqueness. The equation that followed from the virtual work principle had two terms. The first of these was quadratic, obtained by the way of varying the strain energy; it was used to define an inner product in an energy space. The other term, the work of external forces, was a linear functional. Since many problems share this structure of the VWP equation, it makes sense to consider such an equation in general form. So let H be a Hilbert space with inner product $(\cdot,\cdot)_H$ (in this book always an energy space, which can have a complex structure) and consider

$$(u,v)_H = F(v), \qquad (2.13.1)$$

where $F(v)$ is a linear functional in v. A solution of this equation is defined as follows.

Definition 2.13.1. An element $u \in H$ is a solution of (2.13.1) if it satisfies the equation for any $v \in H$.

We formulate

Theorem 2.13.1. *Let $F(u)$ be a linear continuous functional in H. Then (2.13.1) has a solution in the sense of Definition 2.13.1 that is unique in H.*

Proof. By Theorem 2.8.1, we get the unique element $u^* \in H$ such that
$$F(v) = (u^*, v)_H$$
for any $v \in H$. Thus (2.13.1) can be rewritten as
$$(u, v)_H = (u^*, v)_H,$$
and so $(u - u^*, v)_H = 0$. By the arbitrariness of v, we see that $u = u^*$ is the required unique solution. $\square$

This proof parallels our procedure for the string. Theorem 2.13.1 will be used frequently below.

Another simple and useful theorem is equivalent to the minimum total energy principle for many linear equilibrium problems of mechanics. Let us introduce the "total energy" functional
$$\Phi(u) = \frac{1}{2}(u, u)_H - F(u). \tag{2.13.2}$$

Theorem 2.13.2. *Let $F(u)$ be a linear continuous functional in H. The functional $\Phi(u)$ attains its minimum at the unique point $u^* \in H$ that satisfies (2.13.1) in the sense of Definition 2.13.1.*

Proof. By the representation $F(u) = (u^*, u)_H$, we get
$$\Phi(u) = \frac{1}{2}(u, u)_H - (u^*, u)_H = \frac{1}{2}(u - u^*, u - u^*)_H - \frac{1}{2}(u^*, u^*)_H.$$
As $(u^*, u^*)_H$ does not depend on u and $(u - u^*, u - u^*)_H \geq 0$, the minimum is attained when $u = u^*$ and is unique. $\square$

By equating the first variation of $\Phi(u)$ to zero, we obtain (2.13.1). So it is clear why this relationship exists between the minimum point of the total energy functional and (2.13.1).

Remark 2.13.1. The hardest part of proving theorems for particular elastic problems is to obtain conditions under which F — which for all of our problems will be the work of external forces — is a continuous linear functional. Each time we will show that, in the energy space for the problem, the energy norm of F is equivalent to the norm of a Sobolev space $W^{k,m}(\Omega)$; we will then use the Sobolev imbedding theorem to formulate conditions for continuity of F.

We should also add that, in large part, mathematical books that treat questions of existence-uniqueness for similar elliptic problems start with the equation $Au = f$ and introduce the bilinear form (Au, v). This form is typically the product of Au with v in the space $L^2(\Omega)$, and requires one to prove *coerciveness* of A, which in our terms means that the following inequality holds:

$$(Au, u) \geq \|u\|^2_{W^{k,p}}.$$

This is the condition of equivalence of the energy norm to a Sobolev norm. Then they apply the *Lax–Milgram theorem*, which is a consequence of the Riesz representation theorem. So our approach differs from theirs only in form. But the energy approach of this book has a firm mechanical background. □

2.14 Nonlinear Model of a Membrane

Let us turn to another model, one that reduces upon linearization to the familiar Poisson's equation. We consider the physical background for the membrane model. The physical object best described by the model is a soap film. From physics, it is known that the force on a straight edge of such a film is proportional to its length. Thus, the mechanical quantity characterizing the film's equilibrium is given as the force per unit length acting on its edge. Although this "surface tension" depends on temperature and other parameters, in the first approximation it is constant. It should be noted that the surface tension for a soap film is twice that of the liquid from which the film is made.[7] A soap film resists deformation only through surface tension which, on any infinitesimally small line, acts tangent to the film surface in the direction normal to the line. Since its value does not depend on the film's thickness or strain energy, we can model the film with a surface having certain mechanical properties.

The next step is to find the expression for the strain energy due to additional deformation of a soap film. Imagine a film spanning a wire frame. In order to increase the film area, we must apply some force to one side of the frame. This force is infinitesimally greater than that produced by surface tension. The work done during this process is transformed into strain energy

---

[7] By assumption, the surface tension does not depend on the direction of the line and is the same at each point of the film. This reminds us of Pascal's law for the pressure inside a liquid volume. Indeed, in both cases we idealize a mechanical object in such a way that the model lacks certain features of strains possessed by solid bodies.

of the film. Supposing the film is planar, and that the moving side of the frame is straight and moves in the plane in a direction perpendicular to itself, the work is equal to the surface tension multiplied by the increase in area. The same holds for other wire/soap geometries, as long as the surface tension is constant: the strain energy of a soap film due to deformation is proportional to its increase in area. Now a membrane is not a soap film, but there is nothing to stop us from transferring this same property to membrane theory as an *axiom* (or, as mechanicists say, an *assumption*). In fact we will do this, and turn next to the derivation of the equilibrium equation.

When a system is in stable equilibrium, its total energy functional takes a minimum value. If no external forces act and the total and strain energy functionals coincide, a soap film with fixed edge contour has minimum area among all the surfaces spanning this contour. So the film represents a physical solution to Plateau's famous geometric problem: find a surface of minimum area with a given boundary contour. We are interested in solving an extension of Plateau's problem that includes the action of external forces. We assume the latter do not depend on the deformation of the membrane at each point: i.e., that the load is simply a given function $f = f(x, y)$. This "dead load" assumption is natural for linear problems, but becomes questionable for some nonlinear problems.

For simplicity we suppose that the initial state of the membrane is a compact planar domain Ω with a piecewise smooth boundary Γ, and that its deformation is characterized only by a normal deflection $u = u(x, y)$. The corresponding surface area is

$$A = \int_\Omega \sqrt{1 + u_x{}^2 + u_y{}^2}\, dx\, dy, \quad (2.14.1)$$

where the subscripts indicate partial differentiation. Multiplying this by the surface tension coefficient a, we obtain the strain energy of the membrane. For the plane membrane when $u = 0$, this energy is

$$a \int_\Omega dx\, dy;$$

thus the accumulated strain energy is

$$\mathcal{E}_s(u) = a \int_\Omega \left[\sqrt{1 + u_x{}^2 + u_y{}^2} - 1\right] dx\, dy. \quad (2.14.2)$$

By the minimum total energy principle, in the equilibrium state the functional

$$\mathcal{E}_t(u) = a\int_\Omega \left[\sqrt{1 + u_x{}^2 + u_y{}^2} - 1\right] dx\,dy - \int_\Omega fu\,dx\,dy \qquad (2.14.3)$$

takes its minimum over the set of all admissible functions $u = u(x,y)$. The second term on the right is the work of distributed load f over displacements u. If the edge is fixed, admissible functions must satisfy a boundary condition of the form

$$u\big|_\Gamma = \psi(s) \qquad (2.14.4)$$

with given ψ. Thus, when the membrane with fixed edge is in equilibrium, the total energy functional $\mathcal{E}_t(u)$ takes its minimum over all sufficiently smooth functions satisfying (2.14.4). This can be called the Dirichlet problem for a nonlinear membrane. Implementation of the corresponding Euler–Lagrange condition will be left to the reader. A more important point is as follows. Although we derive the equilibrium equations from the principle of minimum total energy, these equations actually turn out to have wider applicability: for nonlinear problems, they can yield solutions that do *not* correspond to points of minimum energy but still describe the membrane equilibrium.

We need not formulate boundary conditions for admissible functions if the membrane edge is free (i.e., the edge is not fixed, but some forces can act on the edge). But we recall that the solution will necessarily satisfy the natural boundary conditions that arise in the variational formulation of the problem. This is the Neumann problem. Note that when the strain energy is zero, we have $u_x = 0$ and $u_y = 0$ at all points so that $u = $ constant. If we add a constant c to some deflection field $u(x,y)$, the strain energy does not change. Mechanically this is clear: translation of the membrane as a rigid whole cannot change its deflection or its strain energy. But any rotation of the membrane as a whole about x or y changes its strain energy by (2.14.2), and this means that in the membrane theory under consideration only the condition that the resultant external force vanishes is necessary. Since the addition of a constant c to the deflection function u does not change the strain energy, the total energy minimum principle is meaningful only if the external forces are self-balanced:

$$c\int_\Omega f\,dx\,dy = 0. \qquad (2.14.5)$$

We obtained a similar condition when considering the equilibrium of a free string; it means that the resultant of the external forces is zero. It is a necessary condition for solvability of this nonlinear problem. Because of the above-mentioned dependence of the energy integral on the rotation of the membrane as a rigid body, we have no other conditions for self-balance of the external forces.

We turn to the simplified membrane theory that is in common use.

2.15 Linear Membrane Theory: Poisson's Equation

The deformation of a planar membrane is described by a deflection function $u = u(x, y)$. We assume any infinitesimal portion of the membrane rotates only through a 'small" angle. Since this angle is characterized by the first derivatives u_x and u_y of u, we must suppose that $|u_x|$ and $|u_y|$ are small as well: i.e., that we can neglect terms of higher order with respect to these quantities. The integrand of the strain energy expression (2.14.2) has the form $\sqrt{1+z} - 1$ where $z = u_x^2 + u_y^2$. Since $|z| \ll 1$, we replace the entire integrand by $z/2 = (u_x^2 + u_y^2)/2$ and obtain the approximate total energy functional

$$\mathcal{E}(u) = \frac{a}{2} \int_\Omega \left[u_x^2 + u_y^2 \right] dx\, dy - \int_\Omega fu\, dx\, dy. \qquad (2.15.1)$$

Supposing we can apply the minimum energy principle, we obtain a new approximate linearized model of the membrane.

Note that it is common practice, when considering mechanical derivations of formulas and models, to initially suppose that certain terms are very small or even infinitesimal but later to regard the simplified formulation as independent and holding for very finite quantities — even forgetting that these quantities must be at least "small" in some sense. Moreover, when considering the corresponding mathematical setup for the simplified problem, we even bring infinite values into consideration. Indeed, such a simplified problem can come to live a life of its own.

We will start with the problem for a membrane with fixed edge:

$$u\big|_\Gamma = \psi(s), \qquad (2.15.2)$$

where ψ is a given smooth function.

So let u be a sufficiently smooth function (in this case, twice differentiable on a closed domain Ω) at which $\mathcal{E}(u)$ attains its minimum (we suppose implicitly, as is common in applications, that f is at least integrable). Let v be a sufficiently smooth function vanishing on Γ. The function $u + tv$, where t is an arbitrary real parameter, satisfies (2.15.2); it is sufficiently smooth, hence admissible for use in comparison of functions on the basis of the minimum energy principle.

At any t, we have

$$\mathcal{E}(u) \leq \mathcal{E}(u+tv). \tag{2.15.3}$$

At fixed u and v, the functional $\mathcal{E}(u+tv)$ is a function of the real variable t and takes its minimum at $t = 0$. This means that at $t = 0$ its derivative with respect to t is zero:

$$a \int_\Omega (u_x v_x + u_y v_y)\, dx\, dy - \int_\Omega fv\, dx\, dy = 0. \tag{2.15.4}$$

Note that this is also the virtual work principle for the membrane.

By the arbitrariness of v, using the ordinary tools of the variational calculus, we arrive at Poisson's equation

$$a\left(u_{xx} + u_{yy}\right) + f = 0. \tag{2.15.5}$$

This also follows directly from the mechanical equilibrium equations. The reader is encouraged to construct a derivation, starting with a small rectangular portion of the membrane and keeping in mind that the sides of this portion rotate during deformation.

2.16 Generalized Setup of the Dirichlet Problem for a Linear Membrane

We are limited to a study of certain properties of associated boundary value problems for Poisson's equation. We would like to consider such problems in a generalized setting. There are several reasons for this. First, a classical study of Poisson's equation is tied too closely to this particular equation; the usual potential-based methods are not readily transferable to other mechanics problems. Second, our approach is comparatively simple (now that we have some familiarity with functional analysis) and can handle the presence of nonsmooth forces. Third, the generalized setup is based on the virtual work principle and therefore has solid mechanical underpinnings. Finally, the powerful finite element method is based on the same generalized setup. It should be added that a generalized solution may have a higher degree of smoothness than is specified in the generalized setup (for sufficiently smooth external parameters, of course).

Our experience indicates that we can base a generalized setup of the Dirichlet problem for a membrane on (2.15.4), which expresses the virtual work principle in this case. For simplicity we consider a membrane with fixed edge:

$$u\big|_\Gamma = 0. \tag{2.16.1}$$

We start with equation (2.15.4):

$$a \int_\Omega (u_x v_x + u_y v_y) \, dx \, dy - \int_\Omega fv \, dx \, dy = 0. \tag{2.16.2}$$

We would like to apply the general theorems of § 2.13 to this problem. The first term in (2.16.2) looks like an inner product. So we introduce the set S_2 of functions twice continuously differentiable on a compact set Ω and vanishing on the boundary Γ. The functional

$$(u, v)_M = a \int_\Omega (u_x v_x + u_y v_y) \, dx \, dy \tag{2.16.3}$$

is an inner product on S_2.

Exercise 2.16.1. *Verify this.*

Because the corresponding inner product space is incomplete (why?), we introduce a norm

$$\|u\|_M = (u, u)_M^{1/2} \tag{2.16.4}$$

and define an energy space for the problem.

Definition 2.16.1. The completion of S_2 with respect to the norm $\|\cdot\|_M$ is called the energy space E_{MD} for the Dirichlet problem for the membrane.

Definition 2.16.2. An element $u \in E_{MD}$ is a generalized solution of the equilibrium problem with condition (2.16.1) if it satisfies equation (2.16.2) for any $v \in E_{MD}$.

Because u belongs to E_{MD} its first derivatives are square-integrable or, equivalently, members of $L^2(\Omega)$. So we seek solutions of Poisson's equation having finite energy. For the linear theory we imposed the more stringent requirement that the first derivatives of the solution be small. But mathematicians commonly try to vary the conditions imposed on a model. Occasionally such investigation yields physical insight. In any case it provides a reliable basis for future use of the model, clarifying its range of applicability and offering justification for numerical solution methods.

Next, according to § 2.13 we should require that the second term of (2.16.2) be a linear continuous functional in the energy space. What does this mean in terms of the distributed load f — i.e., what is the class of admissible forces? Examining the second integral term of (2.16.2), we see that we must know the properties of u, a typical element of E_{MD}. Hence we should establish an imbedding theorem for E_{MD} into another space of functions.

Rather than establishing the imbedding theorem in a sharp form, we will present an older result based on the *Friedrichs inequality*. This states that for any function $u \in S_2$,

$$\int_\Omega u^2 \, dx \, dy \leq c \int_\Omega \left(u_x^2 + u_y^2\right) dx \, dy \qquad (2.16.5)$$

with a constant c that does not depend on u. To obtain it, let us assume Ω lies in the first quadrant of the xy-plane.

Exercise 2.16.2. *Show that the Friedrichs inequality also holds on domains that do not necessarily lie in the first quadrant.*

Since Ω is compact, we can cover it with some square S having side length a and sides parallel to the axes, two of which lie on the axes. We now make the domain of u encompass all of S by setting $u \equiv 0$ on $S \setminus \Omega$ to obtain a new function that we continue to call u. Note that $u(x, y) = 0$ when $x = 0$ or $y = 0$. The representation

$$u(x, y) = \int_0^x u_t(t, y) \, dt$$

holds everywhere in the first quadrant. Let us square both sides and integrate over the domain:

$$\int_\Omega |u(x,y)|^2 \, dx \, dy = \int_0^a \int_0^a \left| \int_0^x u_t(t, y) \, dt \right|^2 dx \, dy.$$

By the Schwarz inequality we have

$$\int_\Omega |u(x,y)|^2 \, dx \, dy = \int_0^a \int_0^a \left| \int_0^x 1 \cdot u_t(t, y) \, dt \right|^2 dx \, dy$$

$$\leq \int_0^a \int_0^a \int_0^x 1^2 \, dt \int_0^x |u_t(t, y)|^2 \, dt \, dx \, dy$$

$$\leq \int_0^a \int_0^a \int_0^a 1^2 \, dt \int_0^a |u_t(t, y)|^2 \, dt \, dx \, dy$$

$$= a^2 \int_0^a \int_0^a |u_t(t, y)|^2 \, dt \, dy,$$

hence
$$\int_\Omega |u|^2 \, dx\, dy \leq a^2 \int_0^a \int_0^a |u_x(x,y)|^2 \, dx\, dy = a^2 \int_\Omega |u_x|^2 \, dx\, dy. \quad (2.16.6)$$
In fact we have the needed estimate, since it is clear that
$$\int_\Omega |u|^2 \, dx\, dy \leq a^2 \int_\Omega \left(|u_x|^2 + |u_y|^2\right) dx\, dy$$
with $c = a^2$. Because the constant c depends only on a, it depends only on Ω (which dictated our original choice of a). Note that we have obtained an even sharper result, inequality (2.16.6).[8]

Applying the Friedrichs inequality to a representative sequence $\{u_n\}$ of an element of E_{MD}, we find that
$$\int_\Omega (u_n - u_m)^2 \, dx\, dy \leq c \int_\Omega \left((u_n - u_m)_x^2 + (u_n - u_m)_y^2\right) dx\, dy. \quad (2.16.7)$$
It follows that $\{u_n\}$ is a Cauchy sequence in $L^2(\Omega)$.

We have introduced $L^p(\Omega)$ as the completion of a set of smooth functions on Ω with respect to the norm of $L^p(\Omega)$. Inequality (2.16.7) means that any representative sequence of an element of E_{MD} is a Cauchy sequence in the norm of $L^2(\Omega)$. Similarly, equivalent Cauchy sequences in the norm E_{MD} stay equivalent in the norm of $L^2(\Omega)$. Hence any element of E_{MD} belongs to $L^2(\Omega)$ and, moreover, Friedrichs inequality yields the following result.

Lemma 2.16.1. *The space E_{MD} imbeds into the space $L^2(\Omega)$. The imbedding operator is continuous and its norm is less than or equal to a.*

The Friedrichs inequality yields conditions on the forces sufficient to make the work functional continuous. For when $f \in L^2(\Omega)$, we have
$$\left|\int_\Omega fv \, dx\, dy\right| \leq \left(\int_\Omega f^2 \, dx\, dy\right)^{1/2} \left(\int_\Omega v^2 \, dx\, dy\right)^{1/2}$$
$$\leq c_1 \left(\int_\Omega f^2 \, dx\, dy\right)^{1/2} \|v\|_M.$$
Thus, applying Theorem 2.13.1, we get

Theorem 2.16.1. *Let $f \in L^2(\Omega)$. The Dirichlet problem for a membrane has a generalized solution in the sense of Definition 2.16.2 that is unique in the space E_{MD}.*

---

[8] In the proof, we estimated u through the values of only one derivative u_x, which indicates that the inequality is not the sharpest possible. A sharper result will be given by Sobolev's imbedding theorem.

In a similar fashion, the reader can reformulate Theorem 2.13.2 for this problem.

Exercise 2.16.3. *Do this.*

Before moving to other questions, let us mention that in this section we simplified the problem by requiring $u = 0$ on Γ. When the membrane edge is fixed by (2.15.2) with $\varphi \neq 0$ at all points of Γ, the treatment is classical. Suppose a function $\tilde{u}$ satisfies (2.15.2). Also suppose $\tilde{u}$ has finite energy so its first derivatives belong to $L^2(\Omega)$. If u satisfies the homogeneous conditions (2.16.1), then $u + \tilde{u}$ satisfies (2.15.2). Replacing u by $u + \tilde{u}$ in the VWP equation (2.15.4), we come to an equation with respect to $u \in E_{MD}$ that formally looks different from (2.16.2):

$$a \int_\Omega (u_x v_x + u_y v_y) \, dx \, dy + a \int_\Omega (\tilde{u}_x v_x + \tilde{u}_y v_y) \, dx \, dy - \int_\Omega f v \, dx \, dy = 0.$$

But the second integral term, under the above conditions for $\tilde{u}$, is also a linear continuous functional in E_{MD}, so we have established the existence theorem in this case as well. The uniqueness theorem for nonhomogeneous Dirichlet boundary conditions should be established separately, since the choice of $\tilde{u}$ is not unique. But this theorem follows from Theorem 2.16.1, since assuming there are two solutions of the problem we find that their difference belongs to the space E_{MD} and satisfies the membrane equation with zero external forces, hence is zero.

We have established solvability in the energy class of Dirichlet's problem when the load belongs to $L^2(\Omega)$. It is possible to weaken this restriction on the load, but we shall require Sobolev's imbedding theorem and related ideas.

Generalized derivatives and Sobolev spaces

We have employed partial derivatives that belong to the space $L^2(\Omega)$. These are not classical in nature and cannot be considered pointwise as ordinary derivatives. An equivalent method of introducing them was advanced by S.L. Sobolev, who pioneered the use of generalized solutions in mechanics and put forth a class of spaces that now bear his name. The Friedrichs inequality represents a particular case of the properties possessed by the elements of one Sobolev space. K.O. Friedrichs proved that in Sobolev spaces, the derivatives obtained from the completion procedure are the same as (more precisely, stand in one-to-one correspondence with) those obtained from Sobolev's definition. Since we use generalized derivatives throughout this book, we should explain Sobolev's initial approach to the concept.

Sobolev's definition of a generalized derivative is based on two things: the formula for integration by parts, and the Main Lemma of the calculus of variations. The former is

$$\int_\Omega u\varphi_x\, dx\, dy = -\int_\Omega u_x \varphi\, dx\, dy, \qquad (2.16.8)$$

which holds for smooth functions u and φ when φ vanishes on the boundary $\partial\Omega$ of Ω. Suppose now that u and another function ω are merely integrable over Ω (in the sense of Riemann or Lebesgue). Let (2.16.8) hold in the form

$$\int_\Omega u\varphi_x\, dx\, dy = -\int_\Omega \omega\varphi\, dx\, dy \qquad (2.16.9)$$

for all smooth functions φ that vanish on $\partial\Omega$. Then ω is called the *generalized derivative* of u in the sense of Sobolev. The generalized derivative, if it exists, is uniquely defined up to a set of "measure zero".[9] It also coincides with the classical derivative of u if the latter exists. With this understanding, we denote it by u_x.

Higher derivatives can be introduced similarly. For example, the generalized derivative $\beta = u_{xy}$ of an integrable function u is uniquely determined by the equality

$$\int_\Omega u\varphi_{xy}\, dx\, dy = \int_\Omega \beta\varphi\, dx\, dy \qquad (2.16.10)$$

provided β is integrable and the equality holds for every smooth φ that vanishes on $\partial\Omega$ along with its first partial derivatives.

In fact, we can introduce generalized derivatives of any order on a finite-dimensional domain. Note that, according to the mode of definition given below, we need not know the previous (i.e., lower-order) derivatives in order to determine a higher-order one. Indeed, we are not guaranteed that other derivatives exist. In Sobolev spaces (of which most of our energy spaces are particular cases) however, all previous derivatives will exist. So the situation is similar to the classical one.

---

[9]Since we have not covered the classical definition of the Lebesgue integral, this remark was directed only toward those who happen to be familiar with that theory. It is worth mentioning, however, that a smooth line in two dimensions has measure zero in this theory; in regards to uniqueness, two functions are considered the same if they are "equal almost everywhere".

Sobolev was able to establish some properties of the elements of Sobolev spaces. Roughly, the imbedding theorems state that derivatives of lower order (and hence the function itself) are "smoother" than those of higher order. Under certain conditions the generalized derivatives (and functions themselves) stand in one-to-one correspondence with continuous functions and therefore can be identified with ordinary functions. In this way we relate the generalized theory to the classical one and obtain results on the existence of classical solutions to particular problems.

But let us return to the introduction of generalized derivatives. Using the completion theorem, we will treat the spaces of functions having generalized derivatives integrable with some degree over the domain. The following *multi-index notation* for partial derivatives is useful:

$$D^\alpha f = \frac{\partial^{|\alpha|} f}{\partial x_1^{\alpha_1} \cdots \partial x_n^{\alpha_n}}, \qquad \alpha = (\alpha_1, \ldots, \alpha_n), \qquad |\alpha| = \alpha_1 + \cdots + \alpha_n.$$

The multi-index α is simply an n-tuple of nonnegative integers where n is the dimension of the space $\mathbb{R}^n$ under consideration. For example, if $n = 3$ and $\alpha = (1, 3, 0)$, then

$$D^\alpha f = \frac{\partial^4 f}{\partial x_1 \partial x_2^3}.$$

Definition 2.16.3. Let Ω be a compact subset of $\mathbb{R}^n$, and let $p \geq 1$ be a fixed integer. Consider the normed space consisting of all l-times continuously differentiable functions $f(\mathbf{x})$ given on Ω and having the norm

$$\|f\| = \left(\int_\Omega \sum_{|\alpha| \leq l} |D^\alpha f|^p \, d\Omega \right)^{1/p}. \qquad (2.16.11)$$

The completion of this space is the *Sobolev space* denoted by $W^{l,p}(\Omega)$.

Sobolev used another definition, in which the norm is given on the set of functions having all generalized derivatives $D^\alpha f$ up to order l integrable to the p-degree over Ω. Using Definition 2.16.3 we get a space equivalent to Sobolev's space: the spaces stand in a one-to-one correspondence under which both the algebraic operations and the norm are preserved, provided Ω is compact and Jordan measurable (the latter means that we can use Riemann integration over Ω). We have confined our attention to such domains so far, and will continue to do so.

We have already met such a space in solving the membrane problem: $W^{1,2}(\Omega)$ for $\Omega \subset \mathbb{R}^2$. It follows from the Friedrichs inequality that, for

Dirichlet boundary conditions, the energy norm for the membrane is equivalent to the norm (2.16.11) with $\alpha = 2$ and $p = 2$.

This equivalence of an energy norm to a Sobolev norm is important. To prove equivalence it is useful to know the forms of various equivalent norms in $W^{l,p}(\Omega)$. This question is treated in Sobolev's book ([Sobolev (1991)], Theorem 2, p. 64). As a consequence of Sobolev's equivalence theorem we have *Poincaré's inequality*

$$\int_\Omega |\varphi|^2\, dV \le M \left[\left\{ \int_\Omega \varphi\, dV \right\}^2 + \int_\Omega \sum_{i=1}^n \left(\frac{\partial \varphi}{\partial x_i} \right)^2 dV \right] \qquad (2.16.12)$$

where $dV = dx_1 \cdots dx_n$. It will be used to treat the Neumann problem for the membrane.

For the string problem we used a one-dimensional energy space that turns out to be equivalent to $W^{1,2}(0,l)$. We found that the elements are in correspondence with continuous functions, and called this correspondence the imbedding operator. Similarly, the elements of a Sobolev space have some additional smoothness properties that are not seen from the definition.

In the one-dimensional case we used the integral representation of a smooth function

$$u(x) = u(x_0) + \int_{x_0}^x \frac{du(s)}{ds}\, ds. \qquad (2.16.13)$$

For the n-dimensional case, Sobolev derived the analogous expression

$$\varphi(\mathbf{x}) = \sum_{\sum \alpha_i \le l-1} x_1^{\alpha_1} \cdots x_n^{\alpha_n} \int_C \zeta_{\alpha_1 \cdots \alpha_n}(\mathbf{y}) \varphi(\mathbf{y})\, dV_\mathbf{y}$$
$$+ \int_\Omega \frac{1}{r^{n-l}} \sum_{\sum \alpha_i = l} w_{\alpha_1 \cdots \alpha_n}(\mathbf{y}, \mathbf{x}) \frac{\partial^l \varphi}{\partial y_1^{\alpha_1} \cdots \partial y_n^{\alpha_n}}\, dV_\mathbf{y}, \qquad (2.16.14)$$

where

$$r = |\mathbf{x} - \mathbf{y}|, \qquad dV_\mathbf{y} = dy_1 \cdots dy_n. \qquad (2.16.15)$$

Here all the ζ and w coefficients are smooth functions determined by the shape of Ω and l. The domain Ω is assumed to be "star-shaped".

Definition 2.16.4. A domain is *star-shaped* if it contains a ball such that any point of the domain can be connected with any point of the ball via a segment lying in the domain.

By studying the properties of the integrals in the representation formula, Sobolev derived a general and rather long imbedding theorem. We will need only special cases of this, which we formulate as separate theorems. Sobolev extended these theorems for finite unions of star-shaped domains. In what follows we shall assume that Ω is a finite union of compact star-shaped domains. Later the imbedding results were extended to domains having the "cone property".

Definition 2.16.5. A domain Ω has the *cone property* if there is a fixed cone such that for any point of Ω we can place the vertex of the cone in such a way that all its points are inside Ω.

It turns out that Ω has the cone property if and only if it is the union of finitely-many star-shaped domains. We assume the cone property holds in what follows.

The degrees q we will encounter below cannot be improved (increased). The significance of this will become clear as we come to understand the role played by the imbedding theorems in proving existence theorems.

The first imbedding theorem we will need applies to the space of functions of two variables.

Theorem 2.16.2. *Let γ be a piecewise differentiable curve in a compact set $\Omega \subset \mathbb{R}^2$. For any finite $q \geq 1$, there are compact (hence continuous) imbeddings*

$$W^{1,2}(\Omega) \hookrightarrow L^q(\Omega), \quad W^{1,2}(\Omega) \hookrightarrow L^q(\gamma). \tag{2.16.16}$$

The next result enables us to discuss plates and shells.

Theorem 2.16.3 (Imbedding of $W^{2,2}$ to C). *Let Ω be a compact subset of $\mathbb{R}^2$. Then there is a continuous imbedding*

$$W^{2,2}(\Omega) \hookrightarrow C(\Omega). \tag{2.16.17}$$

For the first derivatives, the imbedding operators to $L^q(\Omega)$ and $L^q(\gamma)$ are compact for any finite $q \geq 1$.

Theorem 2.16.3 relates the elements of the Sobolev space $W^{2,2}$ to the elements of the space C of continuous functions. Neither Sobolev's definition of $W^{2,2}$ nor the one we have chosen — using equivalent Cauchy sequences of functions — will allow us to say that $u \in W^{2,2}$ is a continuous function on Ω. But these Cauchy sequences have a limit function that is continuous; we will identify this with u and, in this sense, speak of u as a function. Under Sobolev's description, this "function" would be defined uniquely.

For functions of three variables we formulate

Theorem 2.16.4 ($W^{1,2} \hookrightarrow L^q$, **3d case**). *Let γ be a piecewise smooth surface in a compact set $\Omega \subset \mathbb{R}^3$. The imbeddings*

$$W^{1,2}(\Omega) \hookrightarrow L^q(\Omega), \qquad 1 \leq q \leq 6, \tag{2.16.18}$$

$$W^{1,2}(\Omega) \hookrightarrow L^p(\gamma), \qquad 1 \leq p \leq 4, \tag{2.16.19}$$

are continuous. They are compact if $1 \leq q < 6$ or $1 \leq p < 4$, respectively.

Returning to the membrane problem

Lemma 2.16.1 means that E_{MD} has a norm equivalent to that of the space $W^{1,2}(\Omega)$ in the two-dimensional case. Theorem 2.16.2 implies that the elements of the energy space belong to $L^q(\Omega)$ for any finite $q \geq 1$. To prove the existence of a generalized solution to the Dirichlet problem for a membrane, we need only the fact that $\int_\Omega fv \, dx \, dy$ is a linear continuous functional with respect to v in the energy space. This is provided by Hölder's inequality

$$\left| \int_\Omega fv \, dx \, dy \right| \leq \left(\int_\Omega |f|^p \, dx \, dy \right)^{1/p} \left(\int_\Omega |v|^q \, dx \, dy \right)^{1/q}, \quad \frac{1}{p} + \frac{1}{q} = 1. \tag{2.16.20}$$

Hence we obtain an immediate sharpening of the existence result: a generalized solution exists when $f \in L^p(\Omega)$ for some $p > 1$.

Note that part of Theorem 2.16.2 refers to imbedding on a curve. This is also useful for a membrane problem if forces are given on some curve γ. In this case the functional of the work of external forces has a term

$$\int_\gamma \varphi v \, ds. \tag{2.16.21}$$

Exercise 2.16.4. *(a) Show that (2.16.21) is a linear continuous functional in the space $W^{1,2}(\Omega)$ if there exists $p > 1$ such that $\varphi \in L^p(\gamma)$. (b) Using the form of the total energy functional presented below for the Neumann problem, prove the existence-uniqueness theorem for this case.*

Free membrane (Neumann problem)

Let us consider a membrane free from geometrical constraints. Its total energy is given by

$$\mathcal{E}(u) = \frac{a}{2} \int_\Omega \left[u_x^2 + u_y^2 \right] dx \, dy - \int_\Omega fu \, dx \, dy - \int_\gamma \varphi u \, ds, \tag{2.16.22}$$

where φ is a distributed load over a piecewise smooth curve γ, which could be inside Ω or include a part or the whole boundary Γ of Ω.

Equilibrium of the membrane occurs when the total energy takes its minimum value. We have already considered the bar and string free from geometrical constraints. Each time we were led to self-balance conditions for the external forces, and the outcome will be the same here. Let us consider the quadratic part of the strain energy:

$$\frac{a}{2}\int_\Omega [u_x{}^2 + u_y{}^2]\,dx\,dy.$$

Its value does not change if we replace u by $u + c$ for any constant c. However, the linear part of $\mathcal{E}(u)$ — the work of external forces — does change, and can acquire arbitrarily large negative values as c is varied. To prevent this from happening, we require that

$$\int_\Omega f\,dx\,dy + \int_\gamma \varphi\,ds = 0. \qquad (2.16.23)$$

This is the self-balance condition for the external load. It tells us that the resultant force acting on the membrane must be zero in order for the problem of minimum to be meaningful. This is not surprising, since the membrane is free of geometrical constraints and we neglected its inertia. There is one interesting point, however. If we consider the membrane as a rigid body in space, we must require the load to satisfy two other self-balance equations: the two resultant moments of the external load with respect to the x- and y-axes must vanish. These conditions are not needed for the solvability of the problem though. How can we explain this? As in the theory of a stretched string, the membrane is also initially stretched and resists any inclination as a rigid body. So the moment self-balance conditions disappear.

It is worth mentioning that Neumann's problem for Laplace's equation appears in classical mathematical physics. In this case $f = 0$ and the self-balance equation becomes

$$\int_\gamma \varphi\,ds = 0. \qquad (2.16.24)$$

When γ is the boundary contour of Ω, we can use the calculus of variations to show that for a smooth solution of the problem φ/a must be equal to the normal derivative of the solution. Moreover, in mathematical physics they show — and it looks like a strange artificial condition — that (2.16.24) is necessary and sufficient for solvability of the problem. We now see that it has a clear mechanical meaning.

To formulate the equilibrium problem for the free membrane, we calculate the first variation of $\mathcal{E}(u)$. This gives us the equation that we get directly from the use of the virtual work principle:

$$a \int_\Omega (u_x v_x + u_y v_y) \, dx \, dy - \int_\Omega fv \, dx \, dy - \int_\gamma \varphi v \, ds = 0. \qquad (2.16.25)$$

Exercise 2.16.5. Let $\gamma = \Gamma$. Use the calculus of variations to show that the natural boundary condition for $\mathcal{E}(u)$ is

$$a \frac{\partial u}{\partial n}\bigg|_\Gamma = \varphi,$$

where n is the external unit normal to the boundary. Note that up to the factor a, this is the boundary condition that, when taken together with Poisson's equation $\Delta u = -f$, constitutes the classical Neumann problem of mathematical physics.

As for the Dirichlet problem, we can introduce a generalized solution. The energy space should be related to the strain energy of the membrane. We could try to repeat the approach for the Dirichlet problem and attempt to introduce an inner product

$$(u, v)_M = a \int_\Omega (u_x v_x + u_y v_y) \, dx \, dy \qquad (2.16.26)$$

over the set of smooth functions on Ω. But this fails axiom I1 since $(c, c)_M = 0$ for any constant c. Note that $u(x) = c$ is the only smooth function that satisfies the equation $(u, u)_M = 0$. We are reminded of the situation for a free string and, as in that case, there are two ways around the difficulty. One is to announce that the set of all constants forms the zero element of our new space. Then an element of the space is the set of all functions that differ from one another by a constant value. The result is a factor space. Application of the completion theorem allows us to repeat everything that was done for the Dirichlet problem. The self-balance equation (2.16.23) is needed to prove that the functional of external work is linear in the energy space. Unfortunately, we would have to reformulate the Sobolev imbedding theorems in terms of the factor-type energy space. This we avoid by taking another approach. Let us select, from each element of the factor space, a unique representative. We do this in such a way that we obtain an energy space, consisting of the representatives, which is a linear subspace of the Sobolev space $W^{1,2}(\Omega)$. We can then repeat the development for Dirichlet's problem practically without changes.

The condition to select a representative function from the class of functions that differ from one another by a constant follows from Poincaré's inequality (2.16.12):

$$\int_\Omega u \, dx \, dy = 0. \qquad (2.16.27)$$

With this condition, Poincaré's inequality takes the form of Friedrichs's inequality. In the subset of all the functions that have two continuous derivatives in Ω and satisfy condition (2.16.27), $(\cdot, \cdot)_M$ is an inner product. Next, starting with this base set and applying the completion theorem, we can introduce an energy space exactly as is done for the Dirichlet problem for Poisson's equation. Now we can literally repeat everything we did for Dirichlet's problem in order to establish generalized solvability of the equilibrium problem for a free membrane.

Exercise 2.16.6. *Carry out this program in detail; formulate the corresponding existence/uniqueness theorem for the generalized solution.*

It is worth noting that, in order to prove this theorem in the class of representative functions satisfying (2.16.27), we will not need the self-balance condition. We will need only the fact that the forces are in the corresponding spaces L^p. How is this possible after all our emphasis on the self-balance condition? The answer is simple. The condition (2.16.27) that we posed on u is an additional geometric constraint that was not present in the initial formulation of the problem. So here we prove the existence theorem for another problem, for a non-free membrane. If we wish to return to the initial problem for the free membrane, we must reconsider the equilibrium equation not only for representative functions u, but for all functions of the form $u + c$, and so we arrive once again at the self-balance condition as a necessary condition for solvability. This condition, supplemented with the condition $\varphi \in L^p(\gamma)$, is sufficient for existence of a generalized solution for the free membrane.

Exercise 2.16.7. *Formulate the existence-uniqueness theorem for the free membrane. Note that a generalized solution is unique up to an additive constant c.*

2.17 Other Membrane Equilibrium Problems

We have considered two classical boundary value problems for a membrane: Dirichlet's and Neumann's. We could also consider a mixed problem where a portion of the boundary contour is fixed and a force is given over the rest. If the fixed portion contains a piece having finite nonzero length, then considerations involving the existence of a generalized solution are essentially the same as for Dirichlet's problem.

In the definition of the energy space we can consider the set of functions that vanish only on the fixed portion of the boundary. From the results of Sobolev's book on equivalent norms in $W^{1,2}(\Omega)$, it follows that on this set of functions Friedrichs' inequality holds. So on the set, the energy norm is equivalent to the standard norm of $W^{1,2}(\Omega)$. This means that in the energy space the form of the imbedding theorem coincides with the form of the imbedding theorem in $W^{1,2}(\Omega)$. Thus to prove the existence-uniqueness theorem, we can literally repeat everything that was said for Dirichlet's problem.

Existence of a generalized solution does not require a self-balance condition, which is clear from the mechanics of the problem. We leave the details to the reader and proceed to another boundary value problem sometimes encountered for Poisson's equation.

A membrane with Winkler's support on the edge

Mechanically this problem is a bit strange; it involves, along the boundary contour, a distributed support known to civil engineers as *Winkler's foundation*. The reader can imagine that acting at each point of the boundary curve is an elastic support force whose distribution is given by $k(s)u(s)$, where $k(s) \geq 0$ is the elastic coefficient. As the support at a point does not depend on the force at other points of the boundary, Winkler's foundation can be modeled using a continuum of separate springs acting on the boundary. The problem is strange in the sense that it is difficult to imagine such a support for a soap film; however, in other circumstances the idea is quite practical.

The elastic energy of Winkler's foundation is given by the Hooke's law energy formula

$$\frac{1}{2}\int_\Gamma k(s)u^2(s)\,ds.$$

We can imagine some distributed load φ on Γ, so the equilibrium problem for a membrane with elastically supported boundary can be presented as a minimum problem for the total energy functional

$$\frac{1}{2}\int_\Omega (u_x^2 + u_y^2)\, dx\, dy + \frac{1}{2}\int_\Gamma k u^2\, ds - \int_\Omega f u\, dx\, dy - \int_\Gamma \varphi u\, ds.$$

The equilibrium equation, which is the VWP equation as well, is given by the first variation of this functional:

$$a\int_\Omega (u_x v_x + u_y v_y)\, dx\, dy + \int_\Gamma k u v\, ds$$
$$- \int_\Omega f v\, dx\, dy - \int_\Gamma \varphi v\, ds = 0, \qquad (2.17.1)$$

where u is a solution of the problem under consideration when this equality holds for all admissible functions v. Note that u and v need not satisfy any geometrical constraints except to be sufficiently smooth.

It is clear that Winkler's foundation should affect the boundary conditions for the equilibrium function. To see how this happens, we must repeat the variational derivation of the equations and the natural conditions. So let us integrate by parts in the first term of (2.17.1):

$$-a\int_\Omega (u_{xx} + u_{yy}) v\, dx\, dy + \int_\Gamma \left(a\frac{\partial u}{\partial n} + k u\right) v\, ds$$
$$- \int_\Omega f v\, dx\, dy - \int_\Gamma \varphi v\, ds = 0.$$

Taking the set of smooth v that vanish on Γ, we derive Poisson's equation $a\Delta u = -f$. Returning to the equation for v without restrictions on the boundary, we obtain

$$\left(a\frac{\partial u}{\partial n} + k u\right)\bigg|_\Gamma = \varphi. \qquad (2.17.2)$$

This mixed condition, containing the sum of a function and its normal derivative, can be used to pose the *third classical problem* for Poisson's equation.

Let us return to the VWP equation (2.17.1). We can use this to introduce the generalized setup of the problem. The quadratic portion can serve as the inner product in the energy space:

$$(u, v)_{MW} = a\int_\Omega (u_x v_x + u_y v_y)\, dx\, dy + \int_\Gamma k u v\, ds. \qquad (2.17.3)$$

If $k(s) > 0$ on some portion of Γ having non-zero length, and $k(s)$ is piecewise continuous on Γ, it is possible to show [Sobolev (1991)] that

$$\int_\Omega u^2 \, dx \, dy \leq C \, (u,u)_{MW} \tag{2.17.4}$$

for some constant C that does not depend on $u \in C^{(1)}(\Omega)$. So the norm induced by this inner product is equivalent to the norm of $W^{1,2}(\Omega)$. Hence we need not continue to discuss the existence and uniqueness questions for a generalized solution, but can simply refer to § 2.16 since our problem reduces to the following equation in $W^{1,2}(\Omega)$:

$$(u,v)_{MW} - \int_\Omega fv \, dx \, dy - \int_\Gamma \varphi v \, ds = 0, \tag{2.17.5}$$

where $(u,v)_{MW}$ is an inner product of $W^{1,2}(\Omega)$.

We could consider various mixed problems for which the condition (2.17.2) is given only on some portion of Γ. Indeed, (2.17.4) holds if we take the contour integral only along a portion of Γ where $k(s) > 0$.

We could also consider a load φ distributed along a curve *inside* Ω. The results stated above for the generalized solution remain valid. But this is not so for the classical formulation. Over the internal contour where the force is applied, we will have an additional natural "boundary" condition. This happens because a smooth solution u will be continuous together with its derivative in the direction tangent to the contour, while the derivative in the direction normal to the contour will have a jump. Indeed, for solutions that do not depend on y the membrane equation reduces to the string equation, and a force uniformly distributed over a straight line parallel to the y-axis becomes a point force for the corresponding string problem. At the point of application of such a force, we see a jump in the inclination angle of the string. For the membrane, a jump occurs in the normal derivative on the contour. A similar jump in the normal derivative occurs if a Winkler-type support is distributed over a contour inside Ω.

Exercise 2.17.1. *Derive the additional condition that arises at the points of a contour inside of Ω to which a distributed load φ is applied. Repeat for a linearly distributed Winkler support inside Ω.*

A non-classical problem for a membrane

Let us consider a non-classical boundary value problem that is intermediate between the Dirichlet and Neumann problems for a membrane. We suppose the membrane is attached to a contour Γ that maintains its shape but can execute small "rigid-body" motions in space. These motions can be described by linear polynomials $c_1 + c_2 x + c_3 y$ with arbitrary but fixed constants c_1, c_2, c_3. The boundary condition for the membrane is

$$u\big|_\Gamma = \psi(s) + (c_1 + c_2 x + c_3 y)\big|_\Gamma, \qquad (2.17.6)$$

where $\psi(s)$ describes the shape of the rigid contour. The constants c_k are not known in advance.

We assume the existence of a function $\tilde{u} \in W^{1,2}(\Omega)$ that satisfies

$$\tilde{u}\big|_\Gamma = \psi(s).$$

We seek a solution of the VWP equation (2.16.2) in the form

$$u(x,y) + \tilde{u}(x,y) + c_1 + c_2 x + c_3 y$$

where

$$u\big|_\Gamma = 0,$$

i.e., $u \in E_{MD}$ is an unknown function. We also seek c_1, c_2, c_3. This is neither a Dirichlet problem nor a Neumann problem. Furthermore, it is hard to solve without a consideration of mechanical meaning. With such consideration we can reduce it to problems treated earlier.

We begin by defining the class of admissible displacements v in (2.16.2). Since the contour can move as a rigid body and this is described by $c_1 + c_2 x + c_3 y$, the general form of an admissible displacement is $v(x,y) + d_1 + d_2 x + d_3 y$, where $v \in E_{MD}$ (hence $v|_\Gamma = 0$) and d_1, d_2, d_3 are arbitrary constants. Substitution of this admissible displacement into (2.16.2) gives

$$a \int_\Omega (u_x v_x + u_y v_y)\, dx\, dy + a \int_\Omega (u_x d_2 + u_y d_3)\, dx\, dy$$
$$+ a \int_\Omega (\tilde{u}_x v_x + \tilde{u}_y v_y)\, dx\, dy + a \int_\Omega (\tilde{u}_x d_2 + \tilde{u}_y d_3)\, dx\, dy$$
$$+ a \int_\Omega (c_2 v_x + c_3 v_y)\, dx\, dy + a \int_\Omega (c_2 d_2 + c_3 d_3)\, dx\, dy$$
$$- \int_\Omega f v\, dx\, dy - \int_\Omega f(d_1 + d_2 x + d_3 y)\, dx\, dy = 0. \qquad (2.17.7)$$

Next we use the arbitrariness of d_1, d_2, d_3. Setting all these to zero, we get the first necessary equation for equilibrium:

$$a \int_\Omega (u_x v_x + u_y v_y)\, dx\, dy + a \int_\Omega (\tilde{u}_x v_x + \tilde{u}_y v_y)\, dx\, dy$$
$$+ a \int_\Omega (c_2 v_x + c_3 v_y)\, dx\, dy - \int_\Omega fv\, dx\, dy = 0. \qquad (2.17.8)$$

Now we put $d_1 = 1$, $d_2 = 0$, $d_3 = 0$:

$$\int_\Omega f\, dx\, dy = 0. \qquad (2.17.9)$$

Similarly, with $d_1 = 0$, $d_2 = 1$, $d_3 = 0$ we get

$$a \int_\Omega u_x\, dx\, dy + a \int_\Omega \tilde{u}_x\, dx\, dy + a \int_\Omega c_2\, dx\, dy - \int_\Omega fx\, dx\, dy = 0 \qquad (2.17.10)$$

and with $d_1 = 0$, $d_2 = 0$, $d_3 = 1$,

$$a \int_\Omega u_y\, dx\, dy + a \int_\Omega \tilde{u}_y\, dx\, dy + a \int_\Omega c_3\, dx\, dy - \int_\Omega fy\, dx\, dy = 0. \qquad (2.17.11)$$

The set of four equations (2.17.8)–(2.17.11) is equivalent to (2.17.7).

Equation (2.17.9) is the self-balance condition for the external forces f. Since the membrane contour can move as a rigid body, in equilibrium the external forces must be self-balanced. We cannot define c_1 uniquely, since the membrane can freely shift in space through an arbitrary distance c_1.

What can we say about c_2 and c_3? When we try to rotate the membrane, on the contour we immediately see the normal projections of the internal tension forces that resist the rotation. So we see mechanically that c_2 and c_3 should be determined by the remaining equations. How should we construct a procedure that defines these constants?

Let us use the linearity of (2.17.8) in u to define u as a sum of three particular solutions. Supposing for a moment that c_2 and c_3 can be arbitrary, we see that u takes the form

$$u = u_1 + c_2 u_2 + c_3 u_3 + c_1, \qquad (2.17.12)$$

where u_1, u_2, u_3 all belong to E_{MD}; they must satisfy

$$a \int_\Omega [(u_1)_x v_x + (u_1)_y v_y]\, dx\, dy = -a \int_\Omega (\tilde{u}_x v_x + \tilde{u}_y v_y)\, dx\, dy + \int_\Omega fv\, dx\, dy, \qquad (2.17.13)$$

$$a \int_\Omega [(u_2)_x v_x + (u_2)_y v_y]\, dx\, dy = \int_\Omega xv\, dx\, dy, \qquad (2.17.14)$$

and
$$a \int_\Omega [(u_3)_x v_x + (u_3)_y v_y] \, dx \, dy = \int_\Omega yv \, dx \, dy. \qquad (2.17.15)$$

Because these equations are of the form considered in § 2.16, we can immediately state that each has a unique solution in E_{MD}. Note that (2.17.14) and (2.17.15) are particular cases of the Dirichlet problem for "forces" equal to x and y, respectively.

Once u_1, u_2, and u_3 have been determined, we can substitute (2.17.12) into (2.17.10) and (2.17.11). The resulting two equations are sufficient to determine c_2 and c_3 uniquely.

Exercise 2.17.2. *Prove this.*

Note that $\tilde{u}$ can be introduced in a non-unique way, and so we should prove uniqueness of the generalized solution separately. The reader is urged to do this.

Boundary conditions at a corner point

We know that mechanical meaning can help us solve problems. This is not its only advantage, however. When we derive mechanical equations, we often impose conditions that are not reflected in the equations. Commonly these specify that certain values must be small — even infinitesimal. When the equations are applied to real objects, however, such assumptions are often violated. This is definitely the case when, instead of seeking a smooth function u with small derivatives, we seek a generalized solution whose derivatives can be large at points. Many other examples could be given. Whenever we apply an equation to a situation in which its background assumptions are violated, we should provide some justification. This typically requires a clear mechanical understanding of the situation.

For example, let us consider the natural boundary condition for the solution of the Neumann problem for the membrane. In deriving the natural boundary condition an intermediate step was to consider the equation

$$a \int_\Gamma \frac{\partial u}{\partial n} v \, ds - \int_\Gamma \varphi v \, ds = 0. \qquad (2.17.16)$$

Applying the Main Lemma of the calculus of variations to the curvilinear integral and using the arbitrariness of v, we get the natural condition

$$\left. \frac{\partial u}{\partial n} \right|_\Gamma = \varphi(s). \qquad (2.17.17)$$

But suppose Γ has a *corner point* P located at $s = s_0$. To derive the above integral formula, we must apply Green's theorem. At P, however, the normal to Γ is not defined. Here we must consider two one-sided unit normals n^- and n^+. If we suppose that from each side there is some "one-sided limit" by continuity, and that φ is continuous, then

$$\left.\frac{\partial u}{\partial n^-}\right|_{s=s_0^-} = \varphi(s_0) \quad \text{and} \quad \left.\frac{\partial u}{\partial n^+}\right|_{s=s_0^+} = \varphi(s_0).$$

Both of these apply at P, in addition to continuity of $u(s)$ there. The result

$$\left.\frac{\partial u}{\partial n^-}\right|_{s=s_0^-} = \left.\frac{\partial u}{\partial n^+}\right|_{s=s_0^+} \qquad (2.17.18)$$

bears a superficial resemblance to continuity of the normal derivative, but of course it is not. We shall not pursue the issue of corner points further. We merely observe that the resulting problem is nonstandard and requires careful analysis of solution behavior near P. Such an analysis should be based on the idea that, although strange things can happen at P, a valid solution must have finite energy. This occurs not only for the membrane model but for more general models of elastic bodies. Engineers try to avoid corner points when designing elastic structures meant to carry significant loads. We see the possibility of unexpected behavior in mathematical solutions as well.

A great many "strange" mathematical effects are actually borne out somehow in engineering practice — either theoretically or experimentally. So it is not surprising that strange mechanical effects can show a mathematician where to expect difficulties in mathematical investigations with the model. Mathematical studies, in turn, can provide valuable information for the engineering community. The interest in strange effects is mutual.

2.18 Banach's Contraction Mapping Principle

This book presents ideas and models from mechanics, together with modern mathematical tools for their investigation. Banach's contraction mapping principle is used throughout mechanics, from the investigation of numerical tools to the properties of mechanical problems. We interrupt our consideration of mechanical models to consider this powerful principle.

Those who deal with numerical solutions in linear algebra are accustomed to iterative approaches. Such methods can be applied to simultaneous algebraic equations of the form

$$C\mathbf{x} - D\mathbf{x} = \mathbf{b}, \tag{2.18.1}$$

where $\mathbf{x} \in \mathbb{R}^n$ and the $n \times n$ matrices C and D are such that C is easily invertible and $\|C^{-1}D\| = q < 1$. In this case the system is solved by the scheme

$$\mathbf{x}_{n+1} = C^{-1}D\mathbf{x}_n + C^{-1}\mathbf{b}. \tag{2.18.2}$$

For any initial value $\mathbf{x}_0$, which is usually chosen as $\mathbf{x}_0 = \mathbf{0}$, the sequence of iterates $\{\mathbf{x}_n\}$ converges to a unique solution $\mathbf{x}^*$ of the system. Furthermore, the convergence rate is at least as fast as that of a geometric progression having common ratio q. Iterative methods are appropriate for solving large systems, when applicable, since errors do not accumulate during iteration.

In mechanics we encounter the *method of elastic solutions*. It is used to solve the equations of viscoelasticity and plasticity, with given boundary values for the displacement vector $\mathbf{u}$, that take the form

$$C\mathbf{u} = D\mathbf{u} + \mathbf{f}. \tag{2.18.3}$$

Here $\mathbf{f}$ is an external load vector and C is the differential operator of linear elasticity. The operator D characterizes special properties of the body; it can be linear or nonlinear and, in a certain sense, for "small" arguments, is majorized by the linear elastic term. The solution scheme for such equations coincides formally with the above iteration scheme for matrix equations, although it is more complicated.

The convergence of these and similar schemes is covered by a general result known as *Banach's contraction principle*. An attractive feature of this principle is its usefulness not only in justifying numerical methods, but in deriving many qualitative results in the theory of equations — in particular, theorems on existence and uniqueness of solutions. Abstract forms of the implicit function theorem rest on it as well. Banach's principle is one of the few results of nonlinear analysis that holds importance from both theoretical and practical points of view. We should mention its central role in establishing Picard's theorem on existence and uniqueness of solutions to the Cauchy problem for ordinary differential equations. In fact, Banach's principle originated with Picard's theorem.

We will present the contraction mapping theorem in the context of a Banach space B. The classical proof contains two results of practical importance. We begin with a definition that generalizes the above requirement for a matrix to have a norm of value $q < 1$.

Definition 2.18.1. An operator A acting in B is called a *contraction operator* on a subset S of B if there is a constant $q < 1$ such that for any $x, y \in S$ the inequality

$$\|A(x) - A(y)\| \leq q \|x - y\| \qquad (2.18.4)$$

holds.

We see that when A is applied to any pair of points, the images of these points are separated by a factor of q times the distance between the points themselves. The term "contraction" refers to this property of A. It is easily seen that on an open set S the operator A is continuous.

It turns out that many equations of mechanics can be phrased as

$$x = A(x), \qquad (2.18.5)$$

where A acts in some Banach space B. Note that a solution x^* to this equation has the property that $A(x^*) = x^*$; that is, the image of the solution under A must be the solution itself. We call such a point a *fixed point* of the operator A. We are now ready to formulate the first part of Banach's contraction principle.

Theorem 2.18.1. *Let A be a contraction operator on a subset S of a Banach space B, with contraction constant $q < 1$. Then A has no more than one fixed point in S.*

Proof. Supposing the existence of two fixed points x_1, x_2, we have

$$\|x_1 - x_2\| = \|A(x_1) - A(x_2)\| \leq q \|x_1 - x_2\|$$

since A is a contraction operator. Since $q < 1$ we have $x_1 = x_2$. □

The second result concerns the convergence of the iterative scheme

$$x_{n+1} = A(x_n), \quad n = 0, 1, 2, \ldots. \qquad (2.18.6)$$

Theorem 2.18.2. *Suppose*

(i) the operator A in a closed set S of a Banach space B is a contraction with constant $q < 1$:

$$\|A(x) - A(y)\| \leq q \|x - y\| \text{ for any } x, y \in S;$$

(ii) $A(x) \in S$ if $x \in S$.

Then for any initial point $x_0 \in S$

(i) the sequence $\{x_n\}$ with $x_n = A(x_{n-1})$, $n = 1, 2, 3, \ldots$, converges to a fixed point $x^* \in S$ of A, and
(ii) the running approximation error is given by

$$\|x_n - x^*\| \le \frac{q^n}{1-q} \|x_1 - x_0\|. \qquad (2.18.7)$$

Proof. Let us first demonstrate that $\{x_n\}$ is a Cauchy sequence. We begin by writing

$$\begin{aligned}
\|x_{n+1} - x_n\| &= \|A(x_n) - A(x_{n-1})\| \\
&\le q \|x_n - x_{n-1}\| \\
&= q \|A(x_{n-1}) - A(x_{n-2})\| \\
&\le q^2 \|x_{n-1} - x_{n-2}\| \\
&\vdots \\
&\le q^{n-1} \|A(x_1) - A(x_0)\| \\
&\le q^n \|x_1 - x_0\|.
\end{aligned}$$

Using this and the triangle inequality, we get

$$\begin{aligned}
\|x_{n+m} - x_n\| &\le \|x_{n+m} - x_{n+m-1}\| + \|x_{n+m-1} - x_{n+m-2}\| \\
&\quad + \cdots + \|x_{n+1} - x_n\| \\
&\le (q^{n+m-1} + q^{n+m-2} + \cdots + q^n) \|x_1 - x_0\| \\
&= q^n \frac{1 - q^m}{1 - q} \|x_1 - x_0\| \\
&\le \frac{q^n}{1-q} \|x_1 - x_0\|.
\end{aligned}$$

Since $q < 1$, we have

$$\|x_{n+m} - x_n\| \to 0 \quad \text{as } m, n \to \infty.$$

This means that $\{x_n\}$ is a Cauchy sequence; by completeness of B it must converge to a point x^*. Because S is closed, $x^* \in S$. At the same time, the sequence $\{A(x_n)\}$, which is $\{x_{n+1}\}$, is also a Cauchy sequence convergent to the same point x^*. Passing to the limit as $n \to \infty$ in the equality $x_{n+1} = A(x_n)$, we get $x^* = A(x^*)$. So x^* is the needed point. The error estimate (2.18.7) follows from the above estimate for $\|x_{n+m} - x_n\|$ as $m \to \infty$. □

Now we can assert

Theorem 2.18.3. *Let A be a contraction operator in a closed set S of a Banach space B and $A(S) \subseteq S$. Then*

(i) the equation $x = A(x)$ has a unique solution $x^* \in S$;
(ii) for any initial point $x_0 \in S$ the sequence $\{x_n\}$ converges to $x^* \in S$;
(iii) the approximation error at the nth iteration satisfies

$$\|x_n - x^*\| \le \frac{q^n}{1-q} \|x_1 - x_0\|.$$

A particular case of the theorem occurs when A is a contraction operator on all of B. Then A has a unique fixed point in B.

In Chapters 1 and 2 we introduced some principles of continuum mechanics and some methods of pure mathematics with which its problems are now studied. We applied the latter to relatively simple mechanical models which, nonetheless, contain the salient features of more complex models. In Chapter 3 we proceed to a consideration of the main ideas and models of continuum mechanics using the methods we have developed.

Chapter 3

Theory of Elasticity: Statics and Dynamics

3.1 Introduction

Everyday materials exhibit a range of atomic structures. Some are simple and regular; others are complex. So the internal forces acting on an atomic level must differ widely. In continuum mechanics, however, we see a relatively simple type of macro-representation of the internal forces in terms of the stress tensor. How can a simple mathematical structure cover so many possibilities? The answer lies in the way the stress tensor presents an average picture of atomic and molecular interactions. Some authors tell us that continuum mechanics yields results concerning certain "elementary" or "infinitesimal" portions of a material. We are asked to accept that such a portion will contain a typical number of atoms. But while it is interesting to wonder about the difference between real microinteractions and the simplified picture given by the continuum mechanics approach, such considerations properly lie within the realm of the physicist.

Similar comments apply to the problem of describing material deformation. Atomic or molecular structure is complex and differs greatly between materials. So the motion of separate atoms during deformation can be complex. However, in the macro-level integral picture we describe the displacements of points using a relatively simple strain tensor. This yields an average picture of deformation in the body, neglecting the motions of individual atoms.

Engineering experience shows that, to within practical accuracy, the tensor tools of continuum mechanics give good results; we apply these in engineering design despite the fact that they are not valid pointwise. These tools are based on the continuum and solidification principles discussed in Chapter 2. Why are they so good? The reason is that metals and other engineering materials are typically subjected to relative strains on the order of a few thousandths. Clearly, on average, the same changes are seen in the distances between adjacent atoms in the material. Note that in continuum mechanics we deal only with additional quantities that arise during deformation. The relations between changes in interatomic forces and relative changes in interatomic distances are approximated by relatively simple (e.g., linear) functions. Hence the integral quantities we see in practice are given by relatively simple relations as well. These relations we describe with simple tools — stress and strain tensors — to within accuracy sufficient for our purposes.

3.2 An Elastic Bar Under Stretching

In § 1.20 we quoted a few equations related to the model of an elastic bar. We now examine this model in detail. Within its simple framework we will examine certain ideas that apply to all linearly elastic bodies.

Consider the equilibrium of a prismatic bar under an axial force F, as shown in Fig. 3.1.

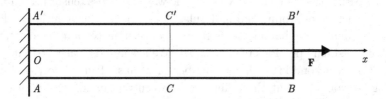

Fig. 3.1 Bar under an axial force.

We cut the bar along CC' (Fig. 3.2) in the direction perpendicular to the axis. To maintain equilibrium of the portion $CC'B'B$ we must apply reaction forces produced by the remainder $AA'C'C$. We expect these to be distributed over the cross section rather than concentrated at a specific point. Although the force distribution could be quite nonuniform in reality, we make the simplifying assumption that it is uniform (in addition to our tacit assumption that it acts only through the cut). Note that this contradicts our original picture where F was applied along Ox. Indeed, we can imagine moving CC' toward BB', eventually encountering a cross section with a distributed load on one side and a concentrated load on the other. Hence we should take F as distributed instead. In accordance with common sense we suppose the distribution σ of F is uniform with $\sigma = F/S$. The solidification principle says that the resultant force $F' = \sigma' S$ on the left should be directed opposite to F and of the same value, which leads to the equality $|\sigma'| = |\sigma|$. Each cross section has two sides and the reactions from each side are equal but opposite in direction. By convention, we take as positive those reactions which place the bar under tension as in Fig. 3.2. Reactions that place the bar under compression are regarded as negative. In the present case we have $\sigma' = \sigma$.

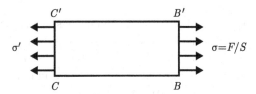

Fig. 3.2 A portion of the bar from Fig. 3.1. F_1 has been changed to an axial uniformly distributed load with density $\sigma = F/S$, where S is the cross-sectional area.

In a sense, the force density σ is not unlike the pressure p in a liquid or gas. The resultant force on a planar area in the latter case can be calculated by multiplying σ by the area S. But the analogy is limited. Pascal's law states that the pressure in a liquid acts normally to any small planar area element inside the liquid and, moreover, its resultant depends only on the element area. Let us see what happens if we cut the bar along DD' at angle α as shown in Fig. 3.3. It is reasonable to suppose that the force remains uniformly distributed, but this means the load still acts along Ox and is therefore *not* normal to DD'. Moreover, because the area of DD' is $S/\sin\alpha$, we must take $\sigma'' = \sigma \sin\alpha$ in order to preserve the resultant; this dependence on cut orientation is not seen in a liquid. We will call σ the *stress*. In the theory of the bar, we will not require the stress values on cross sections that are not perpendicular to the bar axis. Hence we will not make use of quantities such as σ''. But we must understand that in a solid this quantity depends not only on the location of a point but on the orientation of the cut used to define it. This will lead to the notion of a tensor of stress within the body.

So we have introduced the stress in the bar by the formula

$$\sigma = F/S, \tag{3.2.1}$$

and can determine the resultant acting over each cross section. We can also

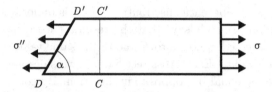

Fig. 3.3 In cross section DD', the density of the uniformly distributed reaction changes to $F\sin\alpha/S$; their direction is not normal to DD'.

calculate the support reaction at AA', which has the same value F. Let us complicate the problem by introducing external forces distributed along the axis. We suppose that such forces would be uniformly distributed over the cross section of application; hence we assume a uniform distribution of stress over any cross section. To get the value of σ at a cross section we will sum up all external forces, lumped and distributed, acting on the right end and divide by the cross-sectional area.

At this point it is not clear why we have introduced the stress instead of using simple reaction forces at a cross section. Indeed, the latter approach is taken in the strength of materials. It might seem that we have overcomplicated things. Since our goal is to discuss the theory of elasticity, however, we are preparing all concepts that will be needed later (in §§ 3.6, 3.8, 3.11, and 3.13).

The above problem seems simple; it suffices to apply an equation of statics. Although we suspect at least some deformation in any bar under load, we have neglected this and treated an undeformed bar. The systems considered by the strength of materials, in which all forces inside the body as well as all support reactions are uniquely determined by the tools of statics only, are called *statically determinate*.

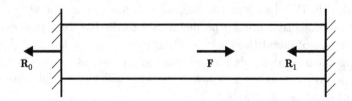

Fig. 3.4 Bar with both ends clamped, under an axial force **F**.

A system that cannot be treated this way is shown in Fig. 3.4. Indeed, the only equation we have from statics is that the force F and the reactions R_0 and R_1 are balanced:

$$F - R_0 - R_1 = 0. \qquad (3.2.2)$$

From this we cannot obtain definite values for R_0 and R_1. A system for which we cannot find all reactions uniquely by the tools of statics alone is said to be *statically indeterminate*. To find the reactions and hence the stresses in the bar, we must make use of additional information. We shall exploit the simplest of all constitutive equations (i.e., the equations that relate stresses to strains): Hooke's law.

For a spring, Hooke's law is formulated as

$$F = k \frac{\Delta l}{l}. \qquad (3.2.3)$$

Here k is the spring constant and Δl is the increase in the length l of the spring under an applied force F. The same idea can be applied to a cable or bar under tension. We characterize the deformation by the relative extension $\Delta l/l$, which is called *strain*. This characteristic can be introduced for a bar if we suppose that, during deformation, any normal cross section shifts uniformly and as a whole along the x-direction as was done above. In this way we can characterize the location of the cross section after deformation by the displacement function $u(x)$ of the points of Ox. For our purposes, we need the local value of the strain when the deformation is not uniform along Ox. We take a small portion of the bar between $[x, x + \Delta x]$, so that $l = \Delta x$, and consider its change under deformation. If the displacements of the segment ends are $u(x)$ and $u(x + \Delta x)$ (Fig. 3.5) then the change in length of the portion is $u(x + \Delta x) - u(x)$ and the strain is

$$\frac{u(x + \Delta x) - u(x)}{\Delta x}.$$

Letting $\Delta x \to 0$ we obtain

$$\varepsilon = \frac{du(x)}{dx}. \qquad (3.2.4)$$

Note that ε is dimensionless. For ordinary structural steel it should not exceed about 0.001, hence it is "very small" from an engineering standpoint. This allows us to ignore the fact that the point x at which we find ε does in fact move during deformation.

We can now generalize Hooke's law as

$$\sigma = E\varepsilon. \qquad (3.2.5)$$

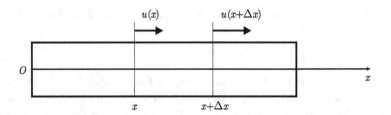

Fig. 3.5 Deformation of a small portion of the bar.

The coefficient E, known as *Young's modulus*, carries units of pressure; its value for ordinary steels is around 2×10^5 MPa. In terms of the reaction N in a bar, Hooke's law takes the form

$$N = ES\varepsilon \tag{3.2.6}$$

where S is the cross-sectional area. Although we have previously considered homogeneous prismatic bars, we can extend the theory by assuming that E and S can depend on x. For this, we assume all the previous assumptions on the uniform distribution of the external load over a cross section must be fulfilled.

With the additional variable ε involved, to define everything uniquely we need the expression relating strain with displacement (3.2.4), the constitutive law (3.2.6), the equilibrium equation which we will derive, and supplementary boundary conditions.

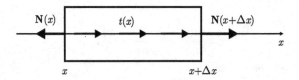

Fig. 3.6 A portion of the bar between x and $x + \Delta x$ under load.

As above, we will isolate a portion of the bar between x and $x + \Delta x$. It is subject to the reactions and the distributed load $q(x)$ shown in Fig. 3.6. The arrow for **N** shows its positive direction on the cross section. When we construct the equilibrium equation in scalar form, we should use the projections onto the axis. Supposing the linear density of the distributed load to be $t(x)$, we get the equilibrium equation:

$$N(x+\Delta x) - N(x) + \int_x^{x+\Delta x} t(s)\,ds = 0.$$

Dividing this by Δx and letting $\Delta x \to 0$, we obtain

$$\frac{dN(x)}{dx} + t(x) = 0. \qquad (3.2.7)$$

This is a static equation that is valid for any constitutive law. Substitution from (3.2.4) and (3.2.6) gives the equilibrium equation for an elastic bar in terms of displacements:

$$\frac{d}{dx}\left(ES\frac{du}{dx}\right) + t(x) = 0. \qquad (3.2.8)$$

This second-order differential equation for $u(x)$ should be supplemented with two boundary conditions, one at each end. Forces or displacements could be specified in any combination.

In the strength of materials, the system of external forces is simple. Normally the load consists of a few lumped forces or a linearly distributed load. Solution can proceed with elementary tools, using linearity of the problem and describing the deformation of finite portions a of the bar by the formula

$$N = ES\frac{\Delta a}{a}. \qquad (3.2.9)$$

The differential equation (3.2.8) is normally used when S depends on x. Lumped forces can be included by incorporating delta-functions into $t(x)$; a term $F_k\,\delta(x - x_k)$ can be used to describe a force F_k applied at x_k.

Let us apply this technique to the simple problem shown in Fig. 3.4. Assume the bar has length l and the force F is applied at $x = a$. We seek the support reactions. We begin by making a cut at the right end and applying a fictitious reaction R_1, which we regard as given. The other reaction is

$$R_0 = F - R_1. \qquad (3.2.10)$$

Supposing R_1 to be given, we transformed the system to a statically determinate one. But this does not change the fact that we still have only one equation for two variables R_0 and R_1. To establish another relationship we will use Hooke's law. Since the portion of the bar that lies to the left of $x = a$ is stretched by a force R_0, its elongation is $\Delta a = aR_0/ES$. The remainder, having length $b = l - a$, is compressed with a force R_1, hence its length decreases by $\Delta b = bR_1/ES$. The total displacement of the right end with respect to the left end is $\Delta a - \Delta b$. Since both ends are fixed, however, we must have $\Delta a - \Delta b = 0$. In the strength of materials this is called a *compatibility condition*. In terms of the reactions, it reads

$$\frac{aR_0}{ES} - \frac{bR_1}{ES} = 0$$

or

$$R_1 = R_0 \frac{a}{b}.$$

The equilibrium equation gives

$$R_0 = \frac{b}{l}F, \qquad R_1 = \frac{a}{l}F.$$

In this case we need not know E and S; it suffices to know that they are constants. To find the displacements we would need numerical values for E and S.

Note that we did not solve the equilibrium equation explicitly. Rather, we used the fact that the tension N is constant in that portion of the bar where no external forces are applied. This is equivalent to solving the equilibrium equation. Because the equation is of second order, we need two boundary conditions: these were that the bar ends are fixed.

This idea — that a statically indeterminate system can be transformed into a determinate one by sectioning, introducing fictitious reaction forces, and supplementing the equilibrium equations with additional equations describing the deformations of the system — is called the *deformation method*. It can be applied to complex problems.

Boundary value problems

We could state results for the bar simply by adapting those for the string, since the equations describing these two objects have the same form. But perhaps we should carry out a few steps independently for the bar, as E and S can vary. We suppose $E(x)S(x)$ is continuously differentiable on $[0, l]$, at least. We start by deriving the virtual work principle for the bar.

We have consistently said that we expect $u(x)$ to be sufficiently smooth. Let us explain what this means in the present case. Suppose $t(x)$ is continuous on $[0, l]$, but that a set of lumped forces F_k also act at points x_k, which can include the endpoints. At points other than the x_k, the equilibrium equation (3.2.7) shows that dN/dx is continuous. If $E(x)S(x)$ is continuously differentiable and nonzero, then $u(x)$ is twice continuously differentiable there.

Although lumped forces were not considered when we derived the equilibrium equation, they are common in engineering practice. It is useful to know what happens near the point x_k of application of a lumped force F_k. We continue to suppose $t(x)$ is continuous. Consider a small portion of the bar corresponding to the segment $[x_k - \Delta, x_k + \Delta]$, $\Delta > 0$. The equilibrium of this portion is determined by the equation

$$-N(x_k - \Delta) + F_k + \int_{x_k - \Delta}^{x_k + \Delta} t(s)\, ds + N(x_k + \Delta) = 0.$$

As $\Delta \to +0$, the value of the integral tends to zero by continuity of $t(x)$. We get

$$N(x_k + 0) - N(x_k - 0) = -F_k, \qquad (3.2.11)$$

where $N(x_k - 0)$ and $N(x_k + 0)$ are the one-sided limits of $N(x)$ as x tends to x_k from the left and right, respectively.

Exercise 3.2.1. *Comment on the smoothness of $u(x)$ at a point where $E(x)S(x)$ has a jump. For simplicity assume there is no lumped force at this point.*

According to (3.2.11), $N(x)$ has a jump at x_k and therefore so does $u'(x)$. This means the equilibrium equation holds only on the intervals (x_k, x_{k+1}) which exclude the points of application of the lumped forces. So it must be considered as the set of equations

$$\frac{d}{dx}\left(ES\frac{du}{dx}\right) + t(x) = 0, \qquad x \in (x_k, x_{k+1}) \quad (k = 0, 1, \ldots, n), \qquad (3.2.12)$$

along with the jump conditions (3.2.11). We should supplement these with the continuity conditions for $u(x)$ at the x_k, which state that the bar cannot be broken:

$$u(x_k - 0) = u(x_k + 0) \qquad (k = 1, \ldots, n-1). \qquad (3.2.13)$$

Finally, we require two conditions at the endpoints x_0 and x_{n+1}.

We have obtained a multipoint boundary value problem: equation (3.2.12) is supplied with boundary conditions at the ends of the bar, but also with conditions at the intermediate points x_k. In engineering practice the functions $t(x)$, $E(x)$, and $S(x)$ are likely to be simple and it should be easy to integrate the equation explicitly over (x_k, x_{k+1}). This allows engineers to solve the problem graphically, using simple rules.

Now we derive the VWP equation and the expression for the total energy. Suppose the virtual displacement function $\delta u(x)$ is continuous together with its first and second derivatives on $[0, l]$. It must also vanish at points where the bar is fixed. We multiply (3.2.8) by $\delta u(x)$ and integrate over the length to obtain

$$\sum_{k=0}^{n} \int_{x_k}^{x_{k+1}} \left[\frac{d}{dx}\left(ES\frac{du}{dx}\right) + t(x) \right] \delta u(x)\, dx = 0.$$

Integrating by parts in the first term, we have

$$\int_0^l ES\frac{du}{dx}\frac{d(\delta u)}{dx}\, dx - \int_0^l t(x)\, \delta u(x)\, dx - \sum_{k=0}^{n} ES\frac{du}{dx}\delta u(x)\bigg|_{x=x_k+0}^{x=x_{k+1}-0} = 0.$$

Taking into account that $N = ESu'$, that $\delta u(x)$ is continuous at x_k, and the jump equation (3.2.11), we finally get

$$\int_0^l ES\frac{du}{dx}\frac{d(\delta u)}{dx}\, dx - \int_0^l t(x)\, \delta u(x)\, dx - \sum_{k=0}^{n} F_k\, \delta u(x_k) = 0. \qquad (3.2.14)$$

Each term is the work of some force over the virtual displacement δu. Equation (3.2.14) expresses the virtual work principle for the bar. A word statement for the equilibrium problem under consideration is as follows.

> Among all functions $u(x)$ that satisfy the geometrical constraints of the clamped ends, the one that satisfies (3.2.14) for all sufficiently smooth virtual displacements that also satisfy the geometrical constraints ($\delta u = 0$ at the clamped ends) is the solution of the equilibrium problem.

In § 2.7 we derived the equilibrium equation for a string under load, which looks like the bar equation. In view of the form taken by the total energy functional for the string, we can try the expression

$$\mathcal{E}(u) = \frac{1}{2}\int_0^l ES\left(\frac{du}{dx}\right)^2 dx - \int_0^l t(x)u(x)\, dx - \sum_{k=0}^{n} F_k\, u(x_k) \qquad (3.2.15)$$

as the total energy for the bar. It should be a functional for which the left-hand side of (3.2.14) is the first variation. Direct calculation confirms this choice: $\mathcal{E}(u)$ is the total energy of the bar-load system, where the first term is the strain energy of the bar.

Thus we can regard the equilibrium problem for the bar as the problem of minimum of the total energy functional. The lumped forces at the ends impose natural boundary conditions on $u(x)$; the remaining conditions (3.2.11) are similar to the natural conditions at the ends, but they arise at intermediate points x_k where the values of u' jump. We can use the integro-differential equation (3.2.14) to introduce generalized setups for the Dirichlet, Neumann, and mixed equilibrium problems for the bar. The steps are strictly analogous to those for the string, and we leave them to the reader.

Exercise 3.2.2. *Derive the natural boundary conditions and the conditions at x_k that follow from the problem of minimum of the total energy functional.*

Exercise 3.2.3. *An initially unstretched spring lies along the axis of the bar treated in Exercise 3.2.2. One end of the spring is attached to the bar at a point $x = c \neq x_k$, while the other end is clamped to an external support. The spring constant is K, hence the constitutive law is $F = Kv$ where v is the displacement of the spring end along the bar. The strain energy of the spring is $\frac{1}{2}Kv^2$. Using the variational procedure, find the equilibrium equations and the natural conditions.*

Exercise 3.2.4. *Formulate the VWP equation for the bar under load ($t(x)$ and F_k for $k = 0, \ldots, n$) when there are also springs attached to the bar. The springs, having elastic coefficients c_k, are attached at points z_k ($k = 1, \ldots, m$) and their other ends are clamped. Consider the cases when, for the unloaded bar, the springs are (1) unstressed, and (2) prestressed. Define a generalized solution, introduce the energy space, and formulate the existence-uniqueness theorem for a generalized solution.*

We finish this section with a few remark on the problem of longitudinal dynamics of the bar. This can be treated simply via d'Alembert's principle. We assume a dependence of the displacement function u on time τ and incorporate inertial forces into the distributed external forces $t = t(x, \tau)$:

$$t(x) \mapsto t(x, \tau) - \rho \frac{\partial^2 u(x, \tau)}{\partial \tau^2},$$

where ρ is the density along the bar. The VWP equation becomes

$$\int_0^l \left(ES \frac{\partial u}{\partial x} \frac{\partial \delta u}{\partial x} + \rho \frac{\partial^2 u(x, \tau)}{\partial \tau^2} \delta u \right) dx$$
$$- \int_0^l t(x, \tau) \, \delta u \, dx - \sum_{k=0}^n F_k \, \delta u(x_k) = 0.$$

Using techniques from the calculus of variations, we can obtain the equation in differential form (which we can obtain from (3.2.8) as well):

$$ES\frac{\partial^2 u(x,\tau)}{\partial x^2} = \rho\frac{\partial^2 u(x,\tau)}{\partial \tau^2} - t(x,\tau).$$

3.3 Bending of a beam

The bending of a beam is one of the most important problems in engineering mechanics. As with the problem of stretching a bar, the mathematical tools needed for its solution are quite elementary. Engineers have even found elementary graphical methods for treating large structures consisting of many beams. Nonetheless, we shall consider the beam problem in order to demonstrate the application of the methods and theorems we have developed.

We consider a straight beam of length l on which act a distributed transverse load $q(x)$ and lumped forces P_k that are perpendicular to the beam. We assume a planar deformation of the beam, hence the forces are parallel to this plane.

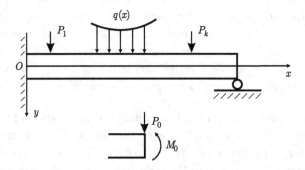

Fig. 3.7 A beam under load. Inset: a cantilever end.

The beam deformation is described with a single function $w(x)$, the deflection of the midline of the beam. In Fig. 3.7 we show a beam along with two ways in which its end can be "fixed". The left end is clamped, which corresponds to the conditions $w = 0$ and $w' = 0$ at $x = 0$. The right end is prevented from moving in the y-direction but is free to rotate; this end is said to be freely supported. The inset shows another type of boundary condition that corresponds to a cantilever beam with a given force and moment.

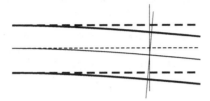

Fig. 3.8 The normals to the midline of a beam before and after deformation.

We will consider the classical theory of beam bending, which starts with hypotheses laid down by D. Bernoulli (1700–1782). We suppose that, during deformation, any straight line segment normal to the midline does not change its length and, moreover, remains normal to the midline after deformation (Fig. 3.8). The midline is defined by the centroids of the beam cross sections; hence, for a rectangular beam it is the set of intersections of the cross-sectional diagonals. Although Bernoulli's hypotheses are only approximate, they provide a good approximation for many civil engineering problems. They allow us to express the strain of the beam at any point in terms of a single function $w(x)$. As a physical model of beam deformation, we can consider the deformation of an elastic midline that resists bending according to certain rules. The resistance of the beam to external forces is described by two quantities: the bending moment M and the shear force Q. When we cut the beam along the cross section, we replace the action of the remainder of the beam by the reactions M and Q. As for the bar, we can regard a cross section — the boundary between the left and right portions of the beam — as "attached" to one portion. The values of M and Q for the two portions must be equal in absolute value but, by Newton's third law, opposite in direction since the cross section itself is in equilibrium. In Fig. 3.9 we show positive directions for M and Q for a segment of beam between cross sections separated by Δx. We recall the solidification principle, which states that in equilibrium under the action of all external forces and the reaction forces from the remainder of the body, any portion of the body must obey the equilibrium law for a rigid body. When we take a section of bar between two cross sections as shown in Fig. 3.9 and suppose the lumped external forces and moments are absent on the segment, we have two equilibrium equations: one for the vertical forces, which is

$$(Q + \Delta Q) - Q + \int_x^{x+\Delta x} q(s)\,ds = 0,$$

and one for the moment with respect to the center point of the left cross section,

$$(M + \Delta M) - M - (Q + \Delta Q)\Delta x - \int_x^{x+\Delta x} sq(s)\,ds = 0.$$

Dividing by Δx and letting $\Delta x \to 0$ we get two equilibrium equations

$$\frac{dM}{dx} = Q, \qquad \frac{dQ}{dx} = -q(x). \tag{3.3.1}$$

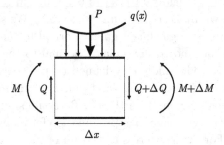

Fig. 3.9 A portion of the beam between x and $x + \Delta x$ under load. The arrows are shown for positive directions of M, Q, q, P.

At point x_k an external lumped force P_k and a lumped moment M_k are given. Supposing that Q and M are continuous from each side at x_k, we obtain

$$M(x_k + 0) - M(x_k - 0) + M_k = 0,$$
$$Q(x_k + 0) - Q(x_k - 0) + P_k = 0. \tag{3.3.2}$$

Hence $M(x)$ and $Q(x)$ have jumps at the point x_k.

For a cantilever beam (i.e., a beam having one end clamped and a force and moment assigned at the other end), equations (3.3.1)–(3.3.2) are sufficient to determine the force characteristics in any cross section; this is an example of a statically determinate problem. But if both ends are clamped, it is impossible to determine the strain state and we must utilize the properties of the beam material. Bernoulli's hypotheses, along with the assumption that the material obeys Hooke law, relate M with w through the formula

$$M = -EI\frac{d^2w}{dx^2}. \qquad (3.3.3)$$

See, for example, [Gere and Timoshenko (1997)] or [Lebedev and Cloud (2004)]. Here E is Hooke's modulus and I is the cross-sectional moment of inertia. Substituting this into (3.3.1), we get the equilibrium equation in terms of the normal deflection $w(x)$:

$$\frac{d^2}{dx^2}\left(EI\frac{d^2w}{dx^2}\right) = q(x). \qquad (3.3.4)$$

This holds at all points excepts those where lumped external forces and moments are applied.

Let us first consider an equilibrium problem for a beam, with constant values of E and I, under the action of a piecewise continuous distributed load $q(x)$. Normally, by the term "classical solution" we mean a solution continuous together with all its derivatives up to the order of the differential equation. We will continue to use this term, however, when a solution has a finite set of singular points at which its behavior can be determined. The problem under consideration is so simple that we can solve it analytically and find conditions for uniqueness. First let us discuss the question of boundary conditions. The general solution to (3.3.4) depends on four independent constants and takes the form

$$w(x) = c_0 + c_1 x + c_2 x^2 + c_3 x^3 + w_p(x), \qquad (3.3.5)$$

where $w_p(x)$ is its particular solution. To determine four constants uniquely, we must formulate four additional conditions. These can be two conditions at $x = 0$ — say $w(0) = 0$ and $w'(0) = 0$ — and two others at $x = l$. Say, for the case of Fig. 3.7, we put $w(l) = 0$ and $w''(l) = 0$. The reader may wonder whether we could place all four conditions at $x = 0$, on w, w', w'', and w''' say. In fact this would give us a Cauchy or initial value problem of the type considered in textbooks on ordinary differential equations. But an engineer must deal with conditions at the other end of the beam. So here we must consider a boundary value problem. It is easy to verify that the four above conditions, two at each end, define the constants uniquely if we know w_p.

What happens to the solution when lumped forces and moments come into play? We previously suggested that the reader investigate the equilibrium equations at such singular points. It is easily seen that if an external couple is applied at some point, then M has a jump at this point. Some books explain that $Q = dM/dx$ has a corresponding δ-function term. The δ-function, which is not a classical function, is a relatively new object in mathematics. Engineers proceeded to use it well before its formalization, however; they would divide a beam into sections by the points of application of lumped forces and couples, while placing some additional conditions at the points (consisting of the two continuity conditions for w and w', and the conditions for the jumps in M and Q). In doing so, they were far ahead of their time; it was only later that so-called generalized solutions would be considered by mathematicians working in the theory of differential equations. These engineers were aided by the use of graphical tools applicable to simplified linear force distributions of the form $q(x) = a + bx$.

Nonetheless, we will consider the beam equilibrium problem in order to explain some peculiarities that arise in more general problems of elasticity where analytic solutions are not available. So we temporarily ignore the solution methods used by engineers and proceed to consider the problem in a complex way — a way that will suffice for more complex problems.

We begin by deriving the VWP equation from the equilibrium equation (3.3.4). We restrict ourselves to the case of a beam with clamped ends:

$$w(0) = 0 = w'(0), \quad w(l) = 0 = w'(l). \tag{3.3.6}$$

Suppose the beam carries a piecewise continuous load $q(x)$, along with lumped forces P_k and moments M_k at the points x_k ($k = 1, 2, \ldots, n-1$). Elementary calculations show that the solution $w(x)$ of the equilibrium problem is continuous together with its first derivative $w'(x)$ on $[0, l]$. The functions $w''(x)$ and $w'''(x)$ are discontinuous only at the points x_k, while $w^{(4)}(x)$ is discontinuous at the x_k and at the jump points of $q(x)$. Let us take a virtual displacement $\delta w(x)$ that belongs to $C^{(2)}(0, l)$ and obeys the geometrical constraints at the ends:

$$\delta w(0) = 0 = \delta w'(0), \quad \delta w(l) = 0 = \delta w'(l). \tag{3.3.7}$$

We recall that the δ-notation is traditional in mechanics; we could just as well write $h(x)$ or $\varphi(x)$ instead of $\delta w(x)$.

We remember that (3.3.4) holds on any interval that does not contain one of the points x_k. We suppose

$$0 = x_0 < x_1 < x_2 < \cdots < x_n = l.$$

Let us multiply (3.3.4) by $\delta w(x)$, integrate the result over each segment

(x_k, x_{k+1}), and then add the equalities for all k. We get

$$\sum_{k=0}^{n-1} \int_{x_k}^{x_{k+1}} \left[\frac{d^2}{dx^2}\left(EI\frac{d^2w}{dx^2}\right) \right] \delta w(x)\, dx - \sum_{k=0}^{n-1} \int_{x_k}^{x_{k+1}} q(x)\, \delta w(x)\, dx = 0.$$

Clearly

$$\sum_{k=0}^{n-1} \int_{x_k}^{x_{k+1}} q(x)\, \delta w(x)\, dx = \int_0^l q(x)\, \delta w(x)\, dx$$

is the work of distributed load $q(x)$ over virtual displacement $\delta w(x)$.

Now let us transform the first sum. We wish to make the integrand symmetric in w and δw, which will lead us to the expression for the strain energy. So we apply integration by parts to each of the integrals twice. Because we are not assuming that $w''(x)$ or the higher-order derivatives are continuous, we use the one-sided notation $x_k + 0$ and $x_k - 0$. We obtain

$$\sum_{k=0}^{n-1} \int_{x_k}^{x_{k+1}} \left[\frac{d^2}{dx^2}\left(EI\frac{d^2w}{dx^2}\right) \right] \delta w(x)\, dx = \int_0^l EI \frac{d^2w}{dx^2} \frac{d^2(\delta w)}{dx^2}\, dx$$

$$+ \sum_{k=0}^{n-1} \frac{d}{dx}\left(EI\frac{d^2w}{dx^2}\right)\delta w \bigg|_{x=x_k+0}^{x=x_{k+1}-0} - \sum_{k=0}^{n-1} EI\frac{d^2w}{dx^2}\frac{d(\delta w)}{dx}\bigg|_{x=x_k+0}^{x=x_{k+1}-0}.$$

As δw and $\delta w'$ are continuous at x_k and vanish at $x = 0$ and $x = l$, we have

$$\sum_{k=0}^{n-1} \frac{d}{dx}\left(EI\frac{d^2w}{dx^2}\right)\delta w \bigg|_{x=x_k+0}^{x=x_{k+1}-0} - \sum_{k=0}^{n-1} EI\frac{d^2w}{dx^2}\frac{d(\delta w)}{dx}\bigg|_{x=x_k+0}^{x=x_{k+1}-0}$$

$$= \sum_{k=1}^{n-1}\left(EIw'''\delta w\bigg|_{x=x_k-0} - EIw'''\delta w\bigg|_{x=x_k+0}\right)$$

$$- \sum_{k=1}^{n-1}\left(EIw''\delta w'\bigg|_{x=x_k-0} - EIw''\delta w'\bigg|_{x=x_k+0}\right)$$

$$= \sum_{k=1}^{n-1}\left(EIw'''\bigg|_{x=x_k-0} - EIw'''\bigg|_{x=x_k+0}\right)\delta w(x_k)$$

$$- \sum_{k=1}^{n-1}\left(EIw''\bigg|_{x=x_k-0} - EIw''\bigg|_{x=x_k+0}\right)\delta w'(x_k).$$

In terms of $w(x)$, equations (3.3.2) are written as

$$-EIw''|_{x=x_k+0} + EIw''|_{x=x_k-0} + M_k = 0,$$
$$-EIw'''|_{x=x_k+0} + EIw'''|_{x=x_k-0} + P_k = 0. \tag{3.3.8}$$

Hence the sum of the non-integrated terms takes the form

$$-\sum_{k=1}^{n-1} Q_k \, \delta w(x_k) + \sum_{k=1}^{n-1} M_k \, \delta w'(x_k).$$

Combining the above formulas, we obtain

$$\int_0^l EI \frac{d^2 w}{dx^2} \frac{d^2(\delta w)}{dx^2} \, dx - \sum_{k=1}^{n-1} Q_k \, \delta w(x_k)$$
$$+ \sum_{k=1}^{n-1} M_k \, \delta w'(x_k) - \int_0^l q(x) \, \delta w(x) \, dx = 0. \qquad (3.3.9)$$

The terms

$$A(\delta w) = \sum_{k=1}^{n-1} Q_k \, \delta w(x_k) - \sum_{k=1}^{n-1} M_k \, \delta w'(x_k) + \int_0^l q(x) \, \delta w(x) \, dx$$

represent the work of the external load over the virtual displacement $\delta w(x)$. If in (3.3.9) we call the first integral term (with a negative sign) the work of the internal forces over virtual displacements $\delta w(x)$, then (3.3.9) expresses the virtual work principle for the beam: the work of all internal and external forces over any virtual displacement of the beam is zero. The reader should consider why the M_k terms differ in sign from the other terms in $A(\delta w)$.

It is easy to verify that the functional for which the left side of (3.3.9) is the first variation is

$$\mathcal{E}(w) = \frac{1}{2} \int_0^l EI {w''}^2(x) \, dx - \sum_{k=1}^{n-1} Q_k w(x_k)$$
$$+ \sum_{k=1}^{n-1} M_k w'(x_k) - \int_0^l q(x) w(x) \, dx. \qquad (3.3.10)$$

By analogy with the theory of the elastic bar, we call this the *total energy* of the beam under load and regard $\frac{1}{2} \int_0^l I {w''}^2(x) \, dx$ as the strain energy of the beam. This analogy is not merely formal: if we calculate the work required to deform the beam from the undeflected state to the one defined by the function $w(x)$, we get this expression for the energy accumulated due to the work. We should note that the expressions for the total energy and other work quantities remain valid when the ends of the bar are not clamped; in that case we simply include the work of the given forces on virtual displacements of the ends in the expression for the complete work $A(\delta w)$.

We will use these relations to introduce and analyze a generalized setup of the equilibrium problem for a beam.

3.4 Generalized Solutions to the Equilibrium Problem for a Beam

For a beam with clamped ends, we have established that the virtual work principle is represented by the equation

$$\int_0^l EI \frac{d^2 w}{dx^2} \frac{d^2(\delta w)}{dx^2} \, dx - \sum_{k=1}^{n-1} Q_k \, \delta w(x_k)$$
$$+ \sum_{k=1}^{n-1} M_k \, \delta w'(x_k) - \int_0^l q(x) \, \delta w(x) \, dx = 0. \qquad (3.4.1)$$

We assumed the existence of a solution $w(x)$, having certain smoothness properties, of the equilibrium problem for a beam with clamped ends. With this $w(x)$, equation (3.4.1) holds for any $\delta w(x)$ from the class described in the previous section. It is easy to see that we can reverse the transformations of that section and demonstrate that, assuming (3.4.1) holds for all admissible virtual displacements on the intervals outside x_k, where $w(x) \in C^{(4)}$, the function $w(x)$ is a solution of the equilibrium equation (3.3.4); moreover, the x_k are singular points of $w(x)$ where the jump relations (3.3.8) hold. It is instructive to derive a condition at those points where $q(x)$ jumps. So in some sense, (3.4.1) is equivalent to the complete formulation of the equilibrium problem for a beam under load.

Now we will tackle this problem without using the common engineering approach of partitioning the beam, etc. Instead we will require that the equilibrium equation hold, in the VWP sense, on the whole segment $[0, l]$. A more classical approach to the problem, with its fourth-order equation, would generate a solution $q(x)$ involving the δ-function and its derivatives at the x_k. Solutions of this type are not normally considered in textbooks on ordinary differential equations. But here we will deal with the generalized approach based on (3.4.1). One advantage is that we will not assume — but rather will prove — the existence of a solution. Clearly there are other advantages and disadvantages, but the principal advantage is that we can extend the approach to equilibrium problems in three-dimensional elasticity and obtain similar results on existence and uniqueness. This will be done in situations where the simple reasoning used in the strength of materials does not apply.

There are two ways in which we can consider this question of how to formulate a generalized setup of the equilibrium problem for a clamped beam. We can start with the virtual work principle as expressed in (3.4.1). We can also start with the problem of minimum of the total energy functional (3.3.10), for which (3.4.1) expresses the equality to zero of its first variation. These approaches are equivalent. In both cases we should delineate the space of the elements in which we will seek a minimizer of (3.3.10) or a solution to (3.4.1). The space is the same in each case. It will be an energy space: that is, a Hilbert space.

First we introduce the set $C_0^{(2)}$ of all functions $w(x)$ that are twice continuously differentiable on $[0, l]$ and satisfy the boundary conditions $w(0) = 0 = w(l)$ and $w'(0) = 0 = w'(l)$. On $C_0^{(2)}$ we define an inner product using the quadratic part of the expression in (3.4.1):

$$(w_1, w_2)_B = \int_0^l EI \frac{d^2 w_1}{dx^2} \frac{d^2 w_2}{dx^2} \, dx. \tag{3.4.2}$$

Of course, the reader should verify satisfaction of the inner product axioms. As is typical of the spaces introduced in the energy approach, our inner product space is incomplete.

Definition 3.4.1. The energy space E_B for the problem of equilibrium of a beam with clamped ends is the completion of $C_0^{(2)}$ with respect to the norm induced by the inner product $(\cdot, \cdot)_B$.

This definition says that the second derivatives of an element of E_B belong to $L^2(0, l)$. In a certain sense then, this element can have a jump in its second derivative as required by (3.3.8). Let us examine some other properties of such an element. For the first derivatives of an element of E_B, the situation is exactly as for an element of the energy space for a string or a bar: the base functions upon which we constructed E_S are continuously differentiable and vanish at the ends. Hence any representative Cauchy sequence $\{w_n(x)\}$ of an element $w \in E_S$ is such that the sequence $\{w_n'(x)\}$ converges uniformly to a continuous function $\phi(x)$. The same thing clearly holds for the sequence $\{w_n(x)\}$, so it converges uniformly to some function $\psi(x)$ on $[0, l]$. This means that we see the convergence of the sequence $\{w_n(x)\}$ in the space $C^{(1)}(0, l)$. By completeness, we have $\psi'(x) = \phi(x)$. We will redenote this function $\psi(x)$ as $w(x)$, the same notation as for the base functions w with which we started constructing E_B. We recall that this continuously differentiable function $w(x)$ does not depend on the choice of representative sequence. Moreover, the imbedding operator from E_B to $C^{(1)}(0,l)$, as it follows from the imbedding result in E_S, is continuous; this is expressed through the inequality

$$\|w(x)\|_{C^{(1)}(0,l)} \leq c \|w(x)\|_B \qquad (3.4.3)$$

where c is a constant that does not depend on $w(x) \in E_B$. We know that despite the appearance of the same notation, the elements on the left- and right-hand sides are really different: on the left $w(x)$ is truly a function; on the right it is a class of equivalent Cauchy sequences that all converge to the function $w(x)$. In the sequel we shall use the notation $w(x)$ in both senses and the reader should bear this in mind.

So now we can introduce two definitions for a generalized solution of the problem under consideration. The first is based on the principle of minimizing the total energy functional.

Definition 3.4.2. A generalized solution for the problem of equilibrium of a beam with clamped ends is an element $w(x) \in E_B$ that minimizes the functional (3.3.10).

The other definition is based on the VWP equation. It can be shown that the two definitions both specify the same element.

Definition 3.4.3. A generalized solution for the problem of equilibrium of a beam with clamped ends is an element $w(x) \in E_B$ that satisfies equation (3.4.1) for all $\delta w(x) \in E_B$.

Let us consider (3.4.1) in light of Definition 3.4.3. The meaning of this equation is as follows. When we substitute the elements $w(x)$ and $\delta w(x)$ of E_B, we then take a representative sequence from each and pass to the limit in each term of (3.4.1). Then we substitute the results into the equation, which must hold in the limiting sense, after the limit passage in each term. Because of (3.4.3), the results of the limit passages in all the non-integrated terms of (3.4.1) are the same as if we had simply substituted the values for the limit functions on the left side of (3.4.3). So we can consider these terms as ordinary functions defined by the imbedding theorems.

Next, for both definitions of a generalized solution, we should establish the properties of the functional describing the work of the external load. It is easy to see that the work of the external load $A(w)$ is a linear functional. If we suppose that $q(x) \in L(0,l)$, then by imbedding inequality (3.4.3) we have

$$|A(w)| \leq c_0 \|w(x)\|_B \qquad (3.4.4)$$

with a constant c_0 that does not depend on $w \in E_B$. Thus $A(w)$ is a linear continuous functional in E_B and so, by the Riesz representation theorem, it can be represented as

$$A(w) = (w, w_0)_B \qquad (3.4.5)$$

with some uniquely determined element $w_0(x) \in E_B$. It is easy to see that $w_0(x)$ is a solution of the problem under consideration in the sense of both Definitions 3.4.2 and 3.4.3. We leave it to the reader to formulate a corresponding uniqueness-existence theorem.

Finally, we should mention the problem of a beam free of geometric constraints. Because the beam can move as a rigid body, the solidification principle requires self-balanced external forces as a necessary condition for equilibrium. In an approximate model, however, the self-balance condition can be modified. In this case there are two conditions: the resultant force in the direction normal to the beam must be zero, and the moment of all external forces and moments with respect to some point must be zero.

For the generalized setup of the free beam, we use the VWP and total energy expressions for the clamped beam; we must, however, add the work of the external load at the beam ends. An arbitrary small displacement of the beam as a rigid body is given as $w = a + bx$ where a and b are constants. If $w(x)$ is a solution of the equilibrium problem for a free beam, then (3.4.1) must hold for $\delta w_0(x) = a + bx$ with any constants a and b. For this $\delta w_0(x)$, the quadratic term in the VWP is zero. Taking $a = 1$, $b = 0$, and then $a = 0$, $b = 1$, we get two equations for the external load; these are the self-balance conditions for the external load as required by the solidification principle.

What follows is merely a sketch of the subsequent procedure. To introduce a generalized solution, we wish to use the same form of the inner product — defined by the internal energy of the beam — as for the beam with clamped ends. However, we find that $\|a + bx\|_B = 0$. So we cannot use this inner product directly. Fortunately, the rigid displacement $a + bx$ is the only smooth function for which $\|w(x)\|_B = 0$. So we can announce that the class of all elements of the form $a + bx$ constitutes the zero element of the new space, and so consider a factor space. This is a mathematically clear way to pose and solve the equilibrium problem. But we can reduce the problem to a simpler one. A solution of the problem under consideration is defined up to the rigid displacement $a + bx$. This suggests that we select from each class a unique function $w(x) + a + bx$. We should select this unique function so that the set of representative functions is a linear space and the energy norm on the new set has all the norm properties. We will do this in such a way that afterwards, in the new energy space, the form of the imbedding theorem for the clamped beam will not change. The conditions for selecting the unique function from the class may be

$$w(0) = 0, \quad w'(0) = 0;$$

they may also be integral conditions such as

$$\int_0^l w(x)\,dx = 0, \quad \int_0^l w'(x)\,dx = 0.$$

After that, on the set of functions twice continuously differentiable on $[0, l]$ and that satisfy one pair of the conditions above, we consider the inner product $(\cdot, \cdot)_B$ and introduce the energy space using the completion theorem. The remaining steps replicate those for the beam with clamped ends. It is interesting that in this beam problem with additional restrictive conditions, formulation of the existence-uniqueness theorem does not require self-balanced external loads. Indeed, we have imposed additional geometric constraints on the beam that were absent in the original problem. When we formulate the existence theorem for the original beam, the self-balance conditions arise necessarily. Uniqueness of the generalized solution for a free beam is guaranteed up to a rigid motion $a + bx$.

3.5 Generalized Setup: Rough Qualitative Discussion

The type of solution we are considering, in addition to being known as a generalized or energy solution, is frequently termed a *weak solution*. When the novice sees integrands containing only derivatives of order less than the order of the corresponding differential equation, he may think that generalized solutions are non-smooth almost everywhere. Consequently he may believe that such solutions should be introduced when non-smooth loading parameters appear in the problem. This is only somewhat correct. Of course, non-smooth loading parameters do yield non-smooth solutions. But in many cases we should solve problems that do not have classical solutions at all. Consider, for example, the following simple problem for a beam under a lumped force P at the center point as shown in Fig. 3.10.

The analytic representation of the solution on each of the intervals $(0, 1)$ and $(1, 2)$ is given by a third-order polynomial, so the solution depends on eight constants. To find these, we need eight equations. The boundary

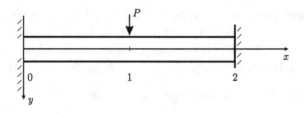

Fig. 3.10 A clamped beam of length 2, loaded with a lumped force P at $x = 1$.

conditions
$$w(0) = 0 = w'(0), \quad w(2) = 0 = w'(2),$$
provide four of them. We also have continuity conditions at $x = 1$:
$$w(1 - 0) = w(1 + 0), \quad w'(1 - 0) = w'(1 + 0).$$
Finally, we have the conditions for the lumped load (3.3.2):
$$EIw''(x_k + 0) = EIw''(x_k - 0),$$
$$-EIw'''(x_k + 0) + EIw'''(x_k - 0) + P = 0.$$
The last equation says that $w'''(x)$ has a jump at $x = x_k$. When we consider the equilibrium problem based on the equation
$$(EIw''(x))'' = q(x),$$
the jump condition is external: it does not follow from the equilibrium equation on the entire beam, but is derived separately via mechanical considerations. Moreover, with a jump in the third derivative of the solution, we cannot consider a classical solution that has fourth derivatives everywhere. So we arrive at the area of generalized solutions. The fact that engineers found a simple way to circumvent this difficulty for beam equilibrium problems does not change the fact that there are many problems for which this is not possible. We should really introduce generalized solutions when a classical setup does not make sense.

The good news is that a generalized solution is not *as* generalized as it might seem at first glance. If a linear mechanical problem based on an elliptic equation or system has a classical solution that possesses all needed derivatives, and if the corresponding energy of the state is finite, then the solution is the generalized solution as well. Indeed, despite the fact that the corresponding generalized solution belongs to an energy space (produced by the completion theorem), it will contain a stationary sequence — each element of which is the classical solution. In this sense the classical solution is also a generalized one.

But there is even better news. Despite the fact that we define the elements of Sobolev and energy spaces as whole entities over some domain S in $\mathbb{R}^n$, they possess certain local properties of additional smoothness in the domain that we can use. For example, from the general theory that lies outside the scope of this book, a generalized solution to the equation $\Delta u = f$ in S, an open bounded domain of $\mathbb{R}^2$, belongs to $W^{2,p}(B)$ for a closed set B in S if $f \in L^p(S)$, where $p > 1$. So the generalized solution has not only the first but the second generalized derivatives on B and these are also in $L^p(B)$. By Sobolev's theorem, this means that the imbedding operator presents a continuous function on B that corresponds to the solution. In all subsequent calculations we can use this function on B instead of the element of the energy space. This situation resembles that for the beam under lumped loads: outside of certain singular portions of S, we can use ordinary continuous functions to describe the solution. On the singular sets and their neighborhoods, we must use the form that is defined as an element of the energy space. For the above problem, the singular set is not the whole boundary of S; rather, it consists only of the points where the boundary is not smooth, such as corner points.

We should add that the equations of the most popular numerical method for the solution of mechanical equilibrium problems — the finite element method — are based on the VWP equation. Hence convergence of the method is always shown as convergence of approximations to the energy solution. The same holds for any variational numerical method applicable to these problems.

We now turn to the description of equilibrium problems for more general deformations. The groundwork was laid by Cauchy, who developed the notion of stress tensor for a solid; this served to generalize the notion of pressure for a liquid or gas.

3.6 Pressure and Stresses

When dealing with material points or rigid bodies, it suffices to consider lumped forces having certain directions and lines of action. For deformable bodies we must consider volume-distributed forces; one cannot shift the point of application of such a force, as this will alter the deformation of the body. As earlier for the bar, we suppose we can describe the action of one portion the body on another by introducing a cut along with suitable reaction forces over this surface. One such reaction is the *pressure*, defined in elementary physics as force per unit area:

$$p = F/S. \tag{3.6.1}$$

This "definition" assumes that F acts normally to the planar area S and is distributed uniformly over that area. Pascal's law for the pressure in a liquid or gas also fits this definition. We can consider a nonuniform distribution of pressure over a surface, introducing a value of p at some point $\mathbf{r}$ via the limit expression

$$p = \lim_{\Delta S \to 0} \frac{\Delta F}{\Delta S}, \tag{3.6.2}$$

where ΔF is the normal force acting on small area ΔS. Here we should use a limit process in which ΔS tends to zero *along with* the maximum distance from $\mathbf{r}$ to all other points of ΔS (i.e., ΔS cannot tend to a segment). Pressure thus defined plays the role of force density over the surface. Knowing this density, we can reconstruct the integral characteristics — resultant force and moment — of the distributed force acting on any finite part of the surface.

Again, because of Pascal's law, pressure is a suitable tool for describing the force distribution inside a gas or liquid. The pressure at any point in a liquid, acting on a small (we could say infinitesimal) planar area, does not depend on the orientation of that area. But consider a point on the surface of a bar stretched by two equal forces F_0 applied at the ends. In an infinitesimal surface tangent to the bar, the normal force, and thus the "pressure", is zero. In the normal cross section it is nonzero and seems close to the average F_0/S_0, where S_0 is the area of the bar cross section. Hence the pressure alone cannot define the strain state in a solid material. We need another tool.

The first idea is to include forces acting on a surface in an arbitrary direction, not just normally. To this end we employ a vectorial version of the above limit construction. Let $\Delta \mathbf{F}$ be the resultant of forces distributed over a small area ΔS of the surface S. Then the "density" of the force distribution at a point is

$$\boldsymbol{\sigma} = \lim_{\Delta S \to 0} \frac{\Delta \mathbf{F}}{\Delta S}. \tag{3.6.3}$$

The new symbol $\boldsymbol{\sigma}$ stands for "stress", an extension of the notion of pressure. The above example of a stretched bar shows that $\boldsymbol{\sigma}$ can depend on the spatial orientation of the infinitesimal surface used to define it. This leads to the notion of stress tensor. Cauchy demonstrated that the stress values acting on any three mutually orthogonal infinitesimal areas that pass through a point of interest uniquely define the "state of stress" at that point.

We have not discussed how the force distribution on a surface can be seen physically. It is clear that no continuous "material" surface can exist inside a body composed of atoms. In fact we have implicitly used the first principle of continuum mechanics: we have employed an ideal model of a continuum without any atoms, where we can draw surfaces and employ any description of the distribution of some characteristic that is given by a smooth function depending on the coordinates. In § 3.10 we will discuss the "coordinates" of the points of a body; this problem is not simple, and for now we shall continue in the footsteps of Cauchy. We will do everything in terms of the Cartesian coordinates (x_1, x_2, x_3) in the volume occupied by the *deformed* body that is in equilibrium. First, we must define the dependence of $\boldsymbol{\sigma}$ on the orientation of the infinitesimal area, which we take to be planar. Let us produce a crosscut inside the body. In doing so, we have "deleted" some internal constraints inside the body. The crosscut is a two-sided surface. Experience in classical mechanics tells us that if we wish to have this cut be an ideal surface and insure that nothing changes in the remainder of the deformed body, we should add appropriate reaction forces distributed over the cut. Here we encounter two questions: (1) Why shouldn't we add the distributed moments that act over the cut? (2) Why shouldn't we consider reactions other than those acting on the sides of the cross section? In both cases the answer is the same: there is no logical reason to ignore these possibilities. Moreover, there are different versions of continuum mechanics in which reactions of these types are included (e.g., Cosserat mechanics). Thus, in making this assumption on the reaction, Cauchy effectively introduced an axiom for continuum mechanics. Cauchy's version of the theory is typically used in engineering practice when dealing with samples of metal, wood, etc, because it agrees with experimental data. So we return to our development. Suppose $\mathbf{n}$ is the "exterior" unit normal to one of the flat cross sections as shown in Fig. 3.11.

We denote the stress on this side of the cut as $\boldsymbol{\sigma}_\mathbf{n}$. The exterior unit normal on the other side is $-\mathbf{n}$, so we can define the corresponding stress on it as $\boldsymbol{\sigma}_{-\mathbf{n}}$. These quantities $\boldsymbol{\sigma}_\mathbf{n}$ and $\boldsymbol{\sigma}_{-\mathbf{n}}$ act on an ideal elementary square that offers no other type of resistance; its equilibrium is assured only if we

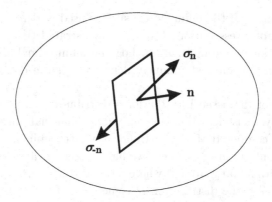

Fig. 3.11 Stresses on the cross section.

suppose that

$$\sigma_\mathbf{n} = -\sigma_{-\mathbf{n}}. \qquad (3.6.4)$$

In some books this equality is called *Cauchy's lemma*.

Note that for a cut perpendicular to the x_k-axis, for which the positive direction gives the direction of the exterior normal, we will use the notation σ_k for the stress.

Following Cauchy, we cut out a small upright pyramid in the deformed body, three faces of which are parallel to the Cartesian coordinate planes; the fourth is a plane with exterior unit normal $\mathbf{n}$ as shown in Fig. 3.12.

We will show that at point (x_1, x_2, x_3), the vertex opposite side $S_\mathbf{n}$, the relation between the stresses is

$$\sigma_\mathbf{n} = \sigma_1 n_1 + \sigma_2 n_2 + \sigma_3 n_3, \qquad (3.6.5)$$

where $\mathbf{n} = (n_1, n_2, n_3)$. Note that this represents a linear dependence $\mathbf{n} \to \sigma_\mathbf{n}$. This transformation can be represented by a matrix composed of the components of σ_k. The vectors $\mathbf{n}$ and $\sigma_\mathbf{n}$ are defined uniquely; they must not depend on the coordinate systems used in the spaces of vectors $\mathbf{n}$ and $\sigma_\mathbf{n}$. So this transformation, termed the stress tensor, must also be frame independent. Also note that, unlike the objects of linear algebra, the matrix components of the transformation have units.

We first apply the solidification principle to the pyramid. Under all the forces applied, the pyramid is in equilibrium. On the face perpendicular to the x_k-axis the resultant reaction force is

$$[-\sigma_k(x_1, x_2, x_3) - \beta_k]\operatorname{mes}(S_k),$$

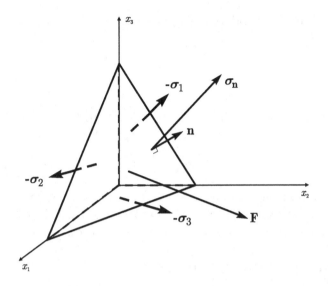

Fig. 3.12 Equilibrium of a pyramid in a deformed body. $-\sigma_k$ is the vector-stress on the upright sides perpendicular to the x_k-axis, and σ_n is the vector-stress on the side having normal **n**. **F** is the density of external forces acting on the pyramid.

where $\text{mes}(S_k)$ is the area of face S_k and the vector β_k (which depends on S_k) tends to zero when $\text{mes}(S_k) \to 0$. The negative sign is a consequence of (3.6.4) and the fact that the unit normal of the face is opposite the direction of the x_k-axis. Similarly, the resultant stress on S_n is

$$[\sigma_n(x'_1, x'_2, x'_3) + \beta_n]\,\text{mes}(S_n),$$

where (x'_1, x'_2, x'_3) is a point of S_n and $\beta_n \to 0$ when $\text{mes}(S_n) \to 0$. The resultant of all other applied forces is proportional to the volume V of the pyramid. These could include gravitational or electrostatic forces, as well as inertia forces. So we believe that all these are given as

$$\mathbf{F} = \gamma V, \tag{3.6.6}$$

where γ is bounded for all small V. The resultant of all the above forces, by the solidification principle, must vanish:

$$[-\sigma_1(x_1, x_2, x_3) - \beta_1]\,\text{mes}(S_1) +$$
$$[-\sigma_2(x_1, x_2, x_3) - \beta_2]\,\text{mes}(S_2) +$$
$$[-\sigma_3(x_1, x_2, x_3) - \beta_3]\,\text{mes}(S_3) +$$
$$[\sigma_n(x'_1, x'_2, x'_3) + \beta_n]\,\text{mes}(S_n) + \gamma V = 0.$$

Dividing by mes($S_\mathbf{n}$) and taking into account that

$$\frac{\text{mes}(S_k)}{\text{mes}(S_\mathbf{n})} = n_k,$$

we get

$$[\boldsymbol{\sigma}_\mathbf{n}(x_1', x_2', x_3') + \boldsymbol{\beta}_\mathbf{n}] = [\boldsymbol{\sigma}_1(x_1, x_2, x_3) + \boldsymbol{\beta}_1]n_1$$
$$+ [\boldsymbol{\sigma}_2(x_1, x_2, x_3) + \boldsymbol{\beta}_2]n_2$$
$$+ [\boldsymbol{\sigma}_3(x_1, x_2, x_3) + \boldsymbol{\beta}_3]n_3 - \gamma \frac{V}{\text{mes}(S_\mathbf{n})}.$$

Now we produce the limit passage when all sides of the pyramid tend to zero proportionally so the direction of $\mathbf{n}$ does not change. We suppose, by the continuum principle, continuity of all functions describing the stresses. During the limit passage, $(x_1', x_2', x_3') \to (x_1, x_2, x_3)$ for all points of V and $V/\text{mes}(S_\mathbf{n}) \to 0$. By the properties of the vector-functions $\boldsymbol{\beta}$, we come to the relation (3.6.5).

So the stress vectors on three mutually orthogonal, infinitesimal, plane areas at a point define the value of the stress on an infinitesimal plane of any orientation at the same point. Each of the stress vectors $\boldsymbol{\sigma}_k$ can be represented in the Cartesian frame using three components $\boldsymbol{\sigma}_k = (\sigma_{k1}, \sigma_{k2}, \sigma_{k3})$. Hence the set of nine components σ_{ij} uniquely defines the stress at a point on an arbitrarily directed infinitesimal area. Note that we could have begun with a pyramid in a non-Cartesian frame. The result would be the same: some nine quantities — components — uniquely define the distribution of stresses at the point. Evidently we have encountered an object whose properties differ from those of a force vector. It is called Cauchy's stress tensor. Now it is time to ask the following question.

Is continuum mechanics an exact mathematical science?

Yes and no. It is an exact science for the models it employs. But it is an approximate science relative to the behavior of real bodies.

In constructing Cauchy's stress tensor, we used a limiting process to obtain the relation (3.6.5) between the stresses acting on infinitesimal areas at a point. At this stage, many books on continuum mechanics contain a discussion regarding the size of the elementary volume V at which we should terminate the limit passage to obtain a reliable result; any body has an atomic structure, and clearly we cannot say anything about the results of the limit passage when V becomes less than the volume of a single atom. But such a procedure contradicts the very definition of limit: *while producing the limit as some quantity tends to zero, we cannot stop at some nonzero finite stage and announce that we have completed the limit passage.* To use the tools of calculus, we must let V tend to zero without restriction. We have done this above. We will encounter the same situation in deriving the equations of equilibrium, the expressions for strains, etc. So what do the equations we derive in continuum mechanics really mean?

Before answering this question we recall a few points from thermodynamics. This science treats various aspects of heat transfer and the behavior of bodies under heating, while trying to maintain the viewpoint that gases and other objects consist of atoms and molecules in motion. Hence thermodynamics is sometimes regarded as a simplified consequence of statistical mechanics. In fact, however, this is not quite the case because statistical mechanics deals with models of gases which do not cover many of the objects of thermodynamics.

But if thermodynamics is so regarded — i.e., as a very imperfect science whose results should be proved by statistical mechanics — then why is this not the case for continuum mechanics as well? Its objects of interest also consist of moving atoms. The answer is simple: we have no analogue of present-day statistical mechanics available to describe solids. Thus we see textbooks present continuous mechanics as though it were an absolute science like geometry or arithmetic. And this is correct in the same sense that geometry provides a description of real objects. The main objects of geometry — points, lines, and surfaces — do not exist in reality. The main objects used in continuum mechanics are also "absolute," but are approximate as representations of real bodies: we cannot talk about the stress on an infinitesimal area that happens to be smaller than an atomic cross section. So in continuum mechanics we deal with idealized objects. But continuum bodies and their relations give us an approximation to phenomena on the macro-level, when the atomic structure of materials can be neglected.

The stress tensor is an idealization: it is the result of some averaging process for the "real" relations between atoms in a body. The same can be said of all the tools and equations we encounter in continuum mechanics.

So the continuum mechanics we find in textbooks is a mathematical science. In some ways it is no less abstract than geometry or arithmetic. It is regarded as part of engineering only because engineers use it in practice. It belongs to the portion of mathematics that considers idealized objects subject to idealized relations.

One additional aspect is rarely mentioned. If continuum mechanics provides only an approximation for real bodies, then is it not inexact? The answer is "yes and no". Continuum mechanics is exact in the way it treats its own models, but these models are only approximate representations of real objects. It is possible to construct finite models of deformable bodies in which limit passages are not used, but which give us some approximations to actual processes. In a certain sense such finite models can be considered as independent models of mechanics even when they are constructed as approximations to other models of continuum mechanics. This is the case, for example, with various versions of the Finite Element Method. Such finite approximations, being constructed with proper understanding, normally obey the main principles of continuum mechanics. The equations of FEM, being a finite model of the deformation of a body, are not "worse" and, in some situations, less exact than the boundary value problem based on the "ideal" differential equations of continuum mechanics.

One consequence of this is the following. When mathematicians study a numerical method for some application, they attempt to prove convergence of the resulting approximations to the exact solution of the problem. If convergence cannot be justified, the reason may be weakness in our technique or simply bad properties of the approximate method. The logical equality of infinitesimal and finite models for mechanical bodies means that each model can be treated independently. If convergence can be justified, on the other hand, then the method has nice properties. However, the question as to which type of model is better — the infinitesimal or the finite — can be answered only through engineering practice.

Before discussing the mechanical aspects of the stress tensor, we would like to introduce tensors in general. We will encounter several of these in our subject, and it will be helpful to understand the properties they share.

3.7 Vectors and Tensors

We have already discussed vectors in this book. It bears repeating that, although the vectors we know from linear algebra reflect the properties of forces, the sets of vectors from classical mechanics sometimes possess additional properties. We must work carefully with such objects. For us, it is essential to regard a vector as an "entity" that can be defined uniquely through specification of its components in any particular coordinate frame. Once its components are known in one frame, they can be calculated in any other frame by certain well-defined rules.

The same holds for tensors, which we will present in a more restrictive way than is done in algebra. A main theme in our introduction of tensors is that these entities are objective like vectors: that is, they do not depend on the choice of basis. For example, second-rank tensors will represent objective linear transformations between spaces of objective vectors, as was the case for the stress tensor above. But we begin with a formal introduction of the second-rank tensors and only later relate them to the set of linear transformations between vectors. Our formal theory of tensors is based on the three-dimensional Euclidean space of vectors (the discussion will hold for two-dimensions as a special case). We then see how the object we have called the "stress tensor" fits into this theory.

We start with the tensor product of two vectors **a** and **b**. Although this is often denoted by $\mathbf{a} \otimes \mathbf{b}$, we will write it simply as **ab** and call it a *dyad*. Dyads are the simplest tensors of the second rank, and are such that $\mathbf{ab} \neq \mathbf{ba}$ if $\mathbf{a} \neq \mathbf{b}$. Now in the Euclidean space of vectors, there are operations of addition and multiplication by a scalar α. We relate these to the tensor product by requiring the following to hold:

$$(\mathbf{a}+\mathbf{b})\mathbf{c} = \mathbf{ac} + \mathbf{bc}, \qquad (3.7.1)$$

$$\mathbf{a}(\mathbf{b}+\mathbf{c}) = \mathbf{ab} + \mathbf{ac}, \qquad (3.7.2)$$

$$\alpha\,\mathbf{ab} = (\alpha\mathbf{a})\mathbf{b} = \mathbf{a}(\alpha\mathbf{b}). \qquad (3.7.3)$$

The reader can verify that if we consider each dyad as a vector (of another space, not the initial three-dimensional Euclidean space) and introduce all the possible algebraic sums of dyads

$$\sum_k \alpha_k \mathbf{a}_k \mathbf{b}_k,$$

we arrive at a new linear space of vectors based on the initial three-dimensional space. This is the *space of second-rank tensors*. It is easy to see that it is nine-dimensional. Moreover, if $(\mathbf{e}_1, \mathbf{e}_2, \mathbf{e}_3)$ is a basis for the three-dimensional space, the set of nine dyads

$$\{\mathbf{e}_1\mathbf{e}_1, \mathbf{e}_1\mathbf{e}_2, \mathbf{e}_1\mathbf{e}_3, \mathbf{e}_2\mathbf{e}_1, \mathbf{e}_2\mathbf{e}_2, \mathbf{e}_2\mathbf{e}_3, \mathbf{e}_3\mathbf{e}_1, \mathbf{e}_3\mathbf{e}_2, \mathbf{e}_3\mathbf{e}_3\}$$

is a basis for the space of second-rank tensors. To represent a dyad **ab** with respect to this basis, we can write

$$\mathbf{a} = \sum_k \alpha^k \mathbf{e}_k, \quad \mathbf{b} = \sum_m \beta^m \mathbf{e}_m,$$

and use the above properties:

$$\mathbf{ab} = \sum_{k=1}^{3}\sum_{m=1}^{3} \alpha^k \beta^m \mathbf{e}_k \mathbf{e}_m. \qquad (3.7.4)$$

So the component of the dyad $\mathbf{ab}$ corresponding to the dyad $\mathbf{e}_k \mathbf{e}_m$ is $\alpha^k \beta^m$. In the same manner, we can obtain the representation of any tensor $\mathbf{A}$ of the new linear space by using the above basis of nine dyads:

$$\mathbf{A} = \sum_{k=1}^{3}\sum_{m=1}^{3} a^{km} \mathbf{e}_k \mathbf{e}_m. \qquad (3.7.5)$$

The a^{km} are called the *components* of $\mathbf{A}$ in the above basis. We can tabulate these in a 3×3 matrix analogous to the component representation of a vector. Tensors play a role in continuum mechanics similar to that played by vectors in classical mechanics. They represent certain objective quantities that do not depend on the choice of coordinate frame.

But if the new space is an ordinary vector space, why must we introduce a new idea? The answer is that we wish to preserve the relationship between the elements of the tensor space and the initial Euclidean space under linear transformations of the coordinate frame of the Euclidean space. The transformation (3.7.4) of a dyad shows that a linear transformation of the coordinate frame leads to a nonlinear dependence on the coefficients of the transformation in the components of a tensor. So the rules of transformation for tensors differ from the rules in ordinary nine-dimensional linear space.

A second-rank tensor can be put in correspondence with a linear transformation from one set of vectors to another. When both kinds of vectors are objective — that is, do not depend on the bases of the spaces — so is the transformation itself. As for the vectors, this allows us to find the relations between the components of a tensor given in different bases. Objectivity is a requirement for tensors of any rank.

It is worth noting that the matrix representation of a tensor allows us to introduce other useful operations with tensors which correspond to the product of matrices and to the product of a matrix by a vector. We will identify a second-rank tensor with a linear transformation acting in three-dimensional vector space. Moreover, the transformation rules for the tensor components allow us to construct the operations in such a way that they are preserved under transformations of the basis of the initial Euclidean space.

Thus our main question is as follows. Suppose a tensor is an objective entity. Given its components with respect to one basis, and given the law describing transformation of bases from one frame to the next, what are its components with respect to a new basis?

Suppose two bases $\mathbf{e}_1, \mathbf{e}_2, \mathbf{e}_3$ and $\mathbf{e}'_1, \mathbf{e}'_2, \mathbf{e}'_3$ are linearly related through the formulas

$$\mathbf{e}_k = \sum_{m=1}^{3} A_k^m \mathbf{e}'_m. \tag{3.7.6}$$

In the first basis a tensor has the representation

$$\mathbf{A} = \sum_{t,s=1}^{3} \alpha^{ts} \mathbf{e}_t \mathbf{e}_s.$$

By (3.7.6) we obtain

$$\mathbf{A} = \sum_{t,s=1}^{3} \alpha^{ts} \sum_{m=1}^{3} A_t^m \mathbf{e}'_m \sum_{n=1}^{3} A_s^n \mathbf{e}'_n = \sum_{m,n=1}^{3} \left(\sum_{t,s=1}^{3} \alpha^{ts} A_t^m A_s^n \right) \mathbf{e}'_m \mathbf{e}'_n.$$

So the components $\alpha^{mn\prime}$ of $\mathbf{A}$ in the new basis $\{\mathbf{e}'_m \mathbf{e}'_n\}$ are

$$\alpha^{mn\prime} = \sum_{t,s=1}^{3} \alpha^{ts} A_t^m A_s^n$$

and

$$\mathbf{A} = \sum_{t,s=1}^{3} \alpha^{ts} \mathbf{e}_t \mathbf{e}_s = \sum_{m,n=1}^{3} \alpha^{mn\prime} \mathbf{e}'_m \mathbf{e}'_n.$$

The dependence between the components of $\mathbf{A}$ in the two bases is quadratic with respect to the transformation coefficients A_i^j. This differs from the formal linear transformations of the nine-dimensional tensor space, which are considered in the theory of vector spaces.

The components of a tensor with respect to a given basis can be represented as a 3×3 matrix. In linear algebra, a matrix represents some linear transformation (operator) of the vector space in a fixed basis. A change of basis of the space is given by a transformation of the type (3.7.6), which can also be represented by a 3×3 matrix. The components of the operator transform by the same formulas as the components of the tensor above. So between the set of linear operators over three-dimensional Euclidean space and the set of tensors as introduced above there is a one-to-one correspondence, and the components of a second-rank tensor in some basis constitute the matrix representation of a linear operator in the same basis. It follows that we can use matrix and tensor techniques interchangeably if we introduce the latter while preserving the correspondence in such operations as matrix multiplication. We saw this for the stress tensor. Although (3.6.5) was derived using a Cartesian basis, we will use it with an arbitrary basis in $\mathbb{R}^3$ that makes sense in continuum mechanics.

Before introducing the tensor apparatus, let us note a few peculiarities in the notation. First, the reader has noted our use of both superscripts and subscripts. In a three-dimensional space we normally introduce a *dual basis* along with any given basis. Its basis vectors, denoted by $\mathbf{e}^m$, satisfy two properties:

(1) $\mathbf{e}^m$ is orthogonal to the two vectors $\mathbf{e}_i$ and $\mathbf{e}_j$ (for $i \neq j$ and $i, j \neq m$);
(2) the length of $\mathbf{e}^m$ is such that $\mathbf{e}^m \cdot \mathbf{e}_m = 1$.

Exercise 3.7.1. *Prove that such a basis exists.*

Any vector
$$\mathbf{x} = \sum_{i=1}^{3} \alpha^i \mathbf{e}_i$$
can be represented using the dual basis: i.e., as
$$\mathbf{x} = \sum_{i=1}^{3} \alpha_i \mathbf{e}^i.$$
The same holds for a tensor $\mathbf{A}$, which can be represented as
$$\mathbf{A} = \sum_{i,j=1}^{3} \alpha^{ij} \mathbf{e}_i \mathbf{e}_j = \sum_{i,j=1}^{3} \alpha_{ij} \mathbf{e}^i \mathbf{e}^j = \sum_{i,j=1}^{3} \alpha^i_{\cdot j} \mathbf{e}_i \mathbf{e}^j = \sum_{i,j=1}^{3} \alpha_i^{\cdot j} \mathbf{e}^i \mathbf{e}_j.$$

So for each basis $\mathbf{e}_1, \mathbf{e}_2, \mathbf{e}_3$ we will use four kinds of dyads.

The use of both superscripts and subscripts provides a convenient formalism and a way to avoid calculation errors. All the indices (in both vectorial and tensorial formulas) come in pairs; note how they occur above: when we have to carry out a summation, they necessarily are one-up and one-down. The superscripted components α^i, α^{ij} are called *contravariant*, while the subscripted components are called *covariant*. The components $\alpha_i^{\cdot j}$ and $\alpha^i_{\cdot j}$ are said to be *mixed*. Note that for the mixed components we insert a dot to avoid ambiguity in the order of indices.

Another formalism is Einstein's rule for summation over repeated indices: when an index is repeated, once as a subscript and once as a superscript, summation over that index is understood and the summation symbol is omitted. Using this, the examples above can be written as

$$\alpha^i \mathbf{e}_i = \sum_{i=1}^{3} \alpha_i \mathbf{e}^i, \quad \alpha^{ij}\mathbf{e}_i\mathbf{e}_j = \alpha_{ij}\mathbf{e}^i\mathbf{e}^j = \alpha^i_{\cdot j}\mathbf{e}_i\mathbf{e}^j = \alpha_i^{\cdot j}\mathbf{e}^i\mathbf{e}_j.$$

Exercise 3.7.2. *Prove that the dual to the dual basis is the initial basis, so the duality is mutual.*

We can introduce the dual basis algebraically and then verify its properties. For this we introduce the *metric coefficients* of the space:

$$g_{ij} = \mathbf{e}_i \cdot \mathbf{e}_j. \tag{3.7.7}$$

As $\mathbf{e}_1, \mathbf{e}_2, \mathbf{e}_3$ are linearly independent, we have $\det(g_{ij}) \neq 0$; hence the matrix (g_{ij}) has an inverse whose components we denote by g^{ij}. The vectors of the dual basis are given by the relation

$$\mathbf{e}^i = g^{ij}\mathbf{e}_j. \tag{3.7.8}$$

Exercise 3.7.3. *Verify this.*

Exercise 3.7.4. *Prove that*

$$g^{ij} = \mathbf{e}^i \cdot \mathbf{e}^j. \tag{3.7.9}$$

The dual basis can help us assign a meaning to the dot product of a tensor with a vector. This provides a way to relate the tensor with a linear operator. We begin by defining the dot product of a dyad with a vector:

$$(\mathbf{ab}) \cdot \mathbf{c} = \mathbf{a}(\mathbf{b} \cdot \mathbf{c}). \tag{3.7.10}$$

The result is a vector **a** multiplied by a numerical factor **b** · **c**. Similarly, we define

$$\mathbf{a} \cdot (\mathbf{bc}) = (\mathbf{a} \cdot \mathbf{b})\mathbf{c}. \tag{3.7.11}$$

So the dot product is defined in such a way that one must dot multiply the two vectors nearest the dot symbol. When we apply the dot product to two dyads, we must retain the order of the vectors left outside the dot product:

$$(\mathbf{ab}) \cdot (\mathbf{cd}) = (\mathbf{b} \cdot \mathbf{c})\mathbf{ad}.$$

The result is a dyad **ad** with numerical coefficient **b** · **c**.

We have extended the notion of dot product so that familiar properties are preserved. We have, for example, the distributive property

$$\mathbf{a}(\mathbf{b}+\mathbf{c})\cdot\mathbf{d} = \mathbf{a}(\mathbf{b}\cdot\mathbf{d}) + \mathbf{a}(\mathbf{c}\cdot\mathbf{d}). \qquad (3.7.12)$$

This lets us introduce the dot product between a vector and an arbitrary tensor, or between two arbitrary tensors. The dot product is introduced so that it mimics matrix multiplication. We formulate this as an exercise.

Exercise 3.7.5. *Let $\mathbf{i}_1, \mathbf{i}_2, \mathbf{i}_3$ be an orthonormal basis. (a) Show that the basis is self-dual. (b) Show that the dot product between a tensor $\mathbf{A} = \alpha^{ij}\mathbf{i}_i\mathbf{i}_j$ and a vector $\mathbf{b} = \beta^k \mathbf{i}_k$ is given by $\sum_{j=1}^{3} \alpha^{ij}\beta^j \mathbf{i}_j$. (c) Show that the product of the matrix (α^{ij}) and a column vector $(\beta^1, \beta^2, \beta^3)^T$ yields a vector with the same components as in part (b). (d) What is the correspondence between a vector-matrix product and the dot product of a vector by a tensor from the left? (e) A tensor is the dot product of two tensors. Show that its components in this basis can be obtained by the corresponding product of two matrices that represent each of the tensors in this basis.*

This clear correspondence between tensor and matrix operations can fail if the basis is not orthonormal. In this case, if we wish to preserve matrix-vector manipulations with the components, we should express one factor in terms of the given basis and the other in terms of the dual basis. This is based on the relation $\mathbf{e}^i \cdot \mathbf{e}_j = \delta^i_j$, which defines the dual basis.

Exercise 3.7.6. *Repeat the previous exercise for an arbitrary basis. Note that when we use only covariant components or contravariant components for the tensors, the formulas for their dot products differ from those obtained by multiplication of the corresponding matrices: they contain g_{ij} or g^{ij}. Only for mixed bases do we get complete correspondence with the formulas of matrix multiplication.*

We could also ask which tensors correspond to the zero and unit operators in a linear space. The zero tensor is simply the tensor whose components in any basis are zero. The situation with the unit tensor is not as obvious. The tensor

$$\mathbf{A}_0 = \mathbf{e}_1\mathbf{e}_1 + \mathbf{e}_2\mathbf{e}_2 + \mathbf{e}_3\mathbf{e}_3,$$

which corresponds to the unit matrix, does not correspond to the unit operator; indeed, the reader can verify that only in a Cartesian basis $(\mathbf{e}_1, \mathbf{e}_2, \mathbf{e}_3)$ do we have $\mathbf{B} \cdot \mathbf{A}_0 = \mathbf{B}$ for any tensor $\mathbf{B}$. In what basis does the unit tensor correspond to the unit matrix? It turns out to be the *mixed basis*, where the unit tensor can be represented as $\mathbf{E} = \mathbf{e}^i \mathbf{e}_i$. The reader should verify that

$$\mathbf{E} = \mathbf{e}^i \mathbf{e}_i = \mathbf{e}_i \mathbf{e}^i = g^{ij}\mathbf{e}_i \mathbf{e}_j = g_{ij}\mathbf{e}^i \mathbf{e}^j. \tag{3.7.13}$$

The *metric tensor* $\mathbf{E}$ plays the role of the unit tensor in the space of tensors:

$$\mathbf{A} \cdot \mathbf{E} = \mathbf{E} \cdot \mathbf{A} = \mathbf{A}. \tag{3.7.14}$$

Exercise 3.7.7. *Verify this fact as well.*

Let us show how to find vector or tensor components using the duality of the bases. For a vector we have

$$\mathbf{a} \cdot \mathbf{e}^k = \alpha^i \mathbf{e}_i \cdot \mathbf{e}^k = \alpha^i \delta_i^k = \alpha^k. \tag{3.7.15}$$

Similarly,

$$\mathbf{a} \cdot \mathbf{e}_k = \alpha_k. \tag{3.7.16}$$

We will also have occasion to employ *double-dot products*. For a tensor $\mathbf{A}$ and a dyad $\mathbf{e}_i \mathbf{e}_j$, we define

$$\mathbf{A} \cdots \mathbf{e}_i \mathbf{e}_j = \alpha_{ij}. \tag{3.7.17}$$

To carry out a double-dot product between higher-order tensors, we (1) dot multiply the two vectors nearest the dot, and then (2) repeat for the two remaining vectors. The result is a product of two scalars. For example, if we double-dot multiply two dyads $\mathbf{ab}$ and $\mathbf{cd}$, we obtain

$$\mathbf{ab} \cdots \mathbf{cd} = (\mathbf{b} \cdot \mathbf{c})(\mathbf{a} \cdot \mathbf{d}).$$

Since any tensor can be expand into dyads, this example will suffice for our purposes.

The metric tensor is symmetric: its components do not change when the indices are swapped. In matrix theory this operation is called *transposition*. It looks like

$$\mathbf{E}^T = \left(g^{ij}\mathbf{e}_i \mathbf{e}_j\right)^T = g^{ji}\mathbf{e}_i \mathbf{e}_j.$$

But in general, tensors are not symmetric: $\mathbf{A}^T \neq \mathbf{A}$.

We have covered only a small portion of tensor analysis. The reader should consult more specialized sources (e.g., [Lebedev and Cloud (2003)]) for additional information. We could add, however, that higher-order tensors can be introduced in the same manner. We will use the tensor of elastic moduli $\{c^{ijkl}\}$, which is a fourth-rank tensor. Construction of such tensors starts with tensor products of four vectors $\mathbf{abcd}$, and parallels our development for second-rank tensors. Results for the dot product operation are similar as well.

3.8 The Cauchy Stress Tensor, Continued

Now we can introduce Cauchy's stress tensor. It is a second-rank tensor given in terms of the Cartesian unit basis $(\mathbf{i}_1, \mathbf{i}_2, \mathbf{i}_3)$ by

$$\begin{aligned}\boldsymbol{\sigma} &= \mathbf{i}_1 \boldsymbol{\sigma}_1 + \mathbf{i}_2 \boldsymbol{\sigma}_2 + \mathbf{i}_3 \boldsymbol{\sigma}_3 \\ &= \sigma^{11} \mathbf{i}_1 \mathbf{i}_1 + \sigma^{12} \mathbf{i}_1 \mathbf{i}_2 + \sigma^{13} \mathbf{i}_1 \mathbf{i}_3 \\ &\quad + \sigma^{21} \mathbf{i}_2 \mathbf{i}_1 + \sigma^{22} \mathbf{i}_2 \mathbf{i}_2 + \sigma^{23} \mathbf{i}_2 \mathbf{i}_3 \\ &\quad + \sigma^{31} \mathbf{i}_3 \mathbf{i}_1 + \sigma^{33} \mathbf{i}_3 \mathbf{i}_2 + \sigma^{33} \mathbf{i}_3 \mathbf{i}_3. \end{aligned} \qquad (3.8.1)$$

Here we can use superscripts and subscripts interchangeably because the Cartesian basis is self-dual. When changing frames, however, we must carefully use the index conventions.

Note that (3.6.5) can be written as

$$\boldsymbol{\sigma_n} = \mathbf{n} \cdot \boldsymbol{\sigma}. \qquad (3.8.2)$$

An important point of tensor theory, often used but seldom stated clearly, is this:

> A formula derived in one coordinate frame but then presented in non-component form remains valid in any frame.

Equation (3.8.2) was derived in a Cartesian frame, but holds in any frame.

Let us return to the problem of equilibrium. By the solidification principle, we have used only the equation for the resultant force. The equation for the resultant moment remains. The reader should pause to solve

Exercise 3.8.1. *Isolate a small cube in the deformed body. Consider the equilibrium equation for the moments of the external forces and the reactions of the cut part as was done for the pyramid. Write out the projections of the resultant moment, taken with respect to the center point of the cube, onto the axes. Produce the limit passage as the side of the cube tends to zero and the cube contracts down to its center. Show that each limit equation yields one equation of the following set: $\sigma_{12} = \sigma_{21}$, $\sigma_{13} = \sigma_{31}$, $\sigma_{23} = \sigma_{32}$.*

The result of the exercise can be rewritten as

$$\boldsymbol{\sigma}^T = \boldsymbol{\sigma}. \qquad (3.8.3)$$

Again, we have derived the symmetry property (3.8.3) in a Cartesian frame. But it holds for any frame, and also takes the same form in any frame:

$$\sigma^{ij} = \sigma^{ji}, \qquad \sigma_{ij} = \sigma_{ji}. \qquad (3.8.4)$$

Exercise 3.8.2. *Write out the symmetry equations for the mixed components of $\boldsymbol{\sigma}$.*

So $\boldsymbol{\sigma}$ is a symmetric tensor. It follows that only six of its components are independent.

Let us return to the problem of stress distribution at a point. Pascal's law for a liquid in equilibrium states that the pressure at a point does not depend on the orientation. Implicit in this formulation is the fact that the pressure in a liquid or gas corresponds to a force that is always perpendicular to the infinitesimal plane area. This is not the case with a moving, nonideal liquid having viscosity. It is clear that we cannot expect such a distribution of stresses at a point in a solid body. What then is the simplest way to represent this distribution in a solid? The answer is this: *at any point, there are three mutually orthogonal directions that define three mutually orthogonal plane infinitesimal areas at which the stress vector is normal to the area element.* For a liquid in equilibrium there are infinitely many such directions.

The proof is simple. The fact that the stress vector is orthogonal to the infinitesimal area with unit normal $\mathbf{n}$ is expressed as $\boldsymbol{\sigma}_\mathbf{n} = p\,\mathbf{n}$, where p gives the value of the stress vector. Using the representation formula for $\boldsymbol{\sigma}_\mathbf{n}$, which by the symmetry of $\boldsymbol{\sigma}$ can be rewritten as

$$\boldsymbol{\sigma}_\mathbf{n} = \boldsymbol{\sigma} \cdot \mathbf{n}, \qquad (3.8.5)$$

we have

$$\boldsymbol{\sigma} \cdot \mathbf{n} = p\,\mathbf{n}. \qquad (3.8.6)$$

But there is a one-to-one correspondence between the set of all tensors over $\mathbb{R}^3$ and the set of linear operators on it. Any fact for operators can be restated for tensors and vice versa. What does (3.8.6) mean in terms of an operator corresponding to $\boldsymbol{\sigma}$? It means that $\mathbf{n}$ is an eigenvector of the operator and p is the corresponding eigenvalue. Now we use a fact about symmetrical operators from linear algebra. It states exactly what we formulated above for the stress tensor: there are three mutually orthonormal vectors, $\mathbf{n}_k$ for $k = 1, 2, 3$, the eigenvectors of the operator. These are the needed normals to the infinitesimal areas on which the stress vector is proportional to the normal. Linear algebra states a bit more: the corresponding eigenvalues p_k are all real; furthermore, when they are distinct there are only three such directions. If the p_k include repeated values, then there are infinitely many such mutually orthonormal directions.

Exercise 3.8.3. *Show that in the frame composed of its eigenvectors,* $\boldsymbol{\sigma}$ *is represented as* $\boldsymbol{\sigma} = p_1 \mathbf{n}_1 \mathbf{n}_1 + p_2 \mathbf{n}_2 \mathbf{n}_2 + p_3 \mathbf{n}_3 \mathbf{n}_3.$

In continuum mechanics eigenvalues and eigenvectors are called, respectively, the *principal stresses* and *principal axes* for the stress tensor at a point. This is explained by the fact that the eigenvalues of a symmetric tensor have extremal properties. One of these may be stated in terms of stresses as

$$|\boldsymbol{\sigma}_\mathbf{n}| \leq \max\{|p_1|, |p_2|, |p_3|\}$$

for any direction **n**.

We shall not elaborate on this or similar statements. We should note, however, the mechanical meaning of the property discussed above. The deformed state at any point of a solid body can be achieved by three simple stretching or compressing actions along three mutually orthogonal directions. Because these directions change from point to point, the viewpoint has practical importance only if the body under consideration (or some portion of it) is homogeneously deformed. Nonetheless, the principal stresses play an important role in many theories of the strength of materials and structures.

Knowledge of stress distribution at a point is important but insufficient to uniquely determine the stress field in a body under load. First, we should derive the equations of equilibrium. Since we have already used the equilibrium equations for an infinitesimal volume on the basis of the solidification principle, it might be surprising that we can still obtain three more equations from it. When we showed how to derive the relation for $\boldsymbol{\sigma}_\mathbf{n}$, we — roughly speaking — approximated the equilibrium equations for an elementary volume to zero order. To obtain differential equations governing the stress tensor, we must approximate the equilibrium equations to higher order. Cauchy considered the equilibrium of a rectangular parallelepiped with faces parallel to the Cartesian coordinate planes. The parallelepiped is cut from the deformed body; for this reason, in addition to the external forces acting on the parallelepiped, we must include the reactions of the rest of the body that act on the faces. Cauchy's approximation of these reactions essentially involved two terms of Taylor's formula, which in one dimension looks like

$$\sigma(x + \delta x) \approx \sigma(x) + \sigma'(x)\, \delta x.$$

Cauchy obtained three differential equations in this way, but we wish to derive them in a slightly different manner. We will use the formula for integration by parts in three dimensions:

$$\iiint_V \frac{\partial f}{\partial x_k} dx_1\, dx_2\, dx_3 = \iint_S f n_k\, dS. \qquad (3.8.7)$$

Here S is the surface bounding a volume V in space, $\mathbf{n}$ is the outward unit normal at a point of S, $\mathbf{i}_k$ is the unit vector along the x_k-axis, and $n_k = \mathbf{n} \cdot \mathbf{i}_k$ is the cosine of the angle between $\mathbf{n}$ and the x_k-axis.

Let us isolate a small volume V, having surface S, within the deformed body. By the solidification principle, the volume should be in equilibrium as a rigid body under the action of all external forces. These include the reaction forces which, at each point of the surface, are given by (3.8.5). Recall that $\boldsymbol{\sigma_n} = \boldsymbol{\sigma} \cdot \mathbf{n}$. Equating the resultant force to zero, we obtain

$$\iint_S \boldsymbol{\sigma} \cdot \mathbf{n}\, dS + \iiint_V \mathbf{F}\, dx_1\, dx_2\, dx_3 = \mathbf{0} \qquad (3.8.8)$$

where $\mathbf{F}$ is the density of external forces at a point of V. Using

$$\mathbf{n} = \sum_{k=1}^{3} n_k \mathbf{i}_k,$$

and (3.8.7), we write the first integral as

$$\iint_S \boldsymbol{\sigma} \cdot \mathbf{n}\, dS = \iint_S \boldsymbol{\sigma} \cdot \sum_{k=1}^{3} n_k \mathbf{i}_k\, dS$$

$$= \sum_{k=1}^{3} \iint_S \boldsymbol{\sigma} n_k\, dS \cdot \mathbf{i}_k$$

$$= \sum_{k=1}^{3} \iiint_V \frac{\partial \boldsymbol{\sigma}}{\partial x_k} dx_1\, dx_2\, dx_3 \cdot \mathbf{i}_k.$$

Hence (3.8.8) takes the form

$$\iiint_V \left(\sum_{k=1}^{3} \frac{\partial \boldsymbol{\sigma}}{\partial x_k} \cdot \mathbf{i}_k + \mathbf{F} \right) dx_1\, dx_2\, dx_3 = \mathbf{0}. \qquad (3.8.9)$$

We assume the integrand is continuous. Since (3.8.9) holds for any V, the integrand is zero:

$$\sum_{k=1}^{3} \frac{\partial \boldsymbol{\sigma}}{\partial x_k} \cdot \mathbf{i}_k + \mathbf{F} = \mathbf{0}. \qquad (3.8.10)$$

We differentiated $\boldsymbol{\sigma}$ while assuming it is written in a fixed Cartesian frame so that only its components can depend on the coordinates x_k. Using

$$\sum_{k=1}^{3} \frac{\partial \boldsymbol{\sigma}}{\partial x_k} \cdot \mathbf{i}_k = \sum_{k=1}^{3} \frac{\partial (\sigma^{ij} \mathbf{i}_i \mathbf{i}_j)}{\partial x_k} \cdot \mathbf{i}_k = \sum_{k=1}^{3} \frac{\partial \sigma^{ik}}{\partial x_k} \mathbf{i}_i = \frac{\partial \sigma^{ik}}{\partial x_k} \mathbf{i}_i,$$

we can rewrite (3.8.10) in component form:

$$\frac{\partial \sigma^{11}}{\partial x_1} + \frac{\partial \sigma^{12}}{\partial x_2} + \frac{\partial \sigma^{13}}{\partial x_3} + F^1 = 0,$$

$$\frac{\partial \sigma^{21}}{\partial x_1} + \frac{\partial \sigma^{22}}{\partial x_2} + \frac{\partial \sigma^{23}}{\partial x_3} + F^2 = 0,$$

$$\frac{\partial \sigma^{31}}{\partial x_1} + \frac{\partial \sigma^{32}}{\partial x_2} + \frac{\partial \sigma^{33}}{\partial x_3} + F^3 = 0. \quad (3.8.11)$$

This system could be obtained by applying the equilibrium equation for forces to an elementary (infinitesimal) parallelepiped. Surprisingly, the equation for the resultant moment yields only the symmetry of $\boldsymbol{\sigma}$.

We can write the same equilibrium equations in non-component form. The reader is aware of the formal vector-operator

$$\boldsymbol{\nabla} = \mathbf{i}_1 \frac{\partial}{\partial x_1} + \mathbf{i}_2 \frac{\partial}{\partial x_2} + \mathbf{i}_3 \frac{\partial}{\partial x_3}. \quad (3.8.12)$$

The formal dot product $\boldsymbol{\nabla} \cdot \boldsymbol{\sigma}$ is

$$\boldsymbol{\nabla} \cdot \boldsymbol{\sigma} = \left(\mathbf{i}_1 \frac{\partial}{\partial x_1} + \mathbf{i}_2 \frac{\partial}{\partial x_2} + \mathbf{i}_3 \frac{\partial}{\partial x_3} \right) \cdot \sigma^{ij} \mathbf{i}_i \mathbf{i}_j = \frac{\partial \sigma^{ij}}{\partial x_i} \mathbf{i}_j = \frac{\partial \sigma^{ji}}{\partial x_i} \mathbf{i}_j$$

by the symmetry of $\boldsymbol{\sigma}$. So the equilibrium equation can be written as

$$\boldsymbol{\nabla} \cdot \boldsymbol{\sigma} + \mathbf{F} = \mathbf{0}. \quad (3.8.13)$$

Later we will discuss how to define $\boldsymbol{\nabla}$ to preserve (3.8.13) in curvilinear coordinates.

Hence we have obtained two forms of the equilibrium equations for a deformed body: (3.8.11) and (3.8.13). They represent three linear equations in the six independent components σ^{ij}, and our experience with linear algebra suggests they cannot determine these unknowns uniquely. Additional equations, known as *constitutive relations*, will characterize the material properties by relating the stress tensor to the deformational characteristics (normally described by the strain tensor). In this book we shall use an extension of Hooke's law to three dimensions.

Before proceeding there is something we should consider. We have obtained the equations of equilibrium. But these equations, and the stress tensor itself, are written in the deformed state of the body. In solving a problem, we normally begin with the undeformed body; we cannot know the deformations (i.e., the final state of the body) in advance. So given the initial state of the body, we must recalculate everything with regard for the deformation (which would be found from the solution). This leads us to nonlinear equilibrium equations. But nonlinear equations are notoriously hard to solve, and we cannot envision Cauchy's contemporaries having much success with this approach. In fact, they understood that the deformations of solids in normal working ranges are quite small. So their approach was to assume that the equilibrium equations and stress tensor are written as if the body were not deformed. We might regard this as a "linearization" process. The result is an approximation to the exact relations of continuum mechanics. Experience shows that it works well for engineering purposes. Of course, a mathematician would prefer to write down the exact relations and then estimate the difference between the "exact" and approximate solutions. But this was never done. Nonlinear problems are difficult, and for real engineering problems the most we can expect is to find some numerical solutions, which are approximate themselves. It is doubtful that, in the near future at least, we will see an analytical comparison between the solutions of nonlinear and linearized versions of the equations for the more difficult problems of mechanics.

It follows that without some discussion we cannot directly apply d'Alembert's principle and write down the dynamical equations for the body. Indeed, if the deformation develops over time, we cannot use the same fixed volume — consisting of the same points — at different time instants. Some points would have left the volume and others would have entered. We cannot add the inertial forces to the external forces, because different instants of time enter the definition of the temporal derivative. This situation can be remedied, but it does lead us to consider how we can and should describe the spatiotemporal state of a body. The two main approaches are named for Lagrange and Euler, respectively. We will discuss them later.

Now we wish to discuss some additional technical points. We have written down the derivatives of the stress tensor. When considered in a Cartesian frame, this amounts to merely differentiating the components of the tensor. In mechanics, however, curvilinear coordinates are often used. In order to properly apply the formulas of calculus in curvilinear systems, we must cover some facts from differential geometry.

3.9 Basic Tensor Calculus in Curvilinear Coordinates

Differentiating a vector-function $f^k \mathbf{e}_k$ with respect to x_i and assuming the basis vectors $\mathbf{e}_k$ are constants, we obtain

$$\frac{\partial f^k}{\partial x_i} \mathbf{e}_k.$$

If the $\mathbf{e}_k$ also depend on x_i, we must apply an analogue of the product rule:

$$\frac{\partial (f^k \mathbf{e}_k)}{\partial x_i} = \frac{\partial f^k}{\partial x_i} \mathbf{e}_k + f^k \frac{\partial \mathbf{e}_k}{\partial x_i}. \tag{3.9.1}$$

We need expressions for the derivatives of the basis vectors. These are clearly needed to differentiate tensors as well. Before addressing this issue, however, we would like to show how a basis may be introduced in curvilinear coordinates. Books on continuum mechanics normally relegate such material to an appendix. Although the reader may be familiar with differential geometry, our presentation will suppose only a knowledge of calculus.

We begin by choosing an origin $\mathbf{0}$ in a three-dimensional space. The position of any other point is determined by a radius vector $\mathbf{r}$ from $\mathbf{0}$ to this point. When a Cartesian frame is given, we can represent $\mathbf{r} = (x_1, x_2, x_3)$ using the Cartesian coordinates of the point. However, we can assume the coordinates of the point are determined through three other quantities q^1, q^2, q^3. Then the x_k become functions of the q^i:

$$x_k = x_k(q^1, q^2, q^3) \qquad (k = 1, 2, 3). \tag{3.9.2}$$

These functions must be invertible in some region so that

$$q^j = q^j(x_1, x_2, x_3) \qquad (j = 1, 2, 3). \tag{3.9.3}$$

We call q^1, q^2, q^3 *curvilinear coordinates* in space. Cylindrical and spherical coordinates are familiar examples. In each of these there are points (like a "pole") where the one-to-one correspondence between coordinates fails. At such points, differential operators typically have singularities. They seldom cause trouble with basic manipulations, however.

So we assume the radius vector of a point is defined by the curvilinear coordinates (q^1, q^2, q^3). Using these, we can introduce three vectors that serve as a local basis at the point. We denote the components of this basis by $\mathbf{r}_k$:

$$\mathbf{r}_k = \frac{\partial \mathbf{r}}{\partial q^k}. \tag{3.9.4}$$

Here the index k in q^k should be regarded as a subscript for purposes of the summation convention. We can easily assign a geometric meaning to $\mathbf{r}_k$. For example, to derive $\mathbf{r}_1$ at a point (q_0^1, q_0^2, q_0^3), we must fix the two other coordinates $q^2 = q_0^2$ and $q^3 = q_0^3$. The end of the radius vector $\mathbf{r} = \mathbf{r}(q_0^1, q_0^2, q_0^3)$ describes a curve in space as q^1 changes. The difference

$$\mathbf{r}(q_0^1 + \delta q^1, q_0^2, q_0^3) - \mathbf{r}(q_0^1, q_0^2, q_0^3)$$

is the vector-chord connecting the points (q_0^1, q_0^2, q_0^3) and $(q_0^1 + \delta q^1, q_0^2, q_0^3)$. A similar structure is used to define the tangent to a curve in calculus. Indeed, the derivative

$$\mathbf{r}_1 = \lim_{\delta q^1 \to 0} \frac{\mathbf{r}(q_0^1 + \delta q^1, q_0^2, q_0^3) - \mathbf{r}(q_0^1, q_0^2, q_0^3)}{\delta q^1}$$

is a vector tangent to the q^1-line at the point (q_0^1, q_0^2, q_0^3). Similarly, $\mathbf{r}_k$ is tangent to the q^k-line.

When the Jacobian of the transformation (3.9.3) is not zero, we obtain a basis $\mathbf{r}_1, \mathbf{r}_2, \mathbf{r}_3$ which, in general, will change from point to point. We introduce the dual basis $\mathbf{r}^1, \mathbf{r}^2, \mathbf{r}^3$ by the relations

$$\mathbf{r}_i \cdot \mathbf{r}^j = \delta_i^j. \tag{3.9.5}$$

In general the $\mathbf{r}^j$ also depend on the coordinates q^i.

Now any vector or tensor field in space can be represented in terms of the two bases. One such tensor field is the metric tensor

$$g^{ij}\mathbf{r}_i\mathbf{r}_j = g_{ij}\mathbf{r}^i\mathbf{r}^j = \delta_i^j \mathbf{r}^i \mathbf{r}_j, \tag{3.9.6}$$

where $g^{ij} = \mathbf{r}^i \cdot \mathbf{r}^j$ and $g_{ij} = \mathbf{r}_i \cdot \mathbf{r}_j$.

Exercise 3.9.1. *Prove that the matrices (g^{ij}) and (g_{ij}) are mutually inverse.*

We can represent a given vector-function in terms of the above bases:

$$\mathbf{f} = f^i \mathbf{r}_i = f_j \mathbf{r}^j. \tag{3.9.7}$$

Here each component and each basis vector can depend on the q^i. Consider the derivative

$$\frac{\partial \mathbf{f}}{\partial q^j} = \frac{\partial (f^i \mathbf{r}_i)}{\partial q^j} = \frac{\partial f^i}{\partial q^j}\mathbf{r}_i + f^i \frac{\partial \mathbf{r}_i}{\partial q^j}. \tag{3.9.8}$$

The last term contains the derivative of the basis vectors. The derivative of a vector is a vector as well. Since $\mathbf{r}_1, \mathbf{r}_2, \mathbf{r}_3$ constitute a basis at each point in space, any vector can be written in terms of this basis. We write

$$\frac{\partial \mathbf{r}_i}{\partial q^j} = \Gamma_{ij}^k \mathbf{r}_k \qquad (3.9.9)$$

where the Γ_{ij}^k are *Christoffel's symbols of the second kind*. They have certain symmetries with respect to the indices ($\Gamma_{ij}^k = \Gamma_{ji}^k$) and can be expressed in terms of the components of the metric tensor. A similar development may be carried out for expansion in terms of the dual basis.

Exercise 3.9.2. *For the cylindrical and spherical coordinate systems, derive expressions for the components of the metric tensor and Christoffel's symbols.*

Returning to the derivative of $\mathbf{f}$, we have

$$\frac{\partial \mathbf{f}}{\partial q^j} = \frac{\partial f^i}{\partial q^j}\mathbf{r}_i + f^i \Gamma_{ij}^k \mathbf{r}_k = \frac{\partial f^i}{\partial q^j}\mathbf{r}_i + f^k \Gamma_{kj}^i \mathbf{r}_i = \left(\frac{\partial f^i}{\partial q^j} + f^k \Gamma_{kj}^i\right)\mathbf{r}_i.$$

The expression in parentheses is called the *covariant* (or *absolute*) *derivative* of the contravariant component of $\mathbf{f}$, and we write

$$\nabla_j f^i = \frac{\partial f^i}{\partial q^j} + \Gamma_{kj}^i f^k. \qquad (3.9.10)$$

Similarly, the covariant derivative of the covariant component of $\mathbf{f}$ is

$$\nabla_j f_i = \frac{\partial f_i}{\partial q^j} - \Gamma_{ij}^k f_k. \qquad (3.9.11)$$

We have

$$\frac{\partial \mathbf{f}}{\partial q^j} = \nabla_j f^i \mathbf{r}_i = \nabla_j f_i \mathbf{r}^i \qquad (3.9.12)$$

(cf., [Lebedev and Cloud (2003), § 4.6]). In the same manner we can obtain the derivatives of a tensor-function with respect to q^k:

$$\frac{\partial}{\partial q^k}\mathbf{A} = \nabla_k a^{ij} \mathbf{r}_i \mathbf{r}_j, \quad \nabla_k a^{ij} = \frac{\partial a^{ij}}{\partial q^k} + \Gamma_{ks}^i a^{sj} + \Gamma_{ks}^j a^{is}. \qquad (3.9.13)$$

In § 3.8 we introduced the $\boldsymbol{\nabla}$-operator in a Cartesian frame. To express it in curvilinear coordinates, we begin with the first differential of a scalar function F depending on q^1, q^2, q^3:

$$dF = \frac{\partial F}{\partial q^i} dq^i.$$

Let us represent it using the differential of the radius vector $\mathbf{r}$, which is

$$d\mathbf{r} = \frac{\partial \mathbf{r}}{\partial q^i} dq^i.$$

We shall use the formula

$$dF = \nabla F \cdot d\mathbf{r},$$

which becomes evident when written in a Cartesian frame:

$$dF = \nabla F \cdot \frac{\partial \mathbf{r}}{\partial q^i} \, dq^i = \nabla F \cdot \mathbf{r}_i dq^i.$$

Because we can obtain this formally by putting

$$\nabla F = \mathbf{r}^k \frac{\partial F}{\partial q^k},$$

the ∇-operator is defined as

$$\nabla = \mathbf{r}^k \frac{\partial}{\partial q^k}. \tag{3.9.14}$$

It is also known as the gradient. It is a formal vector that can be applied to a vector,

$$\nabla \mathbf{f} = \mathbf{r}^k \frac{\partial \mathbf{f}}{\partial q^k}, \tag{3.9.15}$$

and to a second-rank tensor:

$$\nabla \mathbf{A} = \mathbf{r}^k \frac{\partial \mathbf{A}}{\partial q^k}. \tag{3.9.16}$$

Clearly $\nabla \mathbf{f}$ is a second-rank tensor, while $\nabla \mathbf{A}$ is a third-rank tensor.

The ∇-operator plays an essential role in what follows. For example, we derive an important formula

$$\mathbf{g} = \nabla \mathbf{r} = \mathbf{r}^k \frac{\partial \mathbf{r}}{\partial q^k} = \mathbf{r}^k \mathbf{r}_k. \tag{3.9.17}$$

Comparing this with the representation of the metric tensor in terms of the basis $\{\mathbf{e}_k\}$, we see that $\mathbf{g}$ is the metric tensor for the basis $\{\mathbf{r}_k\}$. So we can represent it in any of the forms

$$\mathbf{g} = \mathbf{r}^k \mathbf{r}_k = \mathbf{r}_k \mathbf{r}^k = g^{ij} \mathbf{r}_i \mathbf{r}_j = g_{ij} \mathbf{r}^i \mathbf{r}^j, \tag{3.9.18}$$

where $g^{ij} = \mathbf{r}^i \cdot \mathbf{r}^j$ and $g_{ij} = \mathbf{r}_i \cdot \mathbf{r}_j$. Recall that the metric tensor plays the role of the unit element in the space of second-rank tensors: that is, for any tensor $\mathbf{A}$ we have

$$\mathbf{A} = \mathbf{A} \cdot \mathbf{g} = \mathbf{g} \cdot \mathbf{A}. \tag{3.9.19}$$

We can also introduce a formal dot product of the ∇-operator with a vector $\mathbf{a}$ or tensor $\mathbf{A}$. Known as divergence operations, these are given by

$$\nabla \cdot \mathbf{a} = \mathbf{r}^k \cdot \frac{\partial \mathbf{a}}{\partial q^k}, \qquad \nabla \cdot \mathbf{A} = \mathbf{r}^k \cdot \frac{\partial \mathbf{A}}{\partial q^k}. \tag{3.9.20}$$

The differential of a vector-function **f**, given by

$$d\mathbf{f} = d\mathbf{r} \cdot \boldsymbol{\nabla}\mathbf{f} = (\boldsymbol{\nabla}\mathbf{f})^T \cdot d\mathbf{r}, \tag{3.9.21}$$

plays an important role in the theory of deformation.

We shall use the following version of Green's formula, which we have derived in a Cartesian frame:

$$\int_S \mathbf{n} \cdot \mathbf{f}\, dS = \int_V \boldsymbol{\nabla} \cdot \mathbf{f}\, dV, \tag{3.9.22}$$

where **f** is a vector-function and S is the boundary of the volume V.

We leave the derivation of several analogues of the product rule for differentiation as

Exercise 3.9.3. *Let* **a** *and* **b** *be vectors and* **A** *a second-rank tensor. Prove the following formulas:*

$$\boldsymbol{\nabla} \cdot (\mathbf{a}\mathbf{b}) = \mathbf{b}(\boldsymbol{\nabla} \cdot \mathbf{a}) + \mathbf{a} \cdot (\boldsymbol{\nabla}\mathbf{b}),$$
$$\boldsymbol{\nabla}(\mathbf{a} \cdot \mathbf{b}) = (\boldsymbol{\nabla}\mathbf{a}) \cdot \mathbf{b} + (\boldsymbol{\nabla}\mathbf{b}) \cdot \mathbf{a},$$
$$\boldsymbol{\nabla} \cdot (\mathbf{A} \cdot \mathbf{a}) = \mathbf{a} \cdot (\boldsymbol{\nabla} \cdot \mathbf{A}) + \mathbf{A} \cdot\cdot (\boldsymbol{\nabla}\mathbf{a})^T$$
$$= (\boldsymbol{\nabla} \cdot \mathbf{A}) \cdot \mathbf{a} + \mathbf{A}^T \cdot\cdot (\boldsymbol{\nabla}\mathbf{a}).$$

For a second-rank tensor $\mathbf{A} = a^{ij}\mathbf{e}_i\mathbf{e}_j$ we will need something analogous to the squared magnitude of a vector. It is

$$\mathbf{A} \cdot\cdot \mathbf{A}^T = a^{ij}\mathbf{e}_i\mathbf{e}_j \cdot\cdot a^{km}\mathbf{e}_m\mathbf{e}_k = a^{ij}a^{km}g_{jm}g_{ik} = a^{ij}a_{ij}. \tag{3.9.23}$$

The following characteristic of a transformation $\mathbf{f} = \mathbf{f}(\mathbf{r})$ at a point will be useful:

$$N = \left(\boldsymbol{\nabla}\mathbf{f} \cdot\cdot \boldsymbol{\nabla}\mathbf{f}^T\right)^{1/2}. \tag{3.9.24}$$

In Cartesian coordinates this looks like

$$N = \left[\sum_{i,j=1}^{3} \left(\frac{\partial f_i}{\partial x_j}\right)^2\right]^{1/2}. \tag{3.9.25}$$

It can be viewed as a norm of the Jacobian matrix for the transformation **f**.

3.10 Euler and Lagrange Descriptions of Continua

The stress tensor and equilibrium equations cannot completely specify the state of a medium; we must also describe any motion that takes place. In mechanics, the two principal descriptions of motion are the *Eulerian description* and the *Lagrangian description* (although both were known to Euler). These are intimately related, of course, and full knowledge of one should enable us to obtain the other. But each has advantages depending on the situation.

In the Lagrangian description, the medium points themselves serve as "coordinates"; that is, each individual material point (volume element) receives a label which it retains throughout its motion. Normally the assignment of such labels is done in terms of some initial state of the medium. We might, for example, label a given material point with the Cartesian coordinates (x_1, x_2, x_3) of the spatial point that it happens to occupy in the initial, undeformed state of the medium. This material point carries the "coordinate" label (x_1, x_2, x_3) for all subsequent times as well. If the motion involves deformation, then such coordinates will cease to be Cartesian: initially straight coordinate lines will become curvilinear, and at each point and fixed time instant we will obtain a local, natural basis related to the coordinates in precisely the manner discussed near the beginning of § 3.9. The basis changes with time, and its "origin" moves through absolute space since it is attached to a moving material point. In this description the mass of any fixed "volume" in the coordinates (x_1, x_2, x_3) remains the same; the density of the material, however, can change. Since the main volume forces applied to the material — such as the gravitational or inertial forces — are proportional to the density ρ of the material at a point, they are normally written out as $\mathbf{F} = \rho \mathbf{K}$. Because the reference "coordinates" of the mass points of the medium could be chosen in any way, possibly not even connected with some real spatial coordinates, the coordinate space may possess rather strange properties.

The Lagrangian description is often convenient for theoretical developments. It can be used to derive general statements of the conservation laws, for example. This is so because with any fixed "volume" of material in terms of Lagrange's coordinates, we watch the same points over time. From the equilibrium equations in Lagrangian coordinates, we can write down the dynamical equations immediately using d'Alembert's principle. The Lagrangian approach ideally fits the tools we derived earlier: the Cauchy stress tensor and the equilibrium equations.

The Lagrangian approach is also taken in the linearized theories of solids, where deformations are assumed small. In fact the strains are also assumed small (we could say "infinitesimal", but subsequently the equations are applied to problems with finite strains). As a result we identify the material points with their initial positions. But, unlike the nonlinear theory, we imagine the deformation occurring in such a manner that the material points do not change their positions; the displacements and deformations constitute a sort of "cartoon ghost" picture distributed over the media. In this case we observe displacements under loading, but do not suppose them to affect the positions and the values of external forces (as does happen in nonlinear theories).

The Eulerian description is used for media where displacements may be large and material points may travel over long distances. This is the case for the motion of water in a river. For problems involving gases and liquids, our interest centers on what happens at a certain point in space. It is clear that such a fixed *spatial* point may be occupied by a different *material* point at each successive time instant. Hence the techniques required to implement the Eulerian and Lagrangian approaches differ. But they are equivalent, and we can always choose the best one for the situation at hand.

In summary then, the two approaches differ in the coordinates we use to describe the medium. In the Lagrangian approach, we use coordinates "rigidly" attached to material points — the coordinates of a certain particle do not change during deformation. In the Eulerian approach, the coordinates are attached to spatial points.

3.11 Strain Tensors

We have considered the one-dimensional version of a bar under tension. Placing the bar along the x-axis and denoting by $u = u(x)$ the displacement of a point x on its centerline, we have seen that the strain is given by the derivative du/dx. We would like to extend this to three dimensions. We begin by introducing three-dimensional strains for small deformations, as occurred historically in elasticity theory. In three dimensions, an analogue of the first derivative is the gradient of the displacement vector. Let us introduce this quantity next.

As the reference frame for a continuum medium, we choose Cartesian coordinates erected in the undeformed state of the medium. Hence we introduce an orthonormal basis $\mathbf{i}_1, \mathbf{i}_2, \mathbf{i}_3$. This basis and its dual may be used interchangeably since $\mathbf{i}^k = \mathbf{i}_k$. We describe deformation using Lagrange's approach.

After deformation, a point characterized by the position vector $\mathbf{r} = x_k \mathbf{i}^k$ moves to the spatial point whose position vector is $\mathbf{R}$. The *displacement vector*

$$\mathbf{u} = \mathbf{R} - \mathbf{r} \qquad (3.11.1)$$

describes how the point shifts during deformation. Since we accept (x_1, x_2, x_3) as the coordinates of the point after deformation, the components of $\mathbf{u} = u_i \mathbf{i}^i$ must depend on the coordinates of the point. That is, the u^j are functions of the coordinates:

$$u_i = u_i(x_1, x_2, x_3). \qquad (3.11.2)$$

We take these to be sufficiently smooth and, in particular, continuously differentiable at each point of the medium. Then the gradient of the displacement field is, by definition,

$$\boldsymbol{\nabla} \mathbf{u} = \mathbf{i}^j \frac{\partial}{\partial x_j} u_i \mathbf{i}^i = \frac{\partial u_i}{\partial x^j} \mathbf{i}^j \mathbf{i}^i. \qquad (3.11.3)$$

The components of the gradient contain all first partial derivatives of the u_i. So the tensor $\boldsymbol{\nabla} \mathbf{u}$ is nonzero unless $\mathbf{u}$ is constant. However, we need a characteristic of the deformation itself. This means we expect this characteristic to be zero when the medium moves as a rigid body. But in this case for the gradient we get nonzero values. Therefore we must introduce another characteristic. In linear algebra, any operator can be represented as a sum of symmetric and skew-symmetric operators. This is done by a simple transformation, which we apply to $\boldsymbol{\nabla} \mathbf{u}$:

$$\boldsymbol{\nabla} \mathbf{u} = \frac{1}{2}(\boldsymbol{\nabla} \mathbf{u} + \boldsymbol{\nabla} \mathbf{u}^T) + \frac{1}{2}(\boldsymbol{\nabla} \mathbf{u} - \boldsymbol{\nabla} \mathbf{u}^T). \qquad (3.11.4)$$

The first addend, which we define as

$$\boldsymbol{\varepsilon} = \frac{1}{2}(\boldsymbol{\nabla} \mathbf{u} + \boldsymbol{\nabla} \mathbf{u}^T) \qquad (3.11.5)$$

and call the *strain tensor of small deformation*, is symmetric. The tensor described by the second addend is skew-symmetric.

Let us consider some properties of $\boldsymbol{\varepsilon}$. Its components are

$$\varepsilon_{11} = \frac{\partial u_1}{\partial x^1}, \qquad \varepsilon_{12} = \frac{1}{2}\left(\frac{\partial u_1}{\partial x^2} + \frac{\partial u_2}{\partial x^1}\right), \quad \varepsilon_{13} = \frac{1}{2}\left(\frac{\partial u_1}{\partial x^3} + \frac{\partial u_3}{\partial x^1}\right),$$

$$\varepsilon_{21} = \frac{1}{2}\left(\frac{\partial u_2}{\partial x^1} + \frac{\partial u_1}{\partial x^2}\right), \quad \varepsilon_{22} = \frac{\partial u_2}{\partial x^2}, \qquad \varepsilon_{23} = \frac{1}{2}\left(\frac{\partial u_2}{\partial x^3} + \frac{\partial u_3}{\partial x^2}\right),$$

$$\varepsilon_{31} = \frac{1}{2}\left(\frac{\partial u_3}{\partial x^1} + \frac{\partial u_1}{\partial x^3}\right), \quad \varepsilon_{32} = \frac{1}{2}\left(\frac{\partial u_3}{\partial x^2} + \frac{\partial u_2}{\partial x^3}\right), \quad \varepsilon_{33} = \frac{\partial u_3}{\partial x^3}.$$

Only six of these are independent because $\varepsilon_{12} = \varepsilon_{21}$, $\varepsilon_{13} = \varepsilon_{31}$, and $\varepsilon_{23} = \varepsilon_{32}$. At times we may wish to explicitly show the displacement field $\mathbf{v}$ for which $\boldsymbol{\varepsilon}$ is written out. Then we will write

$$\boldsymbol{\varepsilon} = \boldsymbol{\varepsilon}(\mathbf{v}) = \boldsymbol{\varepsilon}(\mathbf{v}(x^1, x^2, x^3)) = \varepsilon_{ij}(\mathbf{v})\mathbf{i}^i\mathbf{i}^j.$$

We begin to study the properties of $\boldsymbol{\varepsilon}$ with

Exercise 3.11.1. *Prove that for* $\mathbf{u}_0 = \mathbf{a} + \mathbf{b} \times \mathbf{r}$ *with constant vectors* $\mathbf{a}$ *and* $\mathbf{b}$*, we have* $\boldsymbol{\varepsilon}(\mathbf{u}_0) = \mathbf{0}$.

By this exercise, the general form of the solution of the equation $\boldsymbol{\varepsilon} = \mathbf{0}$ is

$$\mathbf{u}_0 = \mathbf{a} + \mathbf{b} \times \mathbf{r}.$$

We call $\mathbf{u}_0$ an *infinitesimal* (or *small*) *rigid motion* of the body. We will use the terms "infinitesimal" and "small" interchangeably. "Infinitesimal" refers to the way in which $\mathbf{u}_0$ can be derived from the formula for the displacements of the body if it moves as a rigid whole in space: we should suppose that the rigid displacement is infinitesimal and take into account only terms of the first order of smallness.

From classical mechanics we know that the infinitesimal displacement vector for a rigid body takes the general form of $\mathbf{u}_0$. So $\boldsymbol{\varepsilon}$, which is linear in the components of the displacement vector, does not change if we add any $\mathbf{u}_0$ to a displacement field $\mathbf{u}$:

$$\boldsymbol{\varepsilon}(\mathbf{u} + \mathbf{u}_0) = \boldsymbol{\varepsilon}(\mathbf{u}).$$

Hence the linear strain tensor of small deformation does not depend on infinitesimal rigid-body motions of the medium.

Now let us examine the components of $\boldsymbol{\varepsilon}$ under certain special deformations. First we stretch the medium uniformly along the x^i-direction. The corresponding displacement vector $\mathbf{u}_i$ has only one nonzero component: that having subscript i, which is cx_i where c is a constant. The only nonzero component of $\boldsymbol{\varepsilon}$ is $\varepsilon_{ii} = \partial u_i/\partial x_i = c$. This value of strain exactly matches that in a stretched bar; in this case the displacement is cx_i.

Next we consider a skew-type displacement. A *pure shear* deformation is characterized by a displacement vector having just one nonzero component given by $u_i = \gamma x_j$, where γ is a constant and $j \neq i$. It is clear that the only nonzero components of $\boldsymbol{\varepsilon}$ are $\varepsilon_{ij} = \varepsilon_{ji} = \frac{1}{2}\gamma$. For small values, γ is approximately equal to the inclination angle of the x_j-axis (Fig. 3.13).

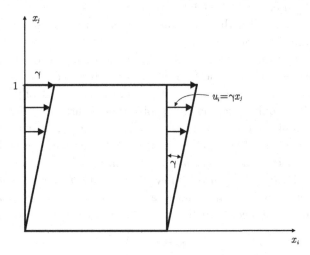

Fig. 3.13 Pure shear in the medium in the (x_i, x_j) plane: $u_i = \gamma x_j$, while the two remaining components of **u** are zero. For small γ, the inclination angle of the lines parallel to the x_j-axis is approximately γ. Here $\varepsilon_{ij} = \gamma/2$.

For these particular types of deformation, we can use the linearity of $\boldsymbol{\varepsilon}$ to see the mechanical sense of each of its components. A component with repeated subscripts shows stretching along the corresponding axis, while a component ε_{ij} with $i \neq j$ characterizes a local pure shear at a point. Any small deformation at a point can be approximated by a superposition of these two elementary deformation types, and we can use $\boldsymbol{\varepsilon}$ to characterize small deformations of the medium.

We have introduced the tensor $\boldsymbol{\varepsilon}$ using Cartesian coordinates in the initial (reference) undeformed state of the medium. With the same success we can use any coordinates of the undeformed state and so, as a consequence of the tensor nature of $\nabla \mathbf{u}$, we see that $\boldsymbol{\varepsilon}$ is a symmetric, second-rank tensor.

Let us discuss some properties of $\boldsymbol{\varepsilon}$ that follow from the general theory. As a symmetric second-rank tensor, $\boldsymbol{\varepsilon}$ has — at each point — three real eigenvalues E_i and three corresponding orthonormal eigenvectors $\mathbf{n}_i$ defined by

$$\boldsymbol{\varepsilon} \cdot \mathbf{n}_i = E_i \mathbf{n}_i.$$

The $\mathbf{n}_i$ are uniquely defined if the E_i are distinct. The deformation at a point can be approximated by special deformations: the elongation or shortening (with no shearing), along the directions $\mathbf{n}_i$ with strains E_i, of a small cube centered at the point and having sides parallel to the $\mathbf{n}_i$. The directions $\mathbf{n}_i$ are called the *principal directions* of the tensor $\boldsymbol{\varepsilon}$ and the E_i are called the *principal strains*.

Other properties of $\boldsymbol{\varepsilon}$ that follow from its tensor nature are discussed in textbooks on linear elasticity. Here we are interested in the accuracy to which $\boldsymbol{\varepsilon}$ approximates the deformation from the viewpoint of the nonlinear theory. We can only touch on this problem of accuracy, as the latter theory is not simple, and the interested reader should consult more advanced books. We suppose that a point of the undeformed medium is defined by the curvilinear coordinates (q^1, q^2, q^3) and employ the notation of § 3.9.

One nonlinear strain tensor, known as the *Cauchy–Green strain tensor* for finite deformation, is introduced as follows. Two infinitesimally separated points of the medium define a vector $d\mathbf{r}$ whose length squared is

$$ds^2 = d\mathbf{r} \cdot d\mathbf{r}.$$

After deformation, the endpoints of $d\mathbf{r}$ move and define a new vector $d\mathbf{R}$ whose length squared is

$$dS^2 = d\mathbf{R} \cdot d\mathbf{R}.$$

Since we deal with infinitesimal quantities, $d\mathbf{R}$ and $d\mathbf{r}$ are related by the formula

$$d\mathbf{R} = d\mathbf{r} \cdot \boldsymbol{\nabla}\mathbf{R} = \boldsymbol{\nabla}\mathbf{R}^T \cdot d\mathbf{r}.$$

Here

$$\boldsymbol{\nabla} = \mathbf{r}^k \frac{\partial}{\partial q^k}.$$

So

$$dS^2 = d\mathbf{R} \cdot d\mathbf{R} = d\mathbf{r} \cdot \boldsymbol{\nabla}\mathbf{R} \cdot \boldsymbol{\nabla}\mathbf{R}^T \cdot d\mathbf{r} = d\mathbf{r} \cdot \mathbf{G}^\times \cdot d\mathbf{r},$$

where

$$\mathbf{G}^\times = \boldsymbol{\nabla}\mathbf{R} \cdot \boldsymbol{\nabla}\mathbf{R}^T \qquad (3.11.6)$$

is called the *Cauchy deformation measure*. Let $\mathbf{u} = \mathbf{R} - \mathbf{r}$. Then
$$\nabla \mathbf{R} = \nabla \mathbf{r} + \nabla \mathbf{u}.$$
Next, we denote $\mathbf{g} = \nabla \mathbf{r}$, the metric tensor in the initial state of the body, which plays the role of the unit operator. We get
$$\begin{aligned}\mathbf{G}^\times &= (\nabla \mathbf{r} + \nabla \mathbf{u}) \cdot (\nabla \mathbf{r} + \nabla \mathbf{u})^T = (\mathbf{g} + \nabla \mathbf{u}) \cdot (\mathbf{g} + \nabla \mathbf{u}^T) \\ &= \mathbf{g} \cdot \mathbf{g} + \nabla \mathbf{u} \cdot \mathbf{g} + \mathbf{g} \cdot \nabla \mathbf{u}^T + \nabla \mathbf{u} \cdot \nabla \mathbf{u}^T \\ &= \mathbf{g} + \nabla \mathbf{u} + \nabla \mathbf{u}^T + \nabla \mathbf{u} \cdot \nabla \mathbf{u}^T.\end{aligned}$$
The Cauchy–Green tensor $\mathcal{E}$ is introduced by the formula
$$\mathbf{G}^\times = \mathbf{g} + 2\mathcal{E} \tag{3.11.7}$$
so that
$$\begin{aligned}\mathcal{E} &= \frac{1}{2}(\nabla \mathbf{u} + \nabla \mathbf{u}^T) + \frac{1}{2}\nabla \mathbf{u} \cdot \nabla \mathbf{u}^T \\ &= \varepsilon + \frac{1}{2}\nabla \mathbf{u} \cdot \nabla \mathbf{u}^T,\end{aligned}$$
where
$$\varepsilon = \frac{1}{2}(\nabla \mathbf{u} + \nabla \mathbf{u}^T) \tag{3.11.8}$$
is the tensor of small deformation. This tensor plays the main role in what follows.

Certain other quantities can be taken as measures of deformation. From knowledge of such a quantity, however, we must be able to recover the displacement field of a body up to rigid motions. The tensor of small deformation has this property. The displacement field is recovered from Cesàro's formula
$$\mathbf{u}(s) = \mathbf{u}_0 + \boldsymbol{\omega}_0 \times (\mathbf{r} - \mathbf{r}_0) + \int_{M_0}^{M} \{\varepsilon(s) + [\mathbf{r}(s) - \mathbf{r}] \times [\nabla \times \varepsilon(s)]\} \cdot d\mathbf{r}(s),$$
where $\mathbf{r}_0$ is the position vector of the initial point M_0, $\mathbf{r}$ is the position vector of the point M, $\mathbf{u}_0$ and $\boldsymbol{\omega}_0$ are the displacement and rotation vectors for M_0, and s defines the position vector $\mathbf{r}(s)$ of an intermediate point of integration between M_0 and M (see textbooks on linear elasticity). If the tensor field ε is expressed through the position vector $\mathbf{r}$, then the condition
$$\nabla \times (\nabla \times \varepsilon) = \mathbf{0}$$
holds. This compatibility condition for ε permits Cesàro's representation to hold.

When does the tensor of small deformations $\boldsymbol{\varepsilon}$ approximate the Cauchy–Green tensor of finite deformation $\boldsymbol{\mathcal{E}}$? The difference between the tensors is

$$\boldsymbol{\mathcal{E}} - \boldsymbol{\varepsilon} = \frac{1}{2}\boldsymbol{\nabla}\mathbf{u} \cdot \boldsymbol{\nabla}\mathbf{u}^T.$$

So $\boldsymbol{\varepsilon} \approx \boldsymbol{\mathcal{E}}$ when

$$\left\|\boldsymbol{\nabla}\mathbf{u} \cdot \boldsymbol{\nabla}\mathbf{u}^T\right\| \ll \|\boldsymbol{\varepsilon}\|.$$

As a rule of thumb engineers may require that $\left\|\boldsymbol{\nabla}\mathbf{u} \cdot \boldsymbol{\nabla}\mathbf{u}^T\right\|$ lie within 1-2% of $\|\boldsymbol{\varepsilon}\|$.

In practical terms, the use of the linear strain tensor requires that the norm $\left\|\boldsymbol{\nabla}\mathbf{u} \cdot \boldsymbol{\nabla}\mathbf{u}^T\right\|$ be small in comparison with $\|\boldsymbol{\varepsilon}\|$. In particular, at any point the portion of the displacement vector that corresponds to the rotation of the medium as a rigid whole must be small as well. Some mechanical structures, such as thin beams or plates, are quite flexible in certain directions. In such cases the angle of rigid rotation can be finite and significant, whereas the deformation remains small. If we wish to preserve the accuracy of the description of deformation, we must supplement the expression for strains with quadratic terms corresponding to the rotation angles. In the most widely used theory of plates and shells, this is done on the basis of an intuitive picture of the deformation, without bringing in the strain tensor of finite deformation.

3.12 The Virtual Work Principle

The Cauchy stress tensor $\boldsymbol{\sigma}$ characterizes the distribution of internal stresses in the deformed state of a medium, whereas the strain tensor $\boldsymbol{\varepsilon}$ of small deformation is given using the initial undeformed state of the medium. If we wish to derive expressions for the components of $\boldsymbol{\sigma}$ and the equilibrium equation in terms of the initial state, we should present these quantities in the same description. In this book we restrict ourselves to the linearized theory of solids, part of which is known as the theory of linear elasticity. Here we consider small deformations of the medium. In engineering practice they are small but finite, and we cannot talk about infinitely small strains, etc. We shall do this, however, in order to obtain the linearized theory: we suppose the strains to be infinitely small, and in this case do not distinguish between the points and frames of the initial undeformed state and the deformed state. We use the terms "infinitely small" or "infinitesimal" only to justify the use of calculus in deriving relations. And, as always, after arriving at a model we shall ignore these issues and apply the model in situations where all quantities are finite. So, in this case, we will consider both the Cauchy stress tensor and the equilibrium equation as if they were introduced in the initial state of the medium. We shall proceed exactly as in numerical analysis when we seek the increment $\Delta f(x)$ of some function $f(x)$ at a point x; we settle for the first differential:

$$\Delta f(x) \approx f'(x)\Delta x.$$

For sufficiently small Δx it is a good approximation. Analogously, the results we obtain in our subject will suffice for much of engineering practice.

Hence, from now on, Cauchy's stress tensor is assumed given in the initial nondeformed state of the medium. Since we will introduce no further stress tensors (and, moreover, since the result is not exactly Cauchy's stress tensor!) we shall refer to it merely as "the stress tensor". As we have said, for this tensor the equilibrium equation should hold:

$$\nabla \cdot \boldsymbol{\sigma} + \mathbf{F} = 0. \tag{3.12.1}$$

This was derived in a Cartesian frame; being written in non-component form, however, it holds in any coordinate frame.

We would like to derive the virtual work principle for continuum media using the above stress and strain tensors. In classical mechanics, the principle reads that for a system in equilibrium, the work of all forces on any virtual displacement equals zero. In continuum mechanics we will consider the displacements $\delta \mathbf{u}$ to be virtual if they are smooth enough and agree with the geometric constraints on the boundary. The symbol δ reflects the idea of smallness of $\delta \mathbf{u}$ as a perturbation of a real displacement field, but in the linearized theory this smallness is not important.

Consider a portion V of the medium, bounded by surface S. Assume it is in equilibrium and recall the solidification principle. A distributed volume force $\mathbf{F}$ is assumed to act on V. If the boundary S is interior to the medium, we must also consider a distributed reaction $\mathbf{f} = \boldsymbol{\sigma}_\mathbf{n}$. As we saw, this is related to $\boldsymbol{\sigma}$ by the formula

$$\mathbf{f} = \mathbf{n} \cdot \boldsymbol{\sigma}, \tag{3.12.2}$$

where $\mathbf{n}$ is the exterior unit normal to S. Equation (3.12.2) must hold at points on the boundary surface of the medium for any given distribution of surface forces $\mathbf{f}$. This type of boundary condition for the body corresponds to the Neumann condition for a membrane. When such a condition holds over the entire boundary of the body, we have what is typically called the first or second problem of elasticity (depending on the author's preference). A problem with prescribed displacements over the entire boundary corresponds to the Dirichlet problem for a membrane. We will discuss this problem later.

The work of all external forces over a virtual displacement $\delta \mathbf{u}$ in the volume V is given by

$$\delta \mathcal{W}_e = \int_V \mathbf{F} \cdot \delta \mathbf{u} \, dV + \int_S \mathbf{f} \cdot \delta \mathbf{u} \, dS. \tag{3.12.3}$$

Let us transform the second integral using (3.12.2), Green's formula (3.9.22), and the formula from Exercise 3.9.3. We get

$$\int_S \mathbf{f} \cdot \delta \mathbf{u} \, dS = \int_S \mathbf{n} \cdot \boldsymbol{\sigma} \cdot \delta \mathbf{u} \, dS$$
$$= \int_V \boldsymbol{\nabla} \cdot (\boldsymbol{\sigma} \cdot \delta \mathbf{u}) \, dV$$
$$= \int_V [(\boldsymbol{\nabla} \cdot \boldsymbol{\sigma}) \cdot \delta \mathbf{u} + \boldsymbol{\sigma} \cdot\cdot (\boldsymbol{\nabla} \delta \mathbf{u})] \, dV.$$

By symmetry of $\boldsymbol{\sigma}$ we have

$$\boldsymbol{\sigma} \cdot\cdot (\boldsymbol{\nabla} \delta \mathbf{u}) = \boldsymbol{\sigma} \cdot\cdot (\boldsymbol{\nabla} \delta \mathbf{u})^T, \tag{3.12.4}$$

hence by definition of the strain tensor of small deformation,

$$\delta \boldsymbol{\varepsilon} = \boldsymbol{\varepsilon}(\delta \mathbf{u}) = \frac{1}{2}(\boldsymbol{\nabla} \delta \mathbf{u} + \boldsymbol{\nabla} \delta \mathbf{u}^T),$$

we have

$$\boldsymbol{\sigma} \cdot\cdot (\boldsymbol{\nabla} \delta \mathbf{u}) = \boldsymbol{\sigma} \cdot\cdot \delta \boldsymbol{\varepsilon}.$$

Therefore

$$\int_S \mathbf{f} \cdot \delta \mathbf{u} \, dS = \int_V [(\boldsymbol{\nabla} \cdot \boldsymbol{\sigma}) \cdot \delta \mathbf{u} + \boldsymbol{\sigma} \cdot\cdot \delta \boldsymbol{\varepsilon}] \, dV.$$

Substituting this into the expression (3.12.3) for $\delta \mathcal{W}_e$, we get

$$\delta \mathcal{W}_e = \int_V [(\mathbf{F} + \boldsymbol{\nabla} \cdot \boldsymbol{\sigma}) \cdot \delta \mathbf{u} + \boldsymbol{\sigma} \cdot\cdot \delta \boldsymbol{\varepsilon}] \, dV = \int_V \boldsymbol{\sigma} \cdot\cdot \delta \boldsymbol{\varepsilon} \, dV,$$

where the last equality holds by the equilibrium equation (3.12.1). Equating this with expression (3.12.3), we finally get

$$\int_V \mathbf{F} \cdot \delta \mathbf{u} \, dV + \int_S \mathbf{f} \cdot \delta \mathbf{u} \, dS - \int_V \boldsymbol{\sigma} \cdot\cdot \delta \boldsymbol{\varepsilon} \, dV = 0. \tag{3.12.5}$$

This is the analytic form of the virtual work principle. It holds for all sufficiently smooth $\delta \mathbf{u}$. The term

$$\delta \mathcal{W}_i = - \int_V \boldsymbol{\sigma} \cdot\cdot \, \delta\boldsymbol{\varepsilon} \, dV \tag{3.12.6}$$

is called the *work of internal forces*. So (3.12.5) expresses the virtual work principle for an arbitrary volume V of the medium. It preserves the word statement of the virtual work principle of classical mechanics: the work of all internal and external forces over any virtual displacement field is zero.

This equation holds even for a continuum body with portions of the boundary clamped. On the clamped portion S_1 we have $\delta\mathbf{u} = \mathbf{0}$ and on the remainder of S the load is given: $\mathbf{f} = \mathbf{n} \cdot \boldsymbol{\sigma}$. However, supposing that on S_1 there is some reaction $\mathbf{f}$, we can repeat the derivation as above but we must take into account that $\delta\mathbf{u} = \mathbf{0}$ holds on S_1 so that (3.12.5) becomes

$$- \int_V \mathbf{F} \cdot \delta\mathbf{u} \, dV - \int_{S\setminus S_1} \mathbf{f} \cdot \delta\mathbf{u} \, dS + \int_V \boldsymbol{\sigma} \cdot\cdot \, \delta\boldsymbol{\varepsilon} \, dV = 0. \tag{3.12.7}$$

Note that during the derivation we did not use the material properties. Hence

The VWP formulas (3.12.5) or (3.12.7) hold for all types of material, whether elastic or not.

Exercise 3.12.1. *Verify (3.12.4).*

Exercise 3.12.2. *Show that $\boldsymbol{\sigma} \cdot\cdot \, \delta\boldsymbol{\varepsilon} = \sigma^{ij} \delta\varepsilon_{ij}$. Write out all the possible representations in components of the tensors of this complete product.*

Exercise 3.12.3. *Show that the virtual work principle implies the equilibrium equation (3.12.1) along with the force boundary condition $\mathbf{n} \cdot \boldsymbol{\sigma}|_S = \mathbf{f}$. This is the natural boundary condition for the VWP equation.*

Later we will see that, for small deformations of a linearly elastic body under load, the left side of (3.12.7) is the first variation of the total energy functional. A consequence is that when a solution $\mathbf{u}$ of the equilibrium problem for a linearly elastic body is sufficiently smooth, the fulfillment of the virtual work principle for all smooth $\delta\mathbf{u}$ implies the equilibrium equations and the natural boundary condition $\mathbf{n} \cdot \boldsymbol{\sigma}|_S = \mathbf{f}$, and conversely.

3.13 Hooke's Law in Three Dimensions

The connection between the stress and strain tensors is called a *constitutive relation* for a material. In elementary physics, such relations are presented in the form of Hooke's law for the wire or spring, or some relation connecting the pressure, volume, and temperature of a gas. In fact, deformation in solids is also accompanied by some temperature change. Practically, however, the influence of temperature is often negligible and need not appear in the constitutive relation. In a more general theory that includes thermodynamics, such problems are called *isothermal*.

We would like to extend Hooke's law to three dimensions. It seems reasonable to seek a linear relation with constant coefficients. So we would like to consider a linear model of an elastic body. We should say at the outset that some materials that can be regarded as solids cannot be described using this law even for small deformations. For example, polymers normally exhibit time-dependent strains when under constant load. The behavior of plastic materials like copper and lead cannot be represented by a linearly elastic constitutive law in principle. But many common solid materials (wood, the steels, various other metals, etc.) are well-described by Hooke's law.

As we said, the linear law is only an approximation. Any textbook on the strength of materials will display a stress-strain diagram for the stretching of a steel bar. This will show a nonlinear elastic portion as well as a portion attributed to plasticity (a regime when an additional stretching force causes strains that remain unrestored when the force is reduced). The diagrams for different materials differ somewhat (even for steels). Fortunately, the portion of a stress-strain diagram near zero, which is nearly linear and corresponds to an elastic material, is fairly large. So the linear elastic model serves well in much of engineering design.

The one-dimensional version of Hooke's law is

$$\sigma = E\varepsilon. \qquad (3.13.1)$$

Its most general linear extension, preserving the tensorial nature of the stress and strain, is

$$\boldsymbol{\sigma} = \mathbf{c} \cdots \boldsymbol{\varepsilon}. \qquad (3.13.2)$$

Here $\boldsymbol{\sigma}$ and $\boldsymbol{\varepsilon}$ are the familiar stress and strain tensors. The quantity $\mathbf{c}$ is a *tensor of elastic moduli*; this is a fourth-rank tensor having components denoted by c^{ijkl}. In component form, (3.13.2) is

$$\sigma^{ij} = c^{ijkl}\varepsilon_{kl}. \tag{3.13.3}$$

Although a fourth-rank tensor has $3^4 = 81$ components, the symmetry of $\boldsymbol{\sigma}$ and $\boldsymbol{\varepsilon}$ implies that we must relate only six components of the former to six components of the latter: in other words, there can be no more than 36 independent elastic constants. Thermodynamic considerations lead to an additional symmetry property,

$$c^{ijkl} = c^{klij}, \tag{3.13.4}$$

which further restricts the number of independent elastic constants to 21.

But 21 is still a huge number of independent elastic constants for the purposes of manual calculation. The success people had prior to the advent of the computer indicates that many elastic materials can be described with fewer constants. Indeed, practical materials in large part have special symmetry properties that reduce the number of constants significantly. The properties of metals and alloys, for instance, do not depend on direction in the material. Such materials are said to be *isotropic* (or to exhibit *isotropy*). If there is some dependence on the direction, the material is *anisotropic*. An example is wood, which exhibits a definite grain structure. Materials can be divided into symmetry classes according to these sorts of properties. The reader can refer to books on engineering elasticity for more information. Our main interest will be in isotropic materials.

For an isotropic material, the components of **c** are expressed in terms of only two elastic constants that are independent. Engineers commonly use Young's modulus E and *Poisson's ratio* ν. The first of these is well known from elementary physics. Let us take a moment to discuss Poisson's ratio. When we stretch an elastic film along some direction so that its strain is ε, the film shortens in the perpendicular direction. In this direction the strain is proportional to ε, that is $-\nu\varepsilon$. Of course, for large deformations the coefficient ν depends on ε, but for small strains experiment shows that it is practically constant. This constant coefficient is called *Poisson's ratio*. For steels its values are close to 0.28.

With Poisson's ratio we can pose a relation between the components of the stress and strain tensors. We will do this in a Cartesian frame. First, when only one stress σ^{11} is present, it should be related to ε^{11} via Hooke's law $\varepsilon^{11} = \sigma^{11}/E$. Applying σ^{22}, which acts perpendicularly, we see an additional contribution $-\nu\sigma^{22}/E$ to ε^{11}. Applying σ^{33}, we see a similar contribution $-\nu\sigma^{33}/E$. By the assumed linearity of the situation we can use the superposition principle; when all three stresses are present we get

$$\varepsilon^{11} = \frac{1}{E}\left(\sigma^{11} - \nu\sigma^{22} - \nu\sigma^{33}\right).$$

Assuming an isotropic material, symmetry leads to the two additional relations

$$\varepsilon^{22} = \frac{1}{E}\left(\sigma^{22} - \nu\sigma^{11} - \nu\sigma^{33}\right),$$

$$\varepsilon^{33} = \frac{1}{E}\left(\sigma^{33} - \nu\sigma^{11} - \nu\sigma^{22}\right).$$

We could ask whether these formulas for ε^{ii} change under the action of some σ^{km}, $k \neq m$. The answer is no: the strains ε^{ii} do not depend on the shear stresses σ^{km}. This can be "proved" experimentally. Here the initial assumption is that for an isotropic material the stress and strain tensors are *coaxial*: that is, the principal directions of the tensors should coincide. Of course, this is only an assumption which, as a consequence, gives us the above relations in which σ^{km} does not appear.

Next we would like to relate the σ^{km} to the ε^{ij}. In the presence of σ^{12}, it is clear that we have the proportionality relationship $\varepsilon^{12} = \sigma^{12}/\mu$ with coefficient $1/\mu$. Again, experiments shows that no other stresses affect ε^{12}. If this statement is unconvincing, we may suppose that the stress and strain tensors are coaxial and obtain the above fact as a direct consequence. So

$$\varepsilon^{12} = \frac{\sigma^{12}}{\mu}, \qquad \varepsilon^{13} = \frac{\sigma^{13}}{\mu}, \qquad \varepsilon^{23} = \frac{\sigma^{23}}{\mu}.$$

Another consequence is the set of relations between the constants E, ν, and μ:

$$E = \frac{\mu(3\lambda + 2\mu)}{\lambda + \mu}, \qquad \nu = \frac{\lambda}{2(\lambda + \mu)}, \qquad \lambda = \frac{(E - 2\mu)\mu}{3\mu - E}. \qquad (3.13.5)$$

In technical books μ may be denoted by G and called the *shear modulus*. Thermodynamic considerations show that $\mu > 0$ and $0 < \nu < 0.5$.

In this book we need the inverse relations between $\boldsymbol{\varepsilon}$ and $\boldsymbol{\sigma}$ where *Lamé's constants* λ and μ appear. In a Cartesian frame they are

$$\sigma^{ij} = \lambda\vartheta\delta^{ij} + 2\mu\varepsilon^{ij}, \qquad \vartheta = \varepsilon^{11} + \varepsilon^{22} + \varepsilon^{33}, \qquad (3.13.6)$$

where δ^{ij} is the Kronecker delta given by

$$\delta^{ij} = \begin{cases} 1, & i = j, \\ 0, & i \neq j. \end{cases}$$

In non-component form (3.13.6) can be written as

$$\boldsymbol{\sigma} = \lambda\vartheta\mathbf{g} + \mu\boldsymbol{\varepsilon}, \qquad (3.13.7)$$

where $\mathbf{g}$ is the metric tensor. In curvilinear coordinates we must take

$$\vartheta = g_{ij}\varepsilon^{ij}, \qquad (3.13.8)$$

which is the trace of the tensor $\boldsymbol{\varepsilon}$ (its *first invariant*). This tensor form of Hooke's law is a consequence of the fact that the stress and strain tensors are coaxial.

3.14 The Equilibrium Equations of Linear Elasticity in Displacements

We are ready to write out the equilibrium equations of linear elasticity in displacements. Substituting Hooke's law (3.13.2) into the equilibrium equation (3.12.1), we obtain

$$\boldsymbol{\nabla} \cdot (\mathbf{c} \cdots \boldsymbol{\varepsilon}) + \mathbf{F} = \mathbf{0}. \qquad (3.14.1)$$

The streamlined tensor notation makes this look simple. But a component representation will be needed for any detailed work. In general curvilinear coordinates, (3.14.1) is cumbersome because the fourth-rank tensor $\mathbf{c}$ depends on the frame vectors; even when it is constant as a tensor, its components become variable along with the basis vectors. Hence, large expressions arise from application of the $\boldsymbol{\nabla}$-operator to $\mathbf{c} \cdots \boldsymbol{\varepsilon}$. In Cartesian coordinates, however, both the basis vectors and the components of $\mathbf{c}$ are constants and the resulting equations are much simpler. For an isotropic elastic body they are called the *Navier* or *Navier–Cauchy equations*. It is easier to derive these directly by substituting (3.13.6) into (3.8.11) given in Cartesian coordinates (where it suffices to use subscripts only). The Navier equations are

$$\mu\left(\frac{1}{1-2\nu}\frac{\partial\vartheta}{\partial x_1} + \Delta u_1\right) + F_1 = 0,$$
$$\mu\left(\frac{1}{1-2\nu}\frac{\partial\vartheta}{\partial x_2} + \Delta u_2\right) + F_2 = 0,$$
$$\mu\left(\frac{1}{1-2\nu}\frac{\partial\vartheta}{\partial x_3} + \Delta u_3\right) + F_3 = 0, \qquad (3.14.2)$$

where
$$\vartheta = \frac{\partial u_1}{\partial x_1} + \frac{\partial u_2}{\partial x_2} + \frac{\partial u_3}{\partial x_3} \qquad (3.14.3)$$
and
$$\Delta = \nabla \cdot \nabla = \frac{\partial^2}{\partial x_1{}^2} + \frac{\partial^2}{\partial x_2{}^2} + \frac{\partial^2}{\partial x_3{}^2}. \qquad (3.14.4)$$

Suppose an elastic body occupies volume V. The three equations (3.14.2) are coupled. We expect that three equations will be sufficient to determine three unknown functions — that they should determine the deformation of the body uniquely given suitable conditions on the boundary S. This is confirmed below. It makes sense to consider the equilibrium problem where the displacement vector is given at each point:

$$\mathbf{u}\big|_S = \mathbf{a}. \qquad (3.14.5)$$

When Navier's system (3.14.2) is supplied with boundary conditions (3.14.5), we have the *first problem of elasticity*. This problem is analogous to Dirichlet's problem for the membrane. When the displacement vector on the boundary is given, we cannot arbitrarily prescribe values for $\mathbf{f}$; similarly, if we are given $\mathbf{f}$ on the boundary we cannot arbitrarily assign $\mathbf{u}$ there. This brings us to the *second problem of elasticity*, in which the surface force $\mathbf{f}$ is given on S:

$$\mathbf{n} \cdot \boldsymbol{\sigma}\big|_S = \mathbf{f}. \qquad (3.14.6)$$

This is analogous to Neumann's problem for the membrane. One can also consider mixed problems for the Navier system. For example, the displacement may be given over part of S and a surface force over the rest. Alternatively, just the normal component of $\mathbf{u}$ may be given along with tangential surface forces. There are no special names for problems in which various elastic supports, etc, may act. In general, the boundary conditions should be such that at each point of the boundary we have three independent scalar conditions.

Anyone familiar with partial differential equations is aware of the role played by Poisson's equation in the theory of second-order elliptic equations: much of the theory attempts to extend the properties of boundary value problems for Poisson's equation to general elliptic equations. It seems that the central role in the theory of second-order elliptic systems in two and three dimensions belongs to the equations of linear elasticity in displacements.

Other boundary value problems are considered in the theory of linear elasticity. For example, we could study a system consisting only of the equilibrium equations in components of the stress tensor along with Hooke's law and the relations for the components of the strain tensor. In this case the system contains many more differential equations and unknown variables. Because it is equivalent, by the way of derivation, to Navier's system, we must supplement it with only three boundary conditions at each boundary point.

Also considered in linear elasticity are the equations in the components of the stress tensor. Here we have three equations of equilibrium in the stresses. However, we know this is not enough to determine the strain state of a body uniquely. The remaining equations come from eliminating **u** from the components of the strain tensor. This yields six nontrivial relations between the components of ε. Using Hooke's law, we produce six additional equations for the components of the stress tensor: these are called *Saint Venant's equations*. Together with the equilibrium equations they constitute a system of nine equations with respect to six unknown independent components of σ. Since this system is a consequence of the complete system of equations of linear elasticity, for unique solvability we should supplement it with three boundary conditions. These boundary value problems normally arise when a load is given on the boundary. We should note that the equilibrium problem for Saint Venant's equations is somewhat strange from the viewpoint of general differential equation theory, despite the fact that it is equivalent to the equilibrium problem in displacements. The system of equations is formulated for six independent components of the stress tensor. It contains three scalar equilibrium equations, each of the first order, and six Saint Venant equations of the second order. Moreover, there are only three boundary conditions at each point of the boundary.

Although common sense is valuable in formulating problems, it should be supported by thorough investigation. We should prove that we have actually formulated uniquely solvable problems — intuition can mislead us otherwise. Our proof for the equilibrium problem in displacements will be based on the fact that the conditions $\mu > 0$ and $0 < \nu < 0.5$ are sufficient to guarantee that the strain energy of the elastic body vanishes only when deformation is absent. The next steps in the investigation of solvability of the problem for an elastic body are similar to those in the equilibrium problem for a membrane.

Before treating the boundary value problems of linear elasticity in more detail, we should mention the following. Pure mathematicians normally begin with the setup of a mechanical problem in final form; they try to avoid interpretation of mechanical meaning whenever possible. This approach is not the best, as mechanical problems — from the simplest problems involving bars to the more complex problems involving three-dimensional bodies — exhibit similar general properties. Knowledge of these can provide an investigator with important cues on how to proceed. For example, the virtual work principle and the principle of minimum total energy dictate the form of the generalized setup of boundary value problems. The generalized solution constitutes a relatively recent tool in pure mathematics, but the principles have been known in mechanics for a long time.

Theoretical investigation of boundary value problems of elasticity centers on Navier's system. We will consider some problems for this system below.

3.15 Virtual Work Principle in Linear Elasticity

In § 3.12, for the state of a body as described by the continuously differentiable stress tensor, we established that the virtual work principle is equivalent to the equilibrium equation and the natural (force) boundary conditions. This holds independently of the constitutive law for the material. So the VWP equation

$$-\int_V \mathbf{F} \cdot \delta\mathbf{u}\, dV - \int_{S\setminus S_1} \mathbf{f} \cdot \delta\mathbf{u}\, dS + \int_V \boldsymbol{\sigma} \cdot\cdot\, \delta\boldsymbol{\varepsilon}\, dV = 0 \qquad (3.15.1)$$

can be used to pose the mixed problem in which the displacement field is given on a portion S_1 of the boundary,

$$\mathbf{u}\big|_{S_1} = \mathbf{a}, \qquad (3.15.2)$$

while a surface load $\mathbf{f}$ acts on the remainder. The virtual displacement field should be sufficiently smooth and satisfy

$$\delta\mathbf{v}\big|_{S_1} = \mathbf{0}. \qquad (3.15.3)$$

Let us apply the VWP equation in the framework of linear elasticity. It turns out to define the setup of the equilibrium problem completely. Certain of its other consequences are physically important as well. Let us invoke Hooke's law. Changing $\delta\mathbf{u}$ to $\mathbf{v}$ and substituting $\sigma^{ij} = c^{ijkl}\varepsilon_{kl}$ into the VWP equation, we get

$$\int_V c^{ijkl}\varepsilon_{kl}(\mathbf{u})\varepsilon_{ij}(\mathbf{v})\,dV - \int_V \mathbf{F}\cdot\mathbf{v}\,dV - \int_{S\setminus S_1} \mathbf{f}\cdot\mathbf{v}\,dS = 0. \qquad (3.15.4)$$

Now we can establish some principal results of linear elasticity. First, the left side of the VWP equation (3.15.4) is the first variation of the functional

$$\mathcal{E}_t(\mathbf{u}) = \frac{1}{2}\int_V c^{ijkl}\varepsilon_{kl}(\mathbf{u})\varepsilon_{ij}(\mathbf{u})\,dV - \int_V \mathbf{F}\cdot\mathbf{u}\,dV - \int_{S\setminus S_1} \mathbf{f}\cdot\mathbf{u}\,dS. \qquad (3.15.5)$$

This follows immediately from symmetry of the quadratic form $c^{ijkl}\varepsilon_{kl}(\mathbf{u})\varepsilon_{ij}(\mathbf{v})$ in $\mathbf{u}$ and $\mathbf{v}$.

Exercise 3.15.1. *Verify this.*

The term

$$\mathcal{E}(\mathbf{u}) = \frac{1}{2}\int_V c^{ijkl}\varepsilon_{kl}(\mathbf{u})\varepsilon_{ij}(\mathbf{u})\,dV \qquad (3.15.6)$$

is called the *strain energy*. Any deformation of the body should be associated with stored energy, which should be positive for an elastic deformation. So we suppose that the functional $\mathcal{E}(\mathbf{u})$ has the following property.

Positiveness assumption. *There is a positive constant c_0 such that for any tensor $\boldsymbol{\varepsilon}$,*

$$\boldsymbol{\varepsilon}\cdot\cdot\mathbf{c}\cdot\cdot\boldsymbol{\varepsilon} \geq c_0\,\boldsymbol{\varepsilon}\cdot\boldsymbol{\varepsilon}. \qquad (3.15.7)$$

This inequality implies positive definiteness of the density of strain energy of the elastic body.

In component form (3.15.7) can be written as

$$c^{ijkl}\varepsilon_{kl}\varepsilon_{ij} \geq c_0\,\varepsilon^{mn}\varepsilon_{mn}. \qquad (3.15.8)$$

Elasticity textbooks normally express this condition in Cartesian coordinates. When written for an isotropic body, it reduces to two ordinary inequalities for the elastic moduli:

$$\lambda + \frac{2}{3}\mu > 0, \qquad \mu > 0. \qquad (3.15.9)$$

The parameter $\lambda + \frac{2}{3}\mu$ is called the *bulk modulus*, and μ the *shear modulus*.

Again, we suppose (3.15.7) holds. The functional $\mathcal{E}(\mathbf{u})$ represents the total energy of the load-elastic body system. So we expect that its minimum is achieved at a solution $\mathbf{u}$ of the mixed equilibrium problem. The fact that the first variation of $\mathcal{E}(\mathbf{u})$ is the left side of the VWP equation (3.15.4) supports this opinion. From the VWP equation we can derive the setup of the equilibrium problem in the form of Navier's differential equations and the natural condition

$$\mathbf{n} \cdot \boldsymbol{\sigma}|_{S \setminus S_1} = \mathbf{f}. \tag{3.15.10}$$

For a complete setup, these should be supplemented with the condition (3.15.2). However, to be convinced that the setup is well-posed we should demonstrate the existence of a unique solution to the problem in the VWP formulation. We start with the next fundamental fact of linear elasticity: the uniqueness theorem due to Kirchhoff.

Theorem 3.15.1. *Let S_1 be a nonempty portion of S. The mixed problem of linear elasticity can have no more than one smooth solution $\mathbf{u}$.*

Proof. Assume the existence of two solutions $\mathbf{u}_1, \mathbf{u}_2$ of (3.15.4). Subtracting the two equations

$$\int_V c^{ijkl} \varepsilon_{kl}(\mathbf{u}_1) \varepsilon_{ij}(\mathbf{v}) \, dV - \int_V \mathbf{F} \cdot \mathbf{v} \, dV - \int_{S \setminus S_1} \mathbf{f} \cdot \mathbf{v} \, dS = 0,$$

$$\int_V c^{ijkl} \varepsilon_{kl}(\mathbf{u}_2) \varepsilon_{ij}(\mathbf{v}) \, dV - \int_V \mathbf{F} \cdot \mathbf{v} \, dV - \int_{S \setminus S_1} \mathbf{f} \cdot \mathbf{v} \, dS = 0,$$

and introducing $\mathbf{u}_0 = \mathbf{u}_2 - \mathbf{u}_1$, we get

$$\int_V c^{ijkl} \varepsilon_{kl}(\mathbf{u}_0) \varepsilon_{ij}(\mathbf{v}) \, dV = 0.$$

But $\mathbf{u}_0 = \mathbf{0}$ on S_1, so we can take $\mathbf{v} = \mathbf{u}_0$ and obtain

$$\int_V c^{ijkl} \varepsilon_{kl}(\mathbf{u}_0) \varepsilon_{ij}(\mathbf{u}_0) \, dV = 0.$$

By the positiveness assumption (3.15.7) we have

$$\varepsilon_{ij}(\mathbf{u}_0) = 0 \quad \text{for all } i, j.$$

So $\boldsymbol{\varepsilon}$ is zero, and consequently $\mathbf{u}_0$ can only take the form of the infinitesimal rigid displacement $\mathbf{u}_0 = \mathbf{a} + \mathbf{b} \times \mathbf{r}$. Because $\mathbf{u}_0 = \mathbf{0}$ on S_1, however, we have $\mathbf{u}_0 = \mathbf{0}$ throughout V. □

We will return to the uniqueness theorem later (Theorems 3.17.1 and 3.18.1). Note that we assumed smoothness in the solution; this could mean the Cartesian components of the displacement vector all belong to $C^{(2)}(V)$, for example.

One simple but important consequence of the VWP equation, used for numerical solution of elasticity problems, is *Betti's duality theorem*. We consider the same body with a portion of the boundary fixed: $\mathbf{u} = \mathbf{0}$ on S_1, say. Let $(\mathbf{F}_1, \mathbf{f}_1)$ and $(\mathbf{F}_2, \mathbf{f}_2)$ be two systems of loads, where the first and second elements of a pair denote the systems of body and surface forces, respectively. If $\mathbf{u}_1$ and $\mathbf{u}_2$ are the solutions of the two problems, then

$$\int_V c^{ijkl}\varepsilon_{kl}(\mathbf{u}_i)\varepsilon_{ij}(\mathbf{v})\,dV - \int_V \mathbf{F}_i \cdot \mathbf{v}\,dV - \int_{S\setminus S_1} \mathbf{f}_i \cdot \mathbf{v}\,dS = 0 \quad (i=1,2).$$

Setting $\mathbf{v} = \mathbf{u}_2$ in the first equation and $\mathbf{v} = \mathbf{u}_1$ in the second, we obtain two equalities. But the first integral is the same in each case by the symmetry of the elastic moduli tensor. It follows that

$$\int_V \mathbf{F}_2 \cdot \mathbf{u}_1\,dV + \int_{S\setminus S_1} \mathbf{f}_2 \cdot \mathbf{u}_1\,dS = \int_V \mathbf{F}_1 \cdot \mathbf{u}_2\,dV + \int_{S\setminus S_1} \mathbf{f}_1 \cdot \mathbf{u}_2\,dS.$$

This equality is a statement of

Theorem 3.15.2 (Betti's duality theorem). *The work of one load system over the displacement field caused by a second load system, on the same body, is equal to the work of the second system over the displacement field caused by the first system.*

Betti's theorem was first established for beam problems.

3.16 Generalized Setup of Elasticity Problems

We will not attempt to treat the classical setup of the boundary value problems of linear elasticity. This may seem strange, but such studies are more complex technically than a study of the generalized approach. So we start to consider the latter, and will try to mimic our earlier treatment of mechanical models.

The VWP equation (3.15.4) has the same quadratic structure as the corresponding equations for all other objects we have considered (bar, beam, string, and membrane). It has a bilinear portion that corresponds to the work of internal forces over admissible displacements, and a linear portion that equals the work of external forces over the same admissible displacements. We are thereby prompted to use (3.15.4) to define a generalized setup. We can then use Theorem 2.13.1 to establish an existence-uniqueness result. Moreover, the theorem will show that our equilibrium formulation for the elastic body is fully satisfactory.

We start by introducing the space in which the problem will be posed. Let C_{20} be the set of all vector-functions defined on the closed and bounded set $V \subset \mathbb{R}^3$. These functions, describing admissible displacements of points of the body, must be of class $C^{(2)}(V)$ and satisfy homogeneous boundary conditions on S_1:

$$\mathbf{u}\big|_{S_1} = \mathbf{0}. \tag{3.16.1}$$

The statement $\mathbf{u} \in C^{(2)}(V)$ means that $\mathbf{u}$ itself has two continuous derivatives on V (in a coordinate-free sense) or that each Cartesian component of $\mathbf{u}$ has all continuous derivatives up to the order two on V.

On C_{20} we introduce the bilinear functional

$$(\mathbf{u}, \mathbf{v})_E = \int_V c^{ijkl} \varepsilon_{kl}(\mathbf{u}_0) \varepsilon_{ij}(\mathbf{v}) \, dV. \tag{3.16.2}$$

This is related to the strain energy of the elastic body that occupies V. By analogy with the membrane and other models we have considered, it should serve as an inner product on C_{20}. It is clearly linear in both $\mathbf{u}$ and $\mathbf{v}$. The symmetry in the indices of the elastic moduli c^{ijkl} gives us the property $(\mathbf{u}, \mathbf{v})_E = (\mathbf{v}, \mathbf{u})_E$. By (3.15.7) we get

$$(\mathbf{u}, \mathbf{u})_E \geq 0.$$

From $(\mathbf{u}, \mathbf{u})_E = 0$ and (3.15.7) it follows that, as we said above,

$$\mathbf{u} = \mathbf{a} + \mathbf{b} \times \mathbf{r}$$

and so by (3.16.1) we have $\mathbf{u} = \mathbf{0}$. Hence (3.16.2) defines an inner product on C_{20}. The resulting inner product space is not complete, however; this is similar to the incompleteness of the space of continuous functions with the norm of $L^p(V)$.

Definition 3.16.1. The energy space E_{E0} for an elastic body occupying volume V, and having a portion S_1 of the boundary surface clamped as in (3.16.1), is the completion of C_{20} in the norm $\|\cdot\|_E$ induced by the inner product (3.16.2).

Now we can define a generalized solution of the problem under consideration. We will use the VWP equation (3.15.4) rewritten via (3.16.2).

Definition 3.16.2. We say that $\mathbf{u} \in E_{E0}$ is a generalized solution to the equilibrium problem for an elastic body occupying volume V and having fixed boundary surface portion S_1 if it satisfies the VWP equation

$$(\mathbf{u}, \mathbf{v})_E - \int_V \mathbf{F} \cdot \mathbf{v} \, dV - \int_{S \setminus S_1} \mathbf{f} \cdot \mathbf{v} \, dS = 0 \tag{3.16.3}$$

for any $\mathbf{v} \in E_{E0}$.

Thus, roughly speaking, this solution satisfies the virtual work principle for the elastic body.

Note that the structure of (3.16.3) is appropriate for the application of Theorem 2.13.1. Of course, we will have to impose conditions on the forces **F** and **f** so that the integral terms in (3.16.3) are linear continuous functionals in $\mathbf{v} \in E_{E0}$. First we should study the properties of the elements of E_{E0}. We wish to show that E_{E0} is a subspace of a Sobolev space. This is a consequence of Korn's inequality, treated next.

Korn's inequality

For vector-functions in E_{E0}, Korn's equality is

$$\int_V \left(|\mathbf{u}|^2 + \boldsymbol{\nabla}\mathbf{u} \cdot\cdot \boldsymbol{\nabla}\mathbf{u}^T\right) dV \leq c \int_V c^{ijkl}\varepsilon_{kl}(\mathbf{u})\varepsilon_{ij}(\mathbf{u}) \, dV, \qquad (3.16.4)$$

where c is a constant that does not depend on $\mathbf{u} \in E_{E0}$ (see (3.9.24)). In Cartesian coordinates this can be rewritten as

$$\int_V \left[\sum_{i=1}^3 u_i^2 + \sum_{i,j=1}^3 \left(\frac{\partial u_i}{\partial x_j}\right)^2\right] dV \leq c \int_V c^{ijkl}\varepsilon_{kl}(\mathbf{u})\varepsilon_{ij}(\mathbf{u}) \, dV. \qquad (3.16.5)$$

In three-dimensional elasticity, this inequality is commonly presented in Cartesian coordinates in the equivalent form

$$\int_V \left[\sum_{i=1}^3 u_i^2 + \sum_{i,j=1}^3 \left(\frac{\partial u_i}{\partial x_j}\right)^2\right] dV \leq c^* \int_V \sum_{i,j=1}^3 \left[\frac{1}{2}\left(\frac{\partial u_i}{\partial x_j} + \frac{\partial u_j}{\partial x_i}\right)\right]^2 dV.$$

(3.16.6)

A general proof of Korn's inequality is beyond the scope of this book. But we can treat the important case for which

$$\mathbf{u}\big|_S = \mathbf{0}. \qquad (3.16.7)$$

The proof depends only on integration by parts. First we prove (3.16.6) for a smooth vector-function $\mathbf{u}$ having second continuous derivatives in V. Denoting the right-hand side of (3.16.6) by A, we get

$$A \geq \int_V \left(\varepsilon_{11}^2 + \varepsilon_{22}^2 + \varepsilon_{33}^2 + \varepsilon_{12}^2 + \varepsilon_{13}^2 + \varepsilon_{23}^2\right) dV = A_1, \text{ say.}$$

Let us expand the integrand of A_1:

$$A_1 = \int_V \left\{ \left(\frac{\partial u_1}{\partial x_1}\right)^2 + \left(\frac{\partial u_2}{\partial x_2}\right)^2 + \left(\frac{\partial u_3}{\partial x_3}\right)^2 \right.$$
$$+ \frac{1}{4}\left[\left(\frac{\partial u_1}{\partial x_2}\right)^2 + \left(\frac{\partial u_1}{\partial x_3}\right)^2 + \left(\frac{\partial u_2}{\partial x_1}\right)^2 + \left(\frac{\partial u_2}{\partial x_3}\right)^2\right.$$
$$\left.+ \left(\frac{\partial u_3}{\partial x_1}\right)^2 + \left(\frac{\partial u_3}{\partial x_2}\right)^2\right]$$
$$\left.+ \frac{1}{2}\left(\frac{\partial u_1}{\partial x_2}\frac{\partial u_2}{\partial x_1} + \frac{\partial u_1}{\partial x_3}\frac{\partial u_3}{\partial x_1} + \frac{\partial u_2}{\partial x_3}\frac{\partial u_3}{\partial x_2}\right)\right\} dV.$$

We integrate by parts twice in each of the last three terms of the integrand:

$$\int_V \frac{\partial u_i}{\partial x_j}\frac{\partial u_j}{\partial x_i} dV = \int_V \frac{\partial u_i}{\partial x_i}\frac{\partial u_j}{\partial x_j} dV.$$

Applying the elementary inequality $|ab| \leq a^2/2 + b^2/2$, we estimate

$$\left|\int_V \frac{\partial u_i}{\partial x_i}\frac{\partial u_j}{\partial x_j} dV\right| \leq \frac{1}{2}\int_V \left[\left(\frac{\partial u_i}{\partial x_i}\right)^2 + \left(\frac{\partial u_j}{\partial x_j}\right)^2\right] dV$$

and so

$$A_1 \geq \int_V \left\{ \left(\frac{\partial u_1}{\partial x_1}\right)^2 + \left(\frac{\partial u_2}{\partial x_2}\right)^2 + \left(\frac{\partial u_3}{\partial x_3}\right)^2 \right.$$
$$+ \frac{1}{4}\left[\left(\frac{\partial u_1}{\partial x_2}\right)^2 + \left(\frac{\partial u_1}{\partial x_3}\right)^2 + \left(\frac{\partial u_2}{\partial x_1}\right)^2 + \left(\frac{\partial u_2}{\partial x_3}\right)^2\right.$$
$$\left.+ \left(\frac{\partial u_3}{\partial x_1}\right)^2 + \left(\frac{\partial u_3}{\partial x_2}\right)^2\right]$$
$$\left.- \frac{1}{2}\left(\left(\frac{\partial u_1}{\partial x_1}\right)^2 + \left(\frac{\partial u_2}{\partial x_2}\right)^2 + \left(\frac{\partial u_3}{\partial x_3}\right)^2\right)\right\} dV.$$

Thus

$$A_1 \geq \frac{1}{4}\int_V \sum_{i,j=1}^3 \left(\frac{\partial u_i}{\partial x_j}\right)^2 dV.$$

Because all components of **u** vanish on the boundary, Friedrichs inequality (2.16.5) yields

$$\int_V |\mathbf{u}|^2 dV \leq c_1 \int_V \sum_{i,j=1}^3 \left(\frac{\partial u_i}{\partial x_j}\right)^2 dV$$

with a constant c_1 that does not depend on $\mathbf{u}$. A consequence of this chain of inequalities is the proof of (3.16.6) for smooth $\mathbf{u}$.

When $\mathbf{u} = \mathbf{0}$ on only a portion of S, the proof is much more technical (see, for example, [Ciarlet (1994)]).

Now we should show that Korn's inequality holds for elements of E_{E0}. This is done in a routine way. An element of E_{E0} is the union of equivalent Cauchy sequences of vector-functions $\{\mathbf{u}_n\} \in C_{20}$ in the norm $\|\mathbf{u}\|_E = (\mathbf{u}, \mathbf{u})_E^{1/2}$. Putting $\mathbf{u} = \mathbf{u}_n - \mathbf{u}_m$ in (3.16.3), we see that $\{\mathbf{u}_n\}$ is such that each of its components is a Cauchy sequence in the norm of $W^{1,2}(V)$. By definition of $W^{1,2}(V)$, if $\mathbf{u} \in E_{E0}$ then $\mathbf{u} \in W^{1,2}(V) \times W^{1,2}(V) \times W^{1,2}(V)$ as well, which means that (3.16.3) holds for all $\mathbf{u} \in E_{E0}$. Moreover, on E_{E0} the norms of E_{E0} and $W^{1,2}(V) \times W^{1,2}(V) \times W^{1,2}(V)$ are equivalent.

Consequently we can use the Sobolev imbedding Theorem 2.16.4. We see that each Cartesian component of $\mathbf{u} \in E_{E0}$ belongs to $L^6(V)$ and thus the magnitude, the Cartesian norm of $\mathbf{u}$, is such that $|\mathbf{u}| \in L^6(V)$. Moreover, on any piecewise smooth surface Z lying in V, the Cartesian components of $\mathbf{u}$ and the absolute value of $\mathbf{u}$ all belong to $L^4(Z)$. It suffices for Z to consist of a finite number of pieces, each representable in its own Cartesian frame (x_1, x_2, x_3) as $x_3 = f_k(x_1, x_2)$, where $f_k(x_1, x_2)$ is continuously differentiable in the domain of parameters x_1, x_2 that describe the corresponding piece. Thus the imbedding Theorem 2.16.4 yields the inequalities

$$\left(\int_V |\mathbf{u}|^6 \, dV\right)^{1/6} \leq c_2 \|\mathbf{u}\|_E \tag{3.16.8}$$

and

$$\left(\int_{S \setminus S_1} |\mathbf{u}|^4 \, dV\right)^{1/4} \leq c_3 \|\mathbf{u}\|_E , \tag{3.16.9}$$

with constants c_k that do not depend on $\mathbf{u}$.

3.17 Existence Theorem for an Elastic Body

Now we can return to the equilibrium problem for an elastic body. An application of Theorem 2.13.2 allows us to formulate

Theorem 3.17.1. *Let the Cartesian components of $\mathbf{F}$ and $\mathbf{f}$ belong to $L^{6/5}(V)$ and $L^{4/3}(S \setminus S_1)$, respectively. Then there exists a unique gen-*

eralized solution of the equilibrium problem for an elastic body occupying volume V, in the sense of Definition 3.16.2.

Proof. To apply Theorem 2.13.2, we should verify that the integrals that yield the work of external forces over $\mathbf{v}$ in (3.16.3) define linear continuous functionals in $\mathbf{v} \in E_{E0}$. Linearity with respect to $\mathbf{v}$ is obvious. Continuity follows from Hölder's inequality, when we take into account (3.16.8) and (3.16.9), by Theorem 2.16.4. Indeed

$$\left| \int_V \mathbf{F} \cdot \mathbf{v} \, dV \right| \leq \int_V |\mathbf{F}| |\mathbf{v}| \, dV$$
$$\leq \left(\int_V |\mathbf{F}|^{6/5} \, dV \right)^{5/6} \left(\int_V |\mathbf{v}|^6 \, dV \right)^{1/6}$$
$$\leq c_3 \|\mathbf{v}\|_E$$

and

$$\left| \int_{S \setminus S_1} \mathbf{f} \cdot \mathbf{v} \, dS \right| \leq \int_{S \setminus S_1} |\mathbf{f}| |\mathbf{v}| \, dS$$
$$\leq \left(\int_{S \setminus S_1} |\mathbf{f}|^{4/3} \, dS \right)^{3/4} \left(\int_{S \setminus S_1} |\mathbf{v}|^4 \, dS \right)^{1/4}$$
$$\leq c_4 \|\mathbf{v}\|_E.$$

It follows that the corresponding functionals are continuous. □

Theorem 3.17.1 assumes the homogeneous boundary condition (3.16.1). The case of a nonhomogeneous condition

$$\mathbf{u}\big|_{S_1} = \mathbf{a} \tag{3.17.1}$$

can be considered exactly as for the nonhomogeneous condition for Poisson equation (see p. 136).

Exercise 3.17.1. *Carry out the steps.*

3.18 Equilibrium of a Free Elastic Body

Let us consider the problem of an elastic body under the action of external forces and without geometrical restrictions. Mechanical experience tells us that equilibrium can occur only if the forces are self-balanced. We came to the same conclusion for the other mechanical objects we studied. Let us investigate further. We start with the equilibrium equation in the form of the virtual work principle, (3.16.3), where S_1 is absent:

$$(\mathbf{u}, \mathbf{v})_E - \int_V \mathbf{F} \cdot \mathbf{v} \, dV - \int_S \mathbf{f} \cdot \mathbf{v} \, dS = 0. \quad (3.18.1)$$

We know that for $\mathbf{u}_0 = \mathbf{a} + \mathbf{b} \times \mathbf{r}$, which is the general form of small rigid motions of the elastic body, we have

$$(\mathbf{u}, \mathbf{u}_0)_E = 0$$

for any admissible $\mathbf{u}$, including a solution. Setting $\mathbf{v} = \mathbf{u}_0$ in (3.18.1) we get

$$\int_V \mathbf{F} \cdot \mathbf{u}_0 \, dV + \int_S \mathbf{f} \cdot \mathbf{u}_0 \, dS = 0$$

or, using the properties of the scalar triple product,

$$\mathbf{a} \cdot \left(\int_V \mathbf{F} \, dV + \int_S \mathbf{f} \, dS \right) + \mathbf{b} \cdot \left(\int_V \mathbf{r} \times \mathbf{F} \, dV + \int_S \mathbf{r} \times \mathbf{f} \, dS \right) = 0.$$

Because the constant vectors $\mathbf{a}$ and $\mathbf{b}$ are arbitrary, two conditions are implied:

$$\int_V \mathbf{F} \, dV + \int_S \mathbf{f} \, dS = \mathbf{0}, \quad (3.18.2)$$

$$\int_V \mathbf{r} \times \mathbf{F} \, dV + \int_S \mathbf{r} \times \mathbf{f} \, dS = \mathbf{0}. \quad (3.18.3)$$

The first of these says that the resultant external force is zero. The second means that the moment of all external forces is zero. The external forces are self-balanced as expected.

It is clear that if a solution $\mathbf{u}$ exists for the equilibrium problem of a body free of geometrical constraints, then $\mathbf{u} + \mathbf{u}_0$ with any small rigid motion $\mathbf{u}_0 = \mathbf{a} + \mathbf{b} \times \mathbf{r}$ is also a solution. From this set of solutions, let us choose a unique solution that satisfies the conditions

$$\int_V (\mathbf{u} + \mathbf{u}_0) \, dV = \mathbf{0}, \quad \int_V \mathbf{r} \times (\mathbf{u} + \mathbf{u}_0) \, dV = \mathbf{0}. \quad (3.18.4)$$

We will use the fact that these equations define the vector constants $\mathbf{a}$ and $\mathbf{b}$ uniquely for any integrable vector function $\mathbf{u}$. Uniqueness of the choice of $\mathbf{a}, \mathbf{b}$ can be established by the reader in

Exercise 3.18.1. *Prove that the equations*

$$\int_V \mathbf{u}_0 \, dV = \mathbf{0}, \quad \int_V \mathbf{r} \times \mathbf{u}_0 \, dV = \mathbf{0},$$

imply $\mathbf{a} = \mathbf{0}$ *and* $\mathbf{b} = \mathbf{0}$.

Indeed, in component form the equations of Exercise 3.18.1 constitute a system of six linear equations in the components of **a** and **b**. Since this system has the unique solution zero, its determinant is nonzero. Hence the system of six scalar equations (3.18.4) with respect to the components of **a** and **b**, which has the same determinant, has a unique solution. In this way we reduce the equilibrium problem for the free body to that for the body restricted by the additional conditions (3.18.4). Note that by requiring that displacements of a rigid body to satisfy (3.18.4), we fix a position of this body completely.

Let us sketch out the equilibrium problem for a free elastic body. We start by introducing the energy space. Suppose V is a compact domain in $\mathbb{R}^3$ for which the Sobolev imbedding theorems hold. Let $C_f^{(2)}(V)$ be the set of vector-functions $\mathbf{u}(\mathbf{r})$ having values in $\mathbb{R}^3$ such that their Cartesian components belong to $C^{(2)}(V)$ and satisfy (3.18.4). Clearly, $C_f^{(2)}(V)$ is a linear space.

Exercise 3.18.2. *Verify that $(\cdot, \cdot)_E$ is an inner product on $C_f^{(2)}(V)$.*

Let us introduce the completion of this space.

Definition 3.18.1. The completion of $C_f^{(2)}(V)$ with respect to the norm induced by the inner product $(\cdot, \cdot)_E$ is called the energy space E_{Ef}.

By analogy with Definition 3.16.2 we define a generalized solution.

Definition 3.18.2. We say that $\mathbf{u} + \mathbf{u}_0$ with $\mathbf{u} \in E_{Ef}$ is a generalized solution of the equilibrium problem for a free (of geometrical constraints) elastic body occupying V if $\mathbf{u}$ satisfies

$$(\mathbf{u}, \mathbf{v} + \mathbf{v}_0)_E - \int_V \mathbf{F} \cdot (\mathbf{v} + \mathbf{v}_0)\, dV - \int_S \mathbf{f} \cdot (\mathbf{v} + \mathbf{v}_0)\, dS = 0$$

for any $\mathbf{v} \in E_{Ef}$ and $\mathbf{v}_0 = \mathbf{a}_1 + \mathbf{b}_1 \times \mathbf{r}$ with arbitrary constant vectors $\mathbf{a}_1$ and $\mathbf{b}_1$. Here $\mathbf{u}_0 = \mathbf{a} + \mathbf{b} \times \mathbf{r}$ where $\mathbf{a}$ and $\mathbf{b}$ are arbitrary constant vectors.

From this definition it follows that the external forces must satisfy the self-balance conditions (3.18.2). When they do, the equation for $\mathbf{u}$ takes the form

$$(\mathbf{u}, \mathbf{v})_E - \int_V \mathbf{F} \cdot \mathbf{v}\, dV - \int_S \mathbf{f} \cdot \mathbf{v}\, dS = 0 \qquad (3.18.5)$$

with $\mathbf{u}, \mathbf{v} \in E_{Ef}$.

The form of (3.18.5) is similar to that of the equation in Definition 3.16.2. Moreover, Korn's inequality (3.16.4) holds for $\mathbf{u} \in E_{Ef}$ (see

[Ciarlet (1994)]). By analogy with the considerations of § 3.17, we have the following existence theorem.

Theorem 3.18.1. *Suppose the Cartesian components of* **F** *belong to* $L^{6/5}(V)$ *and those of* **f** *belong to* $L^{4/3}(S)$. *Suppose the self-balance conditions* (3.18.2) *hold. Then there exists a generalized solution* $\mathbf{u}+\mathbf{a}+\mathbf{b}\times\mathbf{r}$, *where* $\mathbf{u} \in E_{Ef}$, *of the equilibrium problem for a free elastic body occupying volume* V, *in the sense of Definition 3.18.2. Here* **u** *is unique and* **a**, **b** *are arbitrary constant vectors.*

3.19 Variational Methods for Equilibrium Problems

Earlier we studied mechanical models based on differential equations. Despite the widespread belief that all the imprecision of such models comes from imprecise geometrical assumptions, they are fundamentally imprecise. This is because all materials have atomic structure, so the limit passages involved in deriving the differential equations are fundamentally imprecise. No model in continuum mechanics can be truly precise as it is constructed using limit passages. The equations that we have entail some rather vague averaging process. Because of this, the finite models used in numerical calculations are — fundamentally speaking — no less precise than the differential ones if they obey the fundamental laws of mechanics in some sense. Indeed, they can and deserve to be studied independently from the differential equation models. As a rule, the starting point for an approximate (finite-dimensional) numerical model is a corresponding differential equation model. Normally the latter satisfies the fundamental laws of mechanics that we just mentioned. The constructed approximate model may or may not, depending on how we derive it. But there are approximate models that obey the fundamental mechanical laws automatically, as a consequence of the fact that they were derived from a differential equation model. This is the case, in particular, for models obtained by variational methods for the approximate solution of equilibrium problems.

We saw that all the linear problems we considered fit the same elementary abstract equation defined by the problem of minimum of a quadratic functional in a Hilbert space H:

$$\Phi(u) = \frac{1}{2}(u,u)_H - F(u), \qquad (3.19.1)$$

where $F(x)$ is a linear continuous functional. We formulate this problem as

Problem 3.19.1. *(Minimum problem) Find the element $u^* \in H$ that minimizes $\Phi(u)$ in the space H.*

Existence and uniqueness of solution to this problem was given by Theorem 2.13.2. We shall consider an approximate solution using variational methods. The results for this abstract problem can be simply reformulated for any particular equilibrium problem considered in this book. Let us start by formulating the approximation equations.

A variational method of approximation for practical purposes was proposed by Walter Ritz (1878–1909). Given some set of elements $e_1, \ldots, e_n \in H$, we can approximate u^* using a linear combination

$$\sum_{k=1}^{n} c_k e_k. \tag{3.19.2}$$

How should we choose the coefficients c_k for a good approximation? Ritz supposed they can be given as a solution to the following minimization problem:

Problem 3.19.2. *Find real coefficients c_k such that $\Phi\left(\sum_{k=1}^{n} c_k e_k\right)$ takes a minimum value.*

A solution of this problem is called an nth-order *Ritz approximation*. We will denote it by u_n. In Ritz's time, when all calculations were done by hand, n was normally no greater than three and engineers had to choose the e_k carefully to obtain reasonable results. Ritz's method became popular among civil engineers and others (see, e.g., [Timoshenko (1959)]). Modern computers permit much larger values of n and have revitalized this old method. The popular finite element method is a modification and extension of Ritz's method. When Ritz's method was restricted to just 3–5 elements, engineers had to base their selection of the e_k on engineering intuition. When the solution behavior is unknown in advance and n is large, however, it is important to know whether the e_k are linearly independent and whether the method converges.

Gram determinant and linear independence

First we wish to state a practical criterion for the linear independence of a system of vectors in a Hilbert space. The definition of linear independence is familiar from linear algebra.

Definition 3.19.1. A system $e_1, \ldots, e_n$ is *linearly independent* if the equation

$$c_1 e_1 + \cdots + c_n e_n = 0 \tag{3.19.3}$$

with respect to the coefficients c_k has only $c_1 = \cdots = c_n = 0$ as a solution.

Theorem 3.19.1. *In a Hilbert space H, the system $e_1, \ldots, e_n$ is linearly independent if and only if its Gram determinant*

$$\begin{vmatrix} (e_1, e_1)_H & (e_2, e_1)_H & \cdots & (e_n, e_1)_H \\ (e_1, e_2)_H & (e_2, e_2)_H & \cdots & (e_n, e_2)_H \\ \vdots & \vdots & \ddots & \vdots \\ (e_1, e_n)_H & (e_2, e_n)_H & \cdots & (e_n, e_n)_H \end{vmatrix} \tag{3.19.4}$$

is not zero.

Proof. By (3.19.3) we can construct a system of simultaneous equations, obtaining the kth equation by multiplying (3.19.3) by e_k:

$$c_1(e_1, e_1)_H + c_2(e_2, e_1)_H + \cdots + c_n(e_n, e_1)_H = 0,$$
$$c_1(e_1, e_2)_H + c_2(e_2, e_2)_H + \cdots + c_n(e_n, e_2)_H = 0,$$
$$\vdots$$
$$c_1(e_1, e_n)_H + c_2(e_2, e_n)_H + \cdots + c_n(e_n, e_n)_H = 0. \tag{3.19.5}$$

Its determinant is the Gram determinant. If it is not zero, then $c_1 = c_2 = \cdots = c_n = 0$ and so the e_k are linearly independent.

Conversely, suppose the e_k are linearly independent so that (3.19.3) has only the trivial solution. Multiply the first equation of the system (3.19.5) by c_1, the second by c_2, etc, and add all the equations. We get

$$(c_1 e_1 + c_2 e_2 + \cdots + c_n e_n, c_1 e_1 + c_2 e_2 + \cdots + c_n e_n)_H = 0,$$

which implies that

$$c_1 e_1 + c_2 e_2 + \cdots + c_n e_n = 0.$$

Because the e_k are linearly independent, it follows that $c_1 = c_2 = \cdots = c_n = 0$. Thus the system (3.19.5) has only the trivial solution. By the general theory of linear systems this means that its determinant, the Gram determinant, is not zero. $\square$

Our proof was for a real space H, but a small modification would accommodate a complex space H. The present version will suffice for our purposes.

So let us return to the problem of minimizing $\Phi(u)$ over the set of linear combinations (3.19.2), which is the central point of Ritz's method. We seek the working formulas of the method.

Formulas of Ritz's method

Let us put
$$u_n = \sum_{k=1}^{n} c_k e_k \qquad (3.19.6)$$
into the expression for $\Phi(u)$:
$$\Phi(u_n) = \frac{1}{2}\left(\sum_{k=1}^{n} c_k e_k, \sum_{m=1}^{n} c_m e_m\right)_H - F\left(\sum_{k=1}^{n} c_k e_k\right).$$

Now $\Phi(u_n)$ depends only on the coefficients c_k and, moreover, quadratically. Its minimum in Ritz's sense, if it exists (we will see that it does), approximates the solution of Problem 3.19.1. The point of minimum is specified by the set of equations
$$\frac{\partial}{\partial c_k}\Phi(u_n) = 0 \qquad (k=1,2,\ldots,n).$$

Writing these out in detail, we obtain
$$c_1(e_1,e_1)_H + c_2(e_2,e_1)_H + \cdots + c_n(e_n,e_1)_H - F(e_1) = 0,$$
$$c_1(e_1,e_2)_H + c_2(e_2,e_2)_H + \cdots + c_n(e_n,e_2)_H - F(e_2) = 0,$$
$$\vdots$$
$$c_1(e_1,e_n)_H + c_2(e_2,e_n)_H + \cdots + c_n(e_n,e_n)_H - F(e_n) = 0. \qquad (3.19.7)$$

This is called Ritz system of equations of the nth approximation. Since its determinant is the Gram determinant for $e_1, e_2, \ldots, e_n$, we have

Theorem 3.19.2. *Let $e_1, e_2, \ldots, e_n$ be linearly independent in H and $F(u)$ a linear continuous functional. The Ritz system of equations for the nth approximation u_n has a unique solution.*

Thus, when solving Problem 3.19.2 we always get a unique result. Whether this will approximate the solution is another matter. Convergence now becomes an issue: clearly, if the linear combinations (3.19.6) cannot approximate an element u_0 that happens to be the needed solution, we cannot expect convergence. Thus we should add a special property for the sequence $e_1, e_2, \ldots, e_n, \ldots$. Under this property, known as *completeness of a system of elements*, finite linear combinations of the e_k can approximate any element of H to within any predetermined accuracy.

Definition 3.19.2. A sequence of vectors $e_1, e_2, \ldots, e_n, \ldots$ is *complete* in a Hilbert space H if for any $\varepsilon > 0$ and any fixed $u \in H$ there is a linear combination $\sum_{k=1}^{n} c_k e_k$ such that

$$\left\| u - \sum_{k=1}^{n} c_k e_k \right\|_H < \varepsilon$$

with coefficients that depend on ε and u.

This is close to the notion of basis, but the e_k need not actually constitute a basis. In $C([0,1])$, for example, the system of monomials x^k is complete: it is not a basis, but can still approximate any continuous function with a polynomial of finite degree to within any preassigned accuracy.

Now we can formulate

Theorem 3.19.3. *Let $e_1, e_2, \ldots, e_n, \ldots$ be a complete system in H, and assume that any finite subset $e_1, e_2, \ldots, e_n$ is linearly independent. Then the sequence of Ritz approximations $\{u_n\}$ converges to a solution u^* of Problem 3.19.1.*

Proof. Recall that $F(u)$ was represented as $F(u) = (u, u^*)_H$, where u^* turned out to be the solution of the minimum problem (see the proof of Theorem 2.13.2). So we can convert Problem 3.19.1 to that of minimizing the functional

$$\Phi(u) = \frac{1}{2}(u, u)_H - (u, u^*).$$

Since the additive constant $\frac{1}{2}(u^*, u^*)_H$ can be ignored when seeking the minimum, we can minimize

$$\Psi(u) = \Phi(u) + \frac{1}{2}(u^*, u^*)_H = \frac{1}{2} \| u - u^* \|_H^2$$

instead. So we reduce the problem of finding Ritz approximations to the problem of minimizing the functional $\Psi(u)$ over linear combinations $\sum_{k=1}^{n} c_k e_k$, whereas the main Problem 3.19.1 reduces to the trivial problem of minimizing $\Psi(u)$. Obviously this latter minimum is zero and is attained

at $u = u^*$. It is also clear that by extending the basis set $e_1, e_2, \ldots, e_n$ we get a better approximation to the minimum. This means that
$$\Psi(u_1) \geq \Psi(u_2) \geq \cdots \geq \Psi(u_n) \geq \cdots$$
or, in terms of norms,
$$\|u_1 - u^*\|_H \geq \|u_2 - u^*\|_H \geq \cdots \geq \|u_n - u^*\|_H \geq \cdots.$$
Convergence of u_n to u^* will be shown if we can find a subsequence of $\{u_n\}$ convergent to u^*. Given a sequence $\varepsilon_k \to 0$, $\varepsilon_k > 0$, by completeness of $e_1, e_2, \ldots, e_n, \ldots$ there is a sequence $\{\hat{u}_k\}$, where
$$\hat{u}_k = \sum_{r=1}^{n_k} \hat{c}_r e_r,$$
such that
$$\|u - \hat{u}_k\|_H < \varepsilon_k.$$
Remembering that the minimum Ritz problem yields the best approximation for the same basis set of $e_1, \ldots, e_{n_k}$ we see that for the n_kth order of approximation
$$\|u - u_{n_k}\|_H \leq \|u - \hat{u}_k\|_H < \varepsilon_k$$
where u_{n_k} is the Ritz approximation of n_k-th order. We have found the required subsequence. □

Although we have established convergence of the method with no restrictions, the reader should be aware that in practice difficulties can occur with numerical roundoff error in higher approximations.

Bubnov–Galerkin method

Recall equation (2.13.1) for finding the minimum point of Problem 3.19.1:
$$(u, v)_H - F(v) = 0, \tag{3.19.8}$$
where v is an arbitrary element of H. Long ago in Russia there was an engineering journal that published the paper of an author but also a review by the person who submitted the paper to the journal. One such paper was written by S. Timoshenko and submitted by shipbuilding engineer I.G. Bubnov (1872–1919). Timoshenko, at that time a young engineer, considered an equilibrium problem using Ritz's method. In his review, Bubnov mentioned that Ritz's equations could alternatively be obtained from (3.19.8).

In modern terminology, if one substitutes the linear combination (3.19.6) for u in (3.19.8), and then successively puts $e_1, e_2, \ldots, e_n$ in place of v, the result will be (3.19.7). This small remark had the effect of extending Ritz's method beyond its original realm of application, and the extension was named after Bubnov. A later modification was called the *Bubnov–Galerkin method*: in this case the substitution for v is made in terms of another "basis" set $e'_1, e'_2, \ldots, e'_n$. For this second basis, it is important that the set of equations obtained consists of n equations with respect to the n variables c_k. The second set must be such that the system has a unique solution. By selecting this set appropriately, we may obtain an algebraic system that is easier to solve than the original Ritz system. In the non-Russian technical literature, Bubnov's name has largely disappeared from this modification, although B.G. Galerkin (1871–1945) was a student of Bubnov and accepted the latter's role in originating the method.

Finally, we note a particular case of this method. If the e_k are orthonormal so that $(e_k, e_n)_H = \delta_{kn}$ where δ_{kn} is the Kronecker symbol, then the system (3.19.7) is easy to solve:

$$c_k = F(e_k).$$

In this way, as an approximation, we obtain a portion of a Fourier series. Such series will be considered in abstract form in § 3.22.

Ritz's method is used when the number of "basis" elements n (and hence the number of equations) is large. If the resulting matrix is not sparse, a typical numerical method will require something on the order of n^2 operations to solve the system. The finite element method uses a special set of basis elements for which the matrix entries are given by simple formulas. More importantly, the nonzero entries are clustered in a band near the main diagonal. In this case something on the order of n operations are required for a solution, which means we can attack much larger problems. The reader can refer to specialized books for thorough coverage of the method, but we provide a brief sketch for the simplest case of a one-dimensional problem for an elastic bar with clamped ends and distributed load $t(x)$.

For the VWP equation (3.2.14) (simpler case), Ritz's equations are

$$\int_0^l ESu'v'\, dx = \int_0^l t(x)v(x)\, dx. \qquad (3.19.9)$$

The nth-order Ritz approximation is

$$u_n(x) = \sum_{k=1}^n c_k \varphi_k(x).$$

As a basis element e_k we take a function $\varphi_k(x)$ constructed as follows. We partition the segment $[0, 1]$ and number its nodes starting with 1. Let $\varphi_k(x)$ be a piecewise linear "tent function" that is nonzero only between nodes $k-1$ and $k+1$ and equal to 1 at node k as shown in Fig. 3.14.

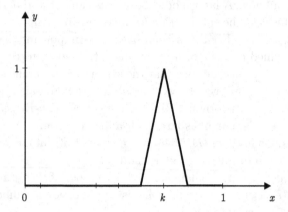

Fig. 3.14   Graph of the kth tent function.

The entry $(e_k, e_m)_H$ in (3.19.7) takes the form

$$\int_0^1 ES\varphi_k'(x)\varphi_m'(x)\,dx,$$

whereas the expression $F(e_k)$ takes the form

$$\int_0^1 t(x)\varphi_k(x)\,dx.$$

Because $(e_k, e_m)_H \neq 0$ only when $|k-m| \leq 1$, the system matrix is tridiagonal. Simple formulas yield the entries for any given partition. Furthermore the partition need not be uniform.

For the two-dimensional case of a membrane, we can partition the domain into triangles and use piecewise-linear functions as the basis set. In this case it is more challenging to number the nodes in such a way that nonzero members of the Gram matrix lie near the main diagonal. Fortunately, algorithms have been developed to handle this numbering automatically.

We will not discuss the difficulties that arise in the application of FEM to various problems. We should note, however, that despite the faith placed by many engineers in "universal" FEM packages, there are many practical problems that such packages cannot solve.

3.20 A Brief but Important Remark

We have studied equilibrium problems for linear elasticity, as well as problems for membranes and other elastic bodies, in Cartesian coordinates. It is important to understand, however, that the results we have obtained are valid in any coordinate system if we can transform between the two systems and the energy integrals make sense. The form of the imbedding theorems will change because, for example, the Jacobian of the transformation will appear in the integrals for the L^p-norm. But that is all. We urge the reader to investigate the form of the existence theorems (the conditions on the external forces) for linear elasticity in polar and cylindrical coordinates.

3.21 Countable Sets and Separable Spaces

We have made use of complete systems of elements in a space, putting aside the question whether such a system — indexed with integers — can exist in principle. Before discussing this further, we must take a brief detour toward the foundations of set theory. Namely, we must ask how the sizes of two given sets can be compared. Which one has "more elements"?

For sets containing just a few elements, we can simply count the elements of each set and compare the numbers. This procedure will not work for infinite sets, however. We can compare two infinite sets as follows. We say that two sets are of *equal power* if a one-to-one correspondence exists between their elements. Of the various infinite sets, we can think of the natural numbers $\mathbb{N} = \{1, 2, 3, \ldots\}$ as having the least power; only a finite set can have a smaller power. Any set that has the same power as $\mathbb{N}$ is said to be *countable*. Because a countable set S can be put into one-to-one correspondence with $\mathbb{N}$, its elements can be indexed using the positive integers: we are permitted to write, say,

$$S = \{s_1, s_2, s_3, \ldots\}.$$

Countable sets possess some interesting properties. We can delete any finite number of elements from a countable set, and the new set will still be countable. We can even delete all the odd integers from the set of all integers and still have a countable set. A particularly useful property is the following.

Theorem 3.21.1. *A countable union of countable sets is countable.*

Proof. We will show how to renumber the elements of the resulting union of sets. Let us denote the elements of the first set by a_{11}, a_{12}, a_{13}, ... This can be done because the set is countable. We denote the elements of the second set by a_{21}, a_{22}, a_{23}, ... The elements of the third set are denoted similarly, and so on. The union of all these sets is the set whose general element can be written in the form a_{ij} for $i,j = 1,2,3,\ldots$. But these elements can be renumbered in many different ways. For example, we could take the first element of the union as a_{11}, but the second and third elements as a_{12} and a_{21} — i.e., as those elements having indices that sum to 3. The fourth, fifth, and sixth elements could be taken as a_{13}, a_{22}, and a_{31} — i.e., as those elements having indices that sum to 4. Continuing in this way (i.e., taking the elements whose indices sum to 5, 6, 7, and so on) we succeed in renumbering the elements of the union in such a way that they stand in one-to-one correspondence with N. □

This theorem has some important consequences. First, the set of rational numbers is countable. Indeed its elements can be labeled using the scheme $a_{ij} = i/j$, where i runs through all integer values and j through all nonzero integer values.

Second, we can apply the theorem in a successive manner any finite number of times. One application shows that the set of polynomials having degree n and rational coefficients is countable. A second application shows that the set of all polynomials having rational coefficients is countable.

Are there sets whose powers exceed that of a countable set? The answer is yes: the set of all real numbers in the segment $[0,1]$ is infinite but not countable. A long proof can be found in any textbook on real analysis. A set having power equal to that of this segment is said to have the *power of the continuum*. The set of points that constitute $\mathbb{R}^3$ is an example.

Using rational numbers, we can approximate any real number to within any desired accuracy. When the accuracy is fixed, we can use a finite set of numbers to approximate a bounded set of numbers. We rely on this fact in ordinary machine computation. It turns out that not all functional spaces have elements that can be approximated in this way, however. Our work will therefore focus on the spaces covered by the following definition.

Definition 3.21.1. A metric space is said to be *separable* if it has a countable dense subset.

So $[0,1]$ is separable as a metric space with the usual metric $d(x,y) = |x-y|$. A much less trivial example is the space $C(a,b)$, in which the set of all polynomials with rational coefficients happens to be countable and dense. This is guaranteed by Weierstrass' theorem, which states that any function continuous on $[a,b]$ can be approximated uniformly to within any fixed accuracy by a polynomial. Clearly, we can select this polynomial with rational coefficients.

Weierstrass' theorem can be extended to the spaces $C^{(k)}(V)$ where V is a compact subset of $\mathbb{R}^n$. It is easy to see that the property of denseness is transitive in a metric space: if A is dense in B, and B is dense in C, then A is dense in C. A consequence of this is that all our energy spaces are separable, because they are constructed on certain base subsets of $C^{(k)}(V)$ using the completion procedure in a countable number of actions.

When we wish to extend the idea of a basis in $\mathbb{R}^n$ to infinite-dimensional spaces in an elementary way, we should do this using separable spaces.

3.22 Fourier Series

We have mentioned orthonormal systems of elements. Any university calculus sequence will contain a treatment of Fourier series. Fourier's initial development in terms of sines and cosines can be extended to a powerful abstract form involving only the property of orthogonality between elements of a Hilbert space. We wish to present this for a complex Hilbert space H, so we take $\mathbb{C}$ as the scalar field over which H is defined. In this case, the Fourier series representation of an element $u \in H$ takes the form

$$u = \sum_{k=1}^{\infty} \alpha_k e_k, \qquad (3.22.1)$$

where the α_k are the Fourier coefficients of u and the e_k are elements of an orthonormal basis. Let us explain this situation in more detail.

Orthonormal system of elements

Definition 3.22.1. The vectors $e_1, e_2, \ldots, e_n$ form an *orthonormal system* if $(e_k, e_m)_H = \delta_{km}$, where δ_{km} is Kronecker's symbol.

An orthonormal sequence $e_1, e_2, \ldots, e_n$ is linearly independent. Indeed, taking the inner product between e_m and both sides of the equation

$$\sum_{k=1}^{n} c_k e_k = 0,$$

we obtain $c_m = 0$ for all m.

Given a linearly independent system of vectors $g_1, g_2, \ldots, g_n$ that is a basis in the n-dimensional linear space of all linear combinations $\sum_{k=1}^{n} c_k g_k$, we can derive another basis of the same space using the *Gram–Schmidt formulas*

$$e_1 = \frac{g_1}{\|g_1\|_H}, \qquad e_k = \frac{\hat{g}_k}{\|\hat{g}_k\|_H} \quad (k = 2, \ldots, n),$$

where

$$\hat{g}_k = g_k - (g_k, e_1)_H e_1 - \cdots - (g_k, e_{k-1})_H e_{k-1} \quad (k = 2, \ldots, n).$$

Because $g_1, g_2, \ldots, g_n$ is a linearly independent system, $\hat{g}_k$ cannot be zero for any k. It can be verified directly that $e_1, e_2, \ldots, e_n$ is an orthonormal system containing n vectors. It is therefore a basis of the same space.

Fourier coefficients

We have seen that if we know a solution u^*, then Ritz's method applied to a minimization problem reduces to the problem of finding the coefficients c_k of the functional

$$\left\| \sum_{k=1}^{n} c_k e_k - u^* \right\|_H^2$$

that minimize this as a function of the c_k. Now suppose u^* is given. Let us rename it u and minimize the functional

$$\left\| u - \sum_{k=1}^{n} c_k e_k \right\|_H^2. \tag{3.22.2}$$

Since H is now complex, our prior technique of differentiation is not appropriate. When $e_1, e_2, \ldots, e_n$ is an orthonormal set, however, we can find the needed coefficients directly.

Theorem 3.22.1. *Let $e_1, e_2, \ldots, e_n$ be an orthonormal set. The best approximation of u by a linear combination $\sum_{k=1}^{n} c_k e_k$ is achieved when*

$$c_k = (u, e_k)_H \quad (k = 1, 2, \ldots, n).$$

We call $\alpha_k = (u, e_k)_H$ the kth *Fourier coefficient of* u.

Proof. Let us represent (3.22.2) as

$$A = \left(u - \sum_{k=1}^{n} c_k e_k,\ u - \sum_{m=1}^{n} c_m e_m\right)_H.$$

Expansion yields

$$A = (u, u)_H - \left(u, \sum_{m=1}^{n} c_m e_m\right)_H - \left(\sum_{k=1}^{n} c_k e_k, u\right)_H$$
$$+ \left(\sum_{k=1}^{n} c_k e_k, \sum_{m=1}^{n} c_m e_m\right)_H$$

or

$$A = (u, u)_H - \sum_{m=1}^{n} \bar{c}_m (u, e_m)_H - \sum_{k=1}^{n} c_k (e_k, u)_H$$
$$+ \sum_{k=1}^{n} c_k \sum_{m=1}^{n} \bar{c}_m (e_k, e_m)_H.$$

Remembering that $(e_k, e_m)_H = \delta_{km}$ and writing $\alpha_k = (u, e_k)_H$, we get

$$A = (u, u)_H - \sum_{m=1}^{n} \bar{c}_m \alpha_m - \sum_{k=1}^{n} c_k \bar{\alpha}_k + \sum_{k=1}^{n} c_k \bar{c}_k.$$

Adding and subtracting $\sum_{k=1}^{n} \alpha_k \bar{\alpha}_k$, we get

$$A = (u, u)_H - \sum_{k=1}^{n} \alpha_k \bar{\alpha}_k + \sum_{k=1}^{n} (c_k - \alpha_k) \overline{(c_k - \alpha_k)}. \qquad (3.22.3)$$

Since

$$\sum_{k=1}^{n} (c_k - \alpha_k) \overline{(c_k - \alpha_k)} = \sum_{k=1}^{n} |c_k - \alpha_k|^2$$

is nonnegative and takes zero as its minimum value when $c_k = \alpha_k$, and the rest of A does not depend on c_k, the result is proved. $\square$

Fourier series

We started with $A \geq 0$. Putting $c_k = \alpha_k$ in (3.22.3), we obtain

$$\sum_{k=1}^{n} |\alpha_k|^2 \leq \|u\|_H^2. \qquad (3.22.4)$$

This is *Bessel's inequality*. It has several important consequences.

(1) *The series $\sum_{k=1}^{\infty} |\alpha_k|^2$ converges.*

Indeed, the terms are nonnegative and for any n the partial sum $\sum_{k=1}^{n} |\alpha_k|^2$ is bounded from above by $\|u\|_H^2$.

(2) *The series $\sum_{k=1}^{\infty} \alpha_k e_k$ converges in H.*

Let $u_n = \sum_{k=1}^{n} \alpha_k e_k$. Then $\{u_n\}$ is a Cauchy sequence in H. Indeed, if $m < n$ then
$$u_n - u_m = \sum_{k=m+1}^{n} \alpha_k e_k$$
and, by direct calculation,
$$\|u_n - u_m\|_H^2 = \sum_{k=m+1}^{n} |\alpha_k|^2 \to 0 \quad \text{as } m \to \infty$$
because $\sum_{k=1}^{\infty} |\alpha_k|^2$ converges. Thus u_n converges as $n \to \infty$. Consequently we get

(3)
$$\left\| \sum_{k=1}^{\infty} \alpha_k e_k \right\|_H^2 = \sum_{k=1}^{\infty} |\alpha_k|^2 \leq \|u\|_H^2.$$

This is all we can say about the Fourier series $\sum_{k=1}^{\infty} \alpha_k e_k$ while assuming only orthonormality of the system $e_1, e_2, \ldots$. Unless this system has basis-like properties, there may exist elements that cannot be approximated by particular Fourier sums u_n.

Theorem 3.22.2. *Suppose the system $e_1, e_2, \ldots$ is complete in H. Then*
$$u = \sum_{k=1}^{\infty} \alpha_k e_k \tag{3.22.5}$$
and Parseval's equality
$$\sum_{k=1}^{\infty} |\alpha_k|^2 = \|u\|_H^2 \tag{3.22.6}$$
holds.

The proof of the first equality, which is the Fourier series representation of u, was given when we considered the convergence of Ritz approximations. Parseval equality is a consequence.

It is difficult to prove directly that $e_1, e_2, \ldots$ is complete in H. In § 3.28 we will examine a class of operators related to eigenvalue problems for elastic structures, and will see that the eigenfunctions of these operators have the necessary property. Now we turn to the problem of vibration.

3.23 Problem of Vibration for Elastic Structures

So far we have considered mostly equilibrium problems. We now turn to vibration problems. These are related to the problem of resonance and have great value in engineering.

Vibration equations can be formulated in the same fashion for all the objects we have considered. First, starting with the equilibrium equation in VWP form and using d'Alembert's principle, we present the dynamical equations. Then, supposing a certain time dependence of the external forces and solutions, we get the equations for forced vibrations.

Let us show how this plan is carried out for a three-dimensional elastic body. We start with the VWP equation (3.15.4):

$$\int_V c^{ijkl} \varepsilon_{kl}(\mathbf{u}) \varepsilon_{ij}(\mathbf{v}) \, dV - \int_V \mathbf{F} \cdot \mathbf{v} \, dV - \int_{S \setminus S_1} \mathbf{f} \cdot \mathbf{v} \, dS = 0. \qquad (3.23.1)$$

D'Alembert's principle says that at each point of V we must replace the external body force $\mathbf{F}$ by

$$\mathbf{F} - \rho \frac{\partial^2 \mathbf{u}}{\partial t^2}$$

where ρ is the density of the material at the point and t is time. In this way we get the equation of motion

$$\int_V c^{ijkl} \varepsilon_{kl}(\mathbf{u}) \varepsilon_{ij}(\mathbf{v}) \, dV - \int_V \left(\mathbf{F} - \rho \frac{\partial^2 \mathbf{u}}{\partial t^2} \right) \cdot \mathbf{v} \, dV - \int_{S \setminus S_1} \mathbf{f} \cdot \mathbf{v} \, dS = 0.$$
$$(3.23.2)$$

To study forced vibrations, we suppose the forces and displacements share the same time dependence: namely, as $e^{i\omega t}$ where ω is the angular frequency. This can be done by the formal replacement (we retain the same notation for the spatially-dependent part of a quantity as for the entire quantity)

$$\mathbf{F} \mapsto e^{i\omega t} \mathbf{F}(\mathbf{r}), \quad \mathbf{f} \mapsto e^{i\omega t} \mathbf{f}(\mathbf{r}), \quad \mathbf{u} \mapsto e^{i\omega t} \mathbf{u}(\mathbf{r}).$$

If we permit ω to take complex values, then $\mathbf{u}$ and $\mathbf{v}$ will have complex components. We shall take $\mathbf{v}$ in conjugate form as $\overline{\mathbf{v}}$. Substituting into (3.23.2) and canceling the common factor $e^{i\omega t}$, we obtain the equation for forced vibrations:

$$\int_V c^{ijkl}\varepsilon_{kl}(\mathbf{u})\varepsilon_{ij}(\overline{\mathbf{v}})\,dV - \int_V (\mathbf{F}+\omega^2\rho\mathbf{u})\cdot\overline{\mathbf{v}}\,dV - \int_{S\setminus S_1} \mathbf{f}\cdot\overline{\mathbf{v}}\,dS = 0. \quad (3.23.3)$$

Now we recast (3.23.3) in operator form in the space E_{E0}. Introducing conjugate values, etc., we come to complex linear spaces. We will continue using our earlier notation for the energy spaces, however. It is clear that the properties we have established for their elements are preserved as well. So the first term of (3.23.3) is

$$\int_V c^{ijkl}\varepsilon_{kl}(\mathbf{u})\varepsilon_{ij}(\overline{\mathbf{v}})\,dV = (\mathbf{u},\mathbf{v})_E. \quad (3.23.4)$$

Assuming $\mathbf{F}$ and $\mathbf{f}$ satisfy the conditions of Theorem 3.17.1, and applying the Riesz representation theorem, we get

$$\int_V \mathbf{F}\cdot\overline{\mathbf{v}}\,dV + \int_{S\setminus S_1} \mathbf{f}\cdot\overline{\mathbf{v}}\,dS = (\mathbf{u}^*,\mathbf{v})_E = 0 \quad (3.23.5)$$

with an element $\mathbf{u}^* \in E_{E0}$ that is uniquely defined by $\mathbf{F}$ and $\mathbf{f}$. Finally, consider the conjugate term

$$\overline{\int_V \rho\mathbf{u}\cdot\overline{\mathbf{v}}\,dV} = \int_V \rho\overline{\mathbf{u}}\cdot\mathbf{v}\,dV.$$

For fixed $\mathbf{u}\in E_{E0}$, this is a linear functional in $\mathbf{v}$. Moreover, because $L^2(V)$ is imbedded to E_{E0} continuously, we get

$$\left|\overline{\int_V \rho\mathbf{u}\cdot\overline{\mathbf{v}}\,dV}\right| \leq c\,\|\mathbf{u}\|_E\,\|\mathbf{v}\|_E \quad (3.23.6)$$

with a constant that does not depend on $\mathbf{u},\mathbf{v}\in E_{E0}$. So this functional is continuous in E_{E0} and we can represent it as an inner product in E_{E0}:

$$\overline{\int_V \rho\mathbf{u}\cdot\overline{\mathbf{v}}\,dV} = (\mathbf{v},\mathbf{u}^{**})_E,$$

which is the same as

$$\int_V \rho\mathbf{u}\cdot\overline{\mathbf{v}}\,dV = (\mathbf{u}^{**},\mathbf{v})_E. \quad (3.23.7)$$

Here $\mathbf{u}^{**}$ depends on the single variable $\mathbf{u}$ in such a way that to $\mathbf{u}\in E_{E0}$ there corresponds a unique $\mathbf{u}^{**}\in E_{E0}$. This is the same as saying that we have defined an operator A:

$$\mathbf{u}^{**} = A(\mathbf{u}).$$

Hence
$$\int_V \rho \mathbf{u} \cdot \overline{\mathbf{v}}\, dV = (A\mathbf{u}, \mathbf{v})_E. \qquad (3.23.8)$$

Exercise 3.23.1. *Show that A is linear.*

This operator is also continuous. Indeed, by (3.23.6) we have
$$|(A\mathbf{u}, \mathbf{v})_E| = \left|\int_V \rho \mathbf{u} \cdot \overline{\mathbf{v}}\, dV\right| \leq c\,\|\mathbf{u}\|_E\, \|\mathbf{v}\|_E\,.$$

Putting $\mathbf{v} = A\mathbf{u}$ we get
$$\|A\mathbf{u}\|_E^2 \leq c\,\|\mathbf{u}\|\, \|A\mathbf{u}\|_E\,,$$
hence $\|A\| \leq c$.

Substituting (3.23.4), (3.23.5), and (3.23.7) into (3.23.3), and denoting
$$\mu = \omega^2, \qquad (3.23.9)$$
we get
$$(\mathbf{u}, \mathbf{v})_E - \mu(A\mathbf{u}, \mathbf{v})_E - (\mathbf{u}^*, \mathbf{v})_E = 0.$$

Since $\mathbf{v} \in E_{E0}$ is arbitrary, we get the necessary operator equation
$$\mathbf{u} - \mu A\mathbf{u} = \mathbf{u}^*. \qquad (3.23.10)$$

A similar equation holds for the forced vibrations of all the mechanical objects we considered earlier.

The eigenvalue problem in finite- and infinite-dimensional space

When $\mathbf{u}^* = \mathbf{0}$ in (3.23.10), the result looks like the eigenvalue problem in linear algebra that is considered in a finite-dimensional space:
$$\mathbf{Mx} = \lambda \mathbf{x}, \qquad (3.23.11)$$
with $\mathbf{M}$ an $n \times n$ matrix and $\lambda = 1/\mu$. There is a difference, however, because E_{E0} is infinite dimensional. Before proceeding, let us recall what is known for the eigenvalue problem in linear algebra. We seek nonzero solutions $\mathbf{x}$ to (3.23.11), known as *eigenvectors*. An eigenvector exists if and only if
$$\det(\mathbf{M} - \lambda \mathbf{I}) = 0. \qquad (3.23.12)$$

The number of eigenvalues does not exceed n. When $\det(\mathbf{M} - \lambda \mathbf{I}) \neq 0$, the inverse $(\mathbf{M} - \lambda \mathbf{I})^{-1}$ exists and the equation

$$\mathbf{M}\mathbf{x} - \lambda \mathbf{x} = \mathbf{b} \qquad (3.23.13)$$

has a solution $\mathbf{x}$ for any $\mathbf{b}$. Linear algebra textbooks rarely mention the class of vectors $\mathbf{b}$ for which a solution exists if λ is an eigenvalue. This question is important in applications. We will discuss it for the general case of equation (3.23.10).

When there is a nontrivial solution x^* of the equation

$$x - \mu B x = 0 \qquad (3.23.14)$$

with a linear operator B in an infinite-dimensional abstract Hilbert space H (that is, x^* is an eigensolution of B), then the equation

$$x - \mu B x = b \qquad (3.23.15)$$

cannot have a solution for any $b \in H$. The same holds for (3.23.10).

Let B be continuous. When (3.23.14) lacks a nontrivial solution, unlike the finite-dimensional case, this does not mean we can solve (3.23.15) for any b. In functional analysis, when it happens that it is possible to find the inverse $(I - \mu B)^{-1}$ but its domain is not the whole space or the inverse is not continuous, we say that μ belongs to the *spectrum* of B. This situation does not arise in finite-dimensional linear algebra. Fortunately, for a bounded elastic object such points do not arise if the body has a "regular" shape (i.e., a shape required by the corresponding imbedding theorems).

First we will establish some properties of the operator A from the elasticity problem of § 3.22. Then we will discuss the spectral properties of a general class of operators to which A belongs.

3.24 Self-Adjointness of A and Its Consequences

We begin with

Definition 3.24.1. A continuous linear operator B acting in a Hilbert space H is *self-adjoint* if the equality

$$(Bx, y) = (x, By) \qquad (3.24.1)$$

holds for all $x, y \in H$.

We noted that all the qualitative properties we can establish for A in the framework of linear elasticity will also apply to the rest of the linear problems of this book. The first is

Theorem 3.24.1. *A is a self-adjoint continuous operator.*

Proof. Continuity was proved earlier. Remember that A is introduced by the equality
$$(A\mathbf{u}, \mathbf{v})_E = \int_V \rho \mathbf{u} \cdot \overline{\mathbf{v}} \, dV.$$
Interchanging $\mathbf{u}$ and $\mathbf{v}$, we get
$$\int_V \rho \mathbf{v} \cdot \overline{\mathbf{u}} \, dV = (A\mathbf{v}, \mathbf{u})_E = \overline{(\mathbf{u}, A\mathbf{v})_E}.$$
Writing
$$\int_V \rho \mathbf{u} \cdot \overline{\mathbf{v}} \, dV \quad \text{as} \quad \overline{\int_V \rho \mathbf{v} \cdot \overline{\mathbf{u}} \, dV}$$
we find that
$$(A\mathbf{u}, \mathbf{v})_E = (\mathbf{u}, A\mathbf{v})_E$$
for all $\mathbf{u}, \mathbf{v} \in E_{E0}$. Hence A is self-adjoint. $\square$

Theorem 3.24.2. *A is strictly positive, which means that $(A\mathbf{u}, \mathbf{u})_E > 0$ whenever $\mathbf{u} \neq \mathbf{0}$.*

Proof. By definition of A we have
$$(A\mathbf{u}, \mathbf{u})_E = \int_V \rho \mathbf{u} \cdot \overline{\mathbf{u}} \, dV$$
and so $(A\mathbf{u}, \mathbf{u})_E \geq 0$. From
$$\int_V \rho \mathbf{u} \cdot \overline{\mathbf{u}} \, dV = 0$$
it follows that $\mathbf{u} = \mathbf{0}$. $\square$

We will refer to a value μ for which the equation
$$\mathbf{u} = \mu A \mathbf{u} \tag{3.24.2}$$
has a nonzero solution as an *eigenvalue* of A, and to the solution itself as an *eigensolution* or *oscillation mode*. Since eigenvalues are related to eigenfrequencies, they are important in the theory of elasticity.

We now prove some consequences of Theorem 3.24.1. Despite their simplicity, the following theorems are valuable in mechanics.

Theorem 3.24.3. *The disk*

$$|\mu| < \frac{1}{\|A\|}$$

contains no eigenvalues of A.

Proof. From (3.24.2) it follows that

$$\|\mathbf{u}\|_E = \|\mu A \mathbf{u}\|_E \leq |\mu|\, \|A\|\, \|\mathbf{u}\|_E .$$

In the disk $|\mu|\, \|A\| = q < 1$ and so

$$\|\mathbf{u}\|_E (1-q) \leq 0.$$

This implies $\mathbf{u} = \mathbf{0}$, hence no μ in the disk can be an eigenvalue. $\square$

By (3.23.9), the eigenfrequencies ω are bounded from below by the value $\sqrt{1/\|A\|}$.

Remark 3.24.1. For a body free of geometrical constraints, the operator A is not strictly positive. It is merely positive, as the rigid infinitesimal displacement $\mathbf{a} + \mathbf{b} \times \mathbf{r}$ is its eigensolution corresponding to $\mu = 0$, an eigenvalue of A. This is an exceptional value for the problem, however, and the remaining eigenvalues that do not correspond to rigid motions of the body do obey the theorem.

Theorem 3.24.4. *The eigenvalues μ are real and can be selected to be positive.*

Proof. An eigenvalue satisfies

$$\mu = (\mathbf{u}, \mathbf{u})_E / (A\mathbf{u}, \mathbf{u})_E$$

and hence is positive by Theorem 3.24.2. $\square$

So the eigenfrequencies ω_k of a bounded elastic body are positive and bounded away from zero.

Theorem 3.24.5. *If $\mu_1 \neq \mu_2$ are two eigenvalues of A, then their corresponding eigensolutions $\mathbf{u}_1$ and $\mathbf{u}_2$ are orthogonal:*

$$(\mathbf{u}_1, \mathbf{u}_2)_E = 0. \qquad (3.24.3)$$

Moreover, the eigenvectors satisfy the generalized orthogonality *property*

$$(A\mathbf{u}_1, \mathbf{u}_2)_E = 0. \qquad (3.24.4)$$

Proof. We have two equalities:
$$\mathbf{u}_1 = \mu_1 A\mathbf{u}_1, \quad \mathbf{u}_2 = \mu_2 A\mathbf{u}_2.$$

We multiply the first by $\mathbf{u}_2$ from the left and the second by $\mathbf{u}_1$ from the right to get
$$(\mathbf{u}_2, \mathbf{u}_1)_E = \mu_1(\mathbf{u}_2, A\mathbf{u}_1)_E,$$
$$(\mathbf{u}_2, \mathbf{u}_1)_E = \mu_2(A\mathbf{u}_2, \mathbf{u}_1)_E.$$

Subtracting, we obtain
$$\mu_1(\mathbf{u}_2, A\mathbf{u}_1)_E - \mu_2(A\mathbf{u}_2, \mathbf{u}_1)_E = 0.$$

Because A is self-adjoint, we have $(\mathbf{u}_2, A\mathbf{u}_1)_E = (A\mathbf{u}_2, \mathbf{u}_1)_E$. Thus
$$(\mu_1 - \mu_2)(\mathbf{u}_2, A\mathbf{u}_1)_E = 0.$$

Since $\mu_1 \neq \mu_2$, it follows that $(\mathbf{u}_2, A\mathbf{u}_1)_E = 0$. This establishes (3.24.4). Orthogonality of $\mathbf{u}_1$ and $\mathbf{u}_2$ follows from the equality $(\mathbf{u}_2, \mathbf{u}_1)_E = \mu_1(\mathbf{u}_2, A\mathbf{u}_1)_E$. □

Later (in Theorem 3.28.2) we will use Theorem 3.24.5 to prove that we can select the set of eigensolutions of A that constitute an orthogonal basis of E_{E0}. This permits application of the separation of variables method to dynamics problems. We will also need the main result of the next section.

3.25 Compactness of A

Before we discuss compactness of an operator, we should introduce the notion of compactness for a set. In elementary calculus we are accustomed to the fact that a closed and bounded subset of $\mathbb{R}^n$ has the following property: from any sequence contained in the set we can select a Cauchy subsequence. In an infinite-dimensional Hilbert space H this property does not hold. For instance, we can take an orthonormal sequence $\{e_k\}$ that definitely exists in H and lies in the unit ball $\|x\|_H \leq 1$; since the distance between any two distinct elements of $\{e_k\}$ is given by
$$\|e_k - e_m\|_H^2 = (e_k - e_m, e_k - e_m)_H = \|e_k\|_H^2 + \|e_m\|_H^2 = 2,$$
this sequence cannot contain a Cauchy subsequence.

Theorem 3.25.1. *Any bounded sequence in a Hilbert space H contains a Cauchy subsequence if and only if H is finite dimensional.*

This theorem holds in a normed space as well. In the following definition, X is a Banach space.

Definition 3.25.1. A set $S \subset X$ is *precompact* if from any of its sequences we can select a Cauchy subsequence. If a precompact set S is closed, it is *compact*.

Note that a sequence $\{x_k\}$ such that $\|x_k\|_X \to \infty$ as $k \to \infty$ does not contain a Cauchy subsequence. So *a precompact subset of a Banach space is bounded*.

To formulate a criterion for compactness, we will need

Definition 3.25.2. A finite set of elements $e_1, e_2, \ldots, e_n$ is a *finite ε-net* of a set $S \subset X$ if for any $x \in S$ there exists e_k such that $\|x - e_k\|_X < \varepsilon$.

In other words, $e_1, e_2, \ldots, e_n$ is a finite ε-net of set S if the union of all balls of radius ε and having centers at these points covers S.

Now let us formulate

Theorem 3.25.2. *(Hausdorff criterion)* $S \subset X$ *is precompact if and only if for any $\varepsilon > 0$ there exists a finite ε-net of S.*

Proof. We will prove the "if" part; the converse can be found in any textbook on functional analysis (e.g., [Lebedev and Vorovich (2002)]). Assume that for any $\varepsilon > 0$ there exists a finite ε-net of S. Let us take an arbitrary sequence $\{x_k\}$ from S and show that it contains a Cauchy subsequence.

(1) Let us construct a finite $\frac{1}{2}$-net of S. S is covered by the union of a finite number of balls of radius $\frac{1}{2}$ with centers at the $\frac{1}{2}$-net. One of these balls, say B_1, contains infinitely many elements of $\{x_k\}$. We denote this subsequence by $\{x_{k_1}\}$. Let us select its first element and call that y_1.

(2) In the same way, a finite union of balls of radius $\frac{1}{2^2}$ that corresponds to a $\frac{1}{2^2}$-net of S covers S, so some ball B_2 contains infinitely many members of $\{x_{k_1}\}$. We denote this subsequence by $\{x_{k_2}\}$. Let us select an element whose position in the initial sequence lies beyond that of y_1 and call it y_2. Because y_1 and y_2 belong to B_1, we have $\|y_1 - y_2\|_X < 2 \cdot \frac{1}{2} = 1$.

(3) Similarly, construct a finite $\frac{1}{2^3}$-net of S and a select corresponding ball B_3 that contains a subsequence $\{x_{k_3}\}$ of the sequence $\{x_{k_2}\}$. Select an element whose position in the initial sequence lies beyond that of y_2

and call it y_3. Because y_2 and y_3 belong to B_2, we have $\|y_2 - y_3\|_X < 2 \cdot \frac{1}{2^2} = \frac{1}{2}$.

We can continue this process indefinitely. On the kth step we construct a finite $\frac{1}{2^k}$-net, then the ball cover, then select a ball B_k that contains an infinite subsequence of the sequence obtained in the previous step. Selecting a corresponding element and denoting it by y_k, we see that both y_k and y_{k-1} lie in B_{k-1} so that

$$\|y_{k-1} - y_k\|_X < 2 \cdot \frac{1}{2^{k-1}} = \frac{1}{2^{k-2}}.$$

We now show that $\{y_k\}$ is the needed Cauchy sequence. For any integer m, we have

$$\|y_{n+m} - y_n\|_X \leq \|(y_{n+m} - y_{n+m-1})\| + \|y_{n+m-1} - y_{n+m-2}\|_X$$
$$+ \cdots + \|y_{n+1} - y_n\|_X$$
$$\leq \frac{1}{2^{n+m-2}} + \frac{1}{2^{n+m-3}} + \cdots + \frac{1}{2^{n-1}}$$
$$< \frac{1}{2^{n-2}} \to 0 \quad \text{as } n \to \infty.$$

This completes the proof. □

A classical result on compactness in infinite-dimensional spaces is

Theorem 3.25.3. *(Arzelà) Let $[a,b]$ be a finite interval. A family of functions $f_\alpha(x)$ is precompact in $C(a,b)$ if and only if it is bounded in $C(a,b)$ and equicontinuous.*

Proof. Let us first explain what is meant by "equicontinuity" of a family of functions. We know that a function $f(x)$ continuous on $[a,b]$ is uniformly continuous: for any $\varepsilon > 0$ there is a $\delta > 0$ such that for any $x, y \in [a,b]$ we have $|f(x) - f(y)| < \varepsilon$. This is called Weierstrass' theorem on uniform continuity of a continuous function on a compact set. If for any $\varepsilon > 0$ we can find $\delta > 0$ having the property that for any $x, y \in [a,b]$ such that $|x - y| < \delta$ we get $|f_\alpha(x) - f_\alpha(y)| < \varepsilon$ for all functions of the family, then the family is said to be *equicontinuous*.

We will present a geometrical proof of the "if" part of the theorem. Assuming the family is bounded and equicontinuous, we construct an ε-net of the family. Let us start with $\varepsilon/3$. By the equicontinuity of the family, we can find $\delta > 0$ such that for any $x, y \in [a,b]$, $|x - y| \leq \delta$, we have $|f_\alpha(x) - f_\alpha(y)| < \varepsilon/3$. Now let us introduce the necessary ε-net.

Since the family is bounded, there exists c such that $|f_\alpha(x)| < c$. Let us depict on the domain $[a,b] \times [-c,c]$ a grid of lines parallel to the axes Ox and Oy (Fig. 3.15). The lines parallel to Oy are separated by a distance δ, and those parallel to Ox by $\varepsilon/3$.

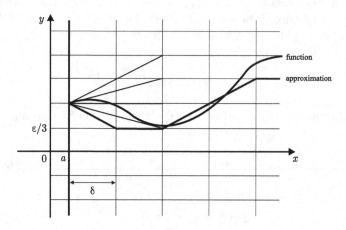

Fig. 3.15 Construction of an ε-net.

Connect all the nodes of this picture (except those in vertical lines) with straight segments. In this way we get a finite number of graphs that represent some piecewise linear continuous functions on $[a,b]$. This is the required ε-net of the family. Indeed, from the picture it is easily seen that any function of the family, because of equicontinuity, can be approximated with one of the piecewise linear functions $\varphi_k(x)$ in such a way that

$$|f_\alpha(x) - \varphi_k(x)| < \varepsilon.$$

We leave the details to the reader. (Hint: On a segment $[x_k, x_k + \delta]$ of the domain, the graph of any function in this family can lie within no more than two vertically-adjacent cells.) □

A version of Arzelà's theorem also holds in $C(V)$, where V is a compact subset of $\mathbb{R}^n$. We advise the reader to adapt the above geometrical proof to this case.

Exercise 3.25.1. *Let $[a,b]$ be a finite interval. Show that a set of functions that is bounded in the space $C^{(1)}(a,b)$ is precompact in $C(a,b)$.*

In § 2.10 we established that each element of the energy space E_{SD} stands in unique correspondence with an element of $C(0,l)$. We called this correspondence an imbedding operator, and found that it is linear and continuous from E_{SD} to $C(0,l)$. Now we wish to prove

Lemma 3.25.1. *The image of the unit ball in E_{SD} under the imbedding operator from E_{SD} to $C(0,l)$ is a precompact set in $C(0,l)$.*

Proof. In order to establish continuity of the imbedding, we started with inequality (2.10.6) for a continuously differentiable function and then, by the limit passage over a representative sequence of an element of E_{SD}, obtained the needed property. We will do the same thing here. For a continuously differentiable function $u(x)$ on $[0,l]$ we have

$$u(y) - u(x) = \int_x^y u'(s)\,ds.$$

Hölder's inequality gives

$$|u(y) - u(x)| = \left|\int_x^y 1 \cdot u'(s)\,ds\right|$$

$$\leq \left|\int_x^y 1^2\,ds\right|^{1/2} \left|\int_x^y |u'(s)|^2\,ds\right|^{1/2}$$

$$= \sqrt{|y-x|}\left|\int_x^y |u'(s)|^2\,ds\right|^{1/2}$$

$$\leq \sqrt{|y-x|}\left(\int_0^l |u'(s)|^2\,ds\right)^{1/2}.$$

Let $U(x)$ be an element of E_{SD} and $\{u_n(x)\}$ a representative sequence. The inequality

$$|u_n(y) - u_n(x)| \leq \sqrt{|y-x|}\left(\int_0^l |u_n'(s)|^2\,ds\right)^{1/2}$$

holds for any $u_n(x)$. Let $u(x)$ be the limit of the sequence $\{u_n(x)\}$, the existence and continuity of which were established in § 2.10. As $n \to \infty$ we obtain

$$|u(y) - u(x)| \leq \sqrt{|y-x|}\lim_{n\to\infty}\left(\int_0^l |u_n'(s)|^2\,ds\right)^{1/2}.$$

But by definition

$$\lim_{n\to\infty}\int_0^l |u_n'(s)|^2\,ds = \int_0^l |U'(s)|^2\,ds$$

and so
$$|u(y) - u(x)| \leq \sqrt{|y-x|}\left(\int_0^l |U'(s)|^2 \, ds\right)^{1/2}.$$

Hence the set of all $u(x)$ that constitute the image of the unit ball of E_{SD} is an equicontinuous and bounded set of functions belonging to $C(0, l)$. An application of Arzelà's theorem completes the proof. □

From the elementary inequality
$$\int_0^l |u(s)|^2 \, ds \leq l \max_{x \in [0,l]} |u(x)|^2,$$

which holds for any function continuous on $[0, l]$, we obtain

Lemma 3.25.2. *If $\{u_n(x)\}$ is a Cauchy sequence in $C(0, l)$, then it is a Cauchy sequence in the norm of $L^2(0, l)$.*

Definition 3.25.3. A linear operator C from a normed space X to a normed space Y is *compact* if the image of the unit ball $\|x\|_X \leq 1$ is precompact in Y.

The reader is asked to work

Exercise 3.25.2. *Show that a linear operator C is compact if and only if any bounded sequence $\{x_k\} \subset X$ contains a subsequence $\{x_{k_1}\}$ such that $\{Cx_{k_1}\}$ is a Cauchy sequence in Y.*

In the exercise, it is sufficient to require $\{x_k\}$ to lie within the ball $\|x\|_X \leq 1$.

A linear compact operator C is continuous. Indeed, the image $C(B)$ of the unit ball $B = \{x\colon \|x\|_X \leq 1\}$ is precompact and therefore bounded; we can define the norm of the operator C as
$$\|C\| = \sup_{\|x\|_X \leq 1} \|Cx\|_Y.$$

Theorem 3.25.4. *The set of linearly independent eigenvectors that correspond to an eigenvalue μ of a compact operator C acting in a Hilbert space H is finite.*

Proof. Let us extend the set of eigenvectors of C that correspond to μ by zero and denote it by N. Then N is a closed subspace of H and therefore a Hilbert space. Let us introduce an operator $C_1 = \mu C$ that is evidently linear and compact as well. But C_1 maps the unit ball of N into itself. By

Definition 3.25.3 this ball is precompact. Thus, by Theorem 3.25.1, N is finite dimensional. □

We now return to our operators for the elastic objects. First we consider the operator A_S for the string. It is given by the representation

$$(A_S w, v)_S = \rho \int_0^l w(x)\overline{v(x)}\,dx, \qquad (3.25.1)$$

where (cf., § 2.9)

$$(w, v)_S = \int_0^l T_0 w'(x)\overline{v'(x)}\,dx \qquad (3.25.2)$$

defined in the complex space E_{SD}. Like the operator A for linear elasticity, A_S is a linear continuous self-adjoint operator and satisfies the inequality

$$|(A_S w, v)_S| \leq c \,\|w\|_{L^2(0,l)} \|v\|_{L^2(0,l)}. \qquad (3.25.3)$$

Now we prove

Theorem 3.25.5. A_S *is a compact operator.*

Proof. From (3.25.3) and the imbedding theorem for E_{SD} we have

$$|(A_S w, v)_S| \leq c_1 \,\|w\|_{L^2(0,l)} \|v\|_S \qquad (3.25.4)$$

with a constant c_1 that does not depend on $w, v \in E_{SD}$. Let us take an arbitrary bounded sequence $\{w_k\}$ in E_{SD}. By Lemma 3.25.2, it contains a subsequence $\{w_{k_1}\}$ that is a Cauchy sequence in $L^2(0, l)$. Let us rename it $\{w_k\}$. To prove the theorem, it suffices to show that the subsequence $\{A_S w_k\}$ is a Cauchy sequence in E_{SD}. Indeed, in (3.25.4) we can substitute $w = w_k - w_m$ and $v = A_S w_k - A_S w_m$ to obtain

$$|(A_S w_k - A_S w_m, A_S w_k - A_S w_m)_S|$$
$$\leq c_1 \,\|w_k - w_m\|_{L^2(0,l)} \|A_S w_k - A_S w_m\|_S.$$

Canceling $\|A_S w_k - A_S w_m\|_S$, we get

$$\|A_S w_k - A_S w_m\|_S \leq c_1 \,\|w_k - w_m\|_{L^2(0,l)} \to 0 \quad \text{as } n, m \to \infty. \quad \square$$

Theorem 3.25.6. *The operator A given by (3.23.8) is compact.*

The proof is based on Rellich's result on the compactness of the imbedding operator in the space $W^{1,2}(V)$. We formulate this as

Theorem 3.25.7. *The elements of the unit ball in the space $W^{1,2}(V)$ constitute a precompact set in $L^2(V)$.*

This is a special case of the Sobolev–Kondrashev theorem on compactness of the imbedding operator to $L^p(V)$. The proof requires techniques that lie outside of the scope of this book.

Having established that operators of the same type as A (equation (3.23.8)) are compact, we shall study the effect of this property on the spectrum of A. The outcome is a collection of results known as the *Fredholm alternative*.

3.26 Riesz–Fredholm Theory for a Linear, Self-Adjoint, Compact Operator in a Hilbert Space

In this section we consider the spectrum of an abstract, linear, self-adjoint, compact operator A in a Hilbert space H. The results will immediately apply to (3.23.8).

We know that the set of eigenvectors of A corresponding to the same eigenvalue, augmented with the zero vector, constitutes a finite-dimensional subspace. Moreover, eigenvectors corresponding to distinct eigenvalues are mutually orthogonal. Now we discuss the distribution of eigenvalues. As we recall, these are all real and bounded away from the origin in the complex plane.

Theorem 3.26.1. *The eigenvalues of A have no finite accumulation point.*

Proof. Suppose to the contrary that μ_0 is an accumulation point of eigenvalues such that $|\mu_0| < \infty$. There is a sequence of eigenvalues μ_k of A such that $\mu_k \to \mu_0$ as $k \to \infty$. For each μ_k let us select a unit eigenvector x_k, $\|x_k\|_H = 1$. Thus $(x_k, x_m) = \delta_{km}$. The sequence $\{\mu_k x_k\}$ is bounded in H. By Exercise 3.25.2 the sequence $\{A(\mu_k x_k)\}$ must contain a Cauchy subsequence. But this is impossible as $\mu_k A x_k = x_k$ and therefore

$$\|A(\mu_k x_k) - A(\mu_m x_m)\|_H^2 = \|x_k - x_m\|_H^2 = \|x_k\|_H^2 + \|x_m\|_H^2 = 2$$

when $k \neq m$. □

Now we wish to study the solvability of the equation

$$x - \mu A x = b. \tag{3.26.1}$$

We denote by N the set of all solutions of the homogeneous equation

$$x - \mu A x = 0. \tag{3.26.2}$$

When μ is an eigenvalue of A, according to Theorem 3.25.4, N is a finite-dimensional subspace of H. By Theorem 1.16.2 there exists an orthogonal complement M of N in H. That is, M is a closed subspace of H, it has only zero in common with N, any element of M is orthogonal to all elements of N, and each element $x \in H$ has the unique representation $x = m + n$ where $n \in N$ and $m \in M$. First we prove

Lemma 3.26.1. *For any $x \in M$ the inequality*

$$\|x - \mu A x\|_H \geq c_0 \|x\|_H \tag{3.26.3}$$

holds with a constant $c_0 > 0$ that does not depend on x.

Proof. Suppose there is no such c_0. Then there is a sequence $\{x_k\}$ such that $\|x_k\|_H = 1$ for all k but $\|x_k - \mu A x_k\|_H \to 0$ as $k \to \infty$. Since A is compact, we can select a Cauchy subsequence $\{A x_{k_1}\}$ from $\{A x_k\}$. Renaming $\{x_{k_1}\}$ as $\{x_k\}$, we find that $\{A x_k\}$ is a Cauchy sequence and $x_k - \mu A x_k = \varepsilon_k$, where $\|\varepsilon_k\|_H \to 0$ as $k \to \infty$. We rewrite the last relation as

$$x_k = \mu A x_k + \varepsilon_k. \tag{3.26.4}$$

Because $\|\varepsilon_k\|_H \to 0$ as $k \to \infty$, the sequence $\{\mu A x_k + \varepsilon_k\}$ has the same limit as $\{\mu A x_k\}$; we denote it by x_0. By (3.26.4), $\lim_{k \to \infty} x_k = x_0$. As M is closed, x_0 belongs to M as well. What we can say about x_0? On one hand, $x_0 \in M$ and

$$\|x_0\|_H = \lim_{k \to \infty} \|x_k\|_H = 1.$$

On the other hand, letting $k \to \infty$ in (3.26.4) we get

$$x_0 = \mu A x_0.$$

Therefore $x_0 \in N$, which contradicts the assertion that x_0 lies in M and is nonzero. $\square$

By Lemma 3.26.1 we can use $\|x - \mu A x\|_H$ as a norm on M. Moreover, because A is continuous, the reverse inequality

$$\|x - \mu A x\|_H \leq c_1 \|x\|_H \tag{3.26.5}$$

holds with a constant c_1 that does not depend on $x \in M$. Taken together with (3.26.3), this means that the expression $\|x - \mu A x\|_H$ constitutes another norm on M that is equivalent to the old norm $\|x\|_H$, and that M is complete in this new norm. The new norm induces an inner product

$$\langle x, y \rangle = (x - \mu A x, y - \mu A y)_H \tag{3.26.6}$$

that we can use on M instead of the initial inner product $(x,y)_H$. With this new inner product M is a Hilbert space as well.

Exercise 3.26.1. *Let B be a continuous linear operator in a normed space X, and suppose the inequality $\|Bx\|_X \geq c\|x\|_X$ holds for all $x \in X$ with some constant $c > 0$ that does not depend on x. Show that the functional $\|Bx\|_X$ constitutes a norm on X equivalent to the initial norm.*

Exercise 3.26.2. *Let B be a continuous linear operator B in a Hilbert space H, and suppose the inequality $\|Bx\|_X \geq c\|x\|_X$ holds for all $x \in H$ with a constant $c > 0$ that does not depend on x. Show that the functional $\langle x, y \rangle = (Bx, By)_X$ is an inner product on H, and that H remains a Hilbert space under the induced norm $\langle x, x \rangle^{1/2}$. Also show that this norm is equivalent to the initial norm on H.*

Now we can formulate

Theorem 3.26.2. *Equation (3.26.1) has a solution if and only if $b \in M$.*

Proof. "Only if" part. Let us first prove that for solvability of (3.26.1), b must be orthogonal to N. Indeed, suppose (3.26.1) has a solution x_0. Let us take the inner product of both sides of (3.26.1) with an arbitrary $n \in N$:

$$(x_0 - \mu A x_0, n)_H = (b, n)_H.$$

Recall that A is self-adjoint and μ is real. Then

$$(x_0 - \mu A x_0, n)_H = (x_0, n - \mu A n)_H = 0$$

and so $(b, n)_H = 0$ for any $n \in N$.

"If" part. Assume $b \in M$. We must prove existence of a solution to equation (3.26.1). Let us start with the functional $(x, b)_H$, which is evidently linear and continuous not only on H but on M as well. As we said above, we can use $\langle x, y \rangle$ instead of $(x, y)_H$ as an inner product on M. So, using the Riesz Representation Theorem 2.8.1, we get a unique element $b_0 \in M$ such that

$$(x, b)_H = \langle x, b_0 \rangle$$

for any $x \in M$. Equivalently,

$$(x, b)_H = (x - \mu A x, b_0 - \mu A b_0)_H. \qquad (3.26.7)$$

Let us show that this holds for any $x \in H$. Indeed, we can uniquely represent any x as $x = n + m$, where $n \in N$ and $m \in M$. Equality (3.26.7)

holds if we put $x = m$. Putting $x = n + m$, we see that
$$(x, b)_H = (n + m, b)_H = (m, b)_H$$
because $(n, b)_H = 0$ (note that $b \in M$ by the condition). On the other hand
$$(x - \mu A x, b_0 - \mu A b_0)_H = ((n - \mu A n) + (m - \mu A m), b_0 - \mu A b_0)_H$$
$$= (m - \mu A m, b_0 - \mu A b_0)_H,$$
because $n - \mu A n = 0$ (n is one of the eigenvectors of A corresponding to μ). So (3.26.7) holds for any $x \in H$ as it is valid for $x = m$. Let us denote $b_0 - \mu A b_0 = x_0$. Then (3.26.7) takes the form
$$(x, b)_H = (x - \mu A x, x_0)_H.$$
Since A is self-adjoint and μ is real, we get
$$(x - \mu A x, x_0)_H = (x, x_0 - \mu A x_0)_H.$$
Hence the identity
$$(x, b)_H = (x, x_0 - \mu A x_0)_H$$
holds for all $x \in H$. But this means that
$$b = x_0 - \mu A x_0,$$
and thus x_0 is the needed solution to (3.26.1). $\square$

Remarks:

(1) The condition for solvability of the equation, $b \in M$, is sufficient for a compact, self-adjoint, linear operator. If the operator is only symmetric and linear, it is necessary for solvability.
(2) We use an abstract form of the theorem of solvability of the equation. For the elastic problems we considered earlier, the abstract condition of solvability can be formulated in a more mechanical form. Here the term $(b, x)_H$ corresponds to the work of external forces and so the solvability condition must be written in terms of work. For example, in linear elasticity it is
$$\int_V \mathbf{F} \cdot \overline{\mathbf{v}}_k \, dV + \int_{S \setminus S_1} \mathbf{f} \cdot \overline{\mathbf{v}}_k \, dS = 0$$
where $\mathbf{v}_k$ is an eigenvector corresponding to μ. In fact, the conjugation of $\mathbf{v}_k$ is not necessary as the eigenfunctions for these problems are real-valued. Mechanicists say that solvability of the problem for a free body requires the external forces to be "orthogonal" to the oscillation modes. This is stated by the above equality.

The reader should understand that we merely proved an existence theorem; the proof does not yield a method of solving (3.26.1). However, the condition that b must be orthogonal to all eigenvectors of A corresponding to an eigenvalue μ is quite practical. Uniqueness of solution of (3.26.1) is related to N; namely, a solution is unique if and only if $N = \{0\}$. Indeed, if $N \neq \{0\}$ and x_0 satisfies (3.26.1), then so does $x_0 + n$ for any $n \in N$. On the other hand, if there are two solutions x_1, x_2 to (3.26.1), then $x_1 - x_2$ is a solution of the homogeneous equation (3.26.1) and thus belongs to N. A direct consequence of this is

Theorem 3.26.3. *Equation* (3.26.1) *has a unique solution for any* $b \in H$ *if and only if* μ *is not an eigenvalue of* A. *Moreover, if* μ *is not an eigenvalue of* A *then there exists a continuous inverse* $(I - \mu A)^{-1}$ *whose domain is* H.

Proof. Suppose μ is not an eigenvalue of A. Then $N = \{0\}$, and thus $M = H$. By Theorem 3.26.2, equation (3.26.1) has a solution for any $b \in H$ and by the above it is unique. Moreover, (3.26.3) shows that $(I - \mu A)^{-1}$ defined on the whole space H is continuous.

Conversely, $M = H$ means that (3.26.1) has a solution for any $b \in H$. But then N, which is the orthogonal complement of M in H, consists only of the zero element. This means that N does not contain eigenvectors and therefore μ is not an eigenvalue. □

It makes sense to collect the properties we have obtained for the spectrum of the operator A.

Theorem 3.26.4. *Let* A *be a linear, self-adjoint, compact operator in a Hilbert space* H. *The spectrum of* A *has the following properties.*

(1) The spectrum of A can contain only eigenvalues of A.
(2) The eigenvalues μ_k of A, if they exist, are real.
(3) The set of eigenvalues has no finite accumulation point.
(4) To an eigenvalue of A there corresponds no more than a finite set of linearly independent eigenvectors.
(5) Eigenvectors x_1, x_2 of A that correspond to distinct eigenvalues are orthogonal: $(x_1, x_2)_H = 0$. They also possess the generalized orthogonality property $(Ax_1, x_2)_H = 0$.
(6) Equation (3.26.1) has a solution if and only if b is orthogonal to all eigenvectors of A that correspond to the eigenvalue μ.
(7) Equation (3.26.1) has a unique solution for any $b \in M$ that depends continuously on b if and only if μ is not an eigenvalue of A.

Additional information about the spectrum can be found from the extremum nature of the eigenvalues. First we need the idea of weak convergence in a Hilbert space.

3.27 Weak Convergence in Hilbert Space

For a sequence of vectors in $\mathbb{R}^n$, norm convergence and componentwise convergence are equivalent. In infinite-dimensional space such as l^2, they are not. We wish to consider an analogue of componentwise convergence in a separable Hilbert space H. The component that defines the orthogonal projection of $x \in H$ onto the axis defined by a unit vector $e \in H$ is the product $(x, e)_H$. We use this fact as follows.

Definition 3.27.1. A sequence $\{x_k\}$ is *weakly convergent* to x_0 if for any $h \in H$ we have

$$\lim_{k \to \infty} (x_k, h)_H = (x_0, h)_H. \tag{3.27.1}$$

We write $x_k \rightharpoonup x_0$ as $k \to \infty$.

Definition 3.27.2. We say that $\{x_k\}$ is a *weak Cauchy sequence* if for any $h \in H$ we have

$$(x_k - x_m, h)_H \to 0 \quad \text{as } k, m \to \infty. \tag{3.27.2}$$

The usual definition of weak convergence in a normed space is based on continuous linear functionals. We based our statements on the Riesz representation theorem to obtain a more geometrical picture of the idea.

From now on, ordinary convergence in norm will be called *strong convergence*. We consider some properties of weak convergence.

Theorem 3.27.1. *A limit of a weakly convergent sequence is unique.*

Proof. Let $\{x_k\}$ have two weak limits x' and x''. That is,

$$\lim_{k \to \infty} (x_k, h)_H = (x', h)_H \quad \text{and} \quad \lim_{k \to \infty} (x_k, h)_H = (x'', h)_H.$$

It follows that $(x', h)_H = (x'', h)_H$ or $(x' - x'', h)_H = 0$. Putting $h = x' - x''$, we get $(x' - x'', x' - x'')_H = 0$ and thus $x' = x''$. $\square$

An orthonormal sequence $\{e_k\}$ cannot converge. Nor can it contain a Cauchy subsequence, because $\|e_k - e_m\|_H = \sqrt{2}$ for $k \neq m$. However, for any $x \in H$ we get

$$\lim_{k \to \infty} (x, e_k)_H = 0$$

because the sequence of Fourier coefficients of x with respect to e_k tends to zero by Bessel's inequality. Hence $\{e_k\}$ has zero as a weak limit in H, so strong convergence and weak convergence are different. Clearly, a strongly convergent sequence converges weakly to the same element. Failure of the converse raises the question whether each weak Cauchy sequence has a weak limit in H. The answer is given by

Theorem 3.27.2. *Any weak Cauchy sequence $\{x_k\}$ has a weak limit in H.*

Proof. In Lemma 3.27.1 we will show that there is a constant c such that $\|x_k\|_H \leq c$. For any $x \in H$, because $\{(x_k, x)_H\}$ is a numerical Cauchy sequence we have

$$\lim_{k \to \infty} \overline{(x_k, x)_H} = \lim_{k \to \infty} (x, x_k)_H = F(x).$$

The reader can verify that $F(x)$ is a linear functional. Next,

$$|F(x)| \leq \sup_k \|x_k\|_H \|x\|_H \leq c \|x\|_H$$

and so by the Riesz representation theorem

$$F(x) = (x, x')_H$$

where $x' \in H$ is uniquely defined by $F(x)$. For x' it follows that

$$\lim_{k \to \infty} (x_k, x)_H = (x', x)_H \quad \text{for all } x \in H$$

and so x' is a weak limit of $\{x_k\}$. □

We still need to prove

Lemma 3.27.1. *Let $\{x_k\}$ be a weak Cauchy sequence in a Hilbert space H. Then it is bounded: there is a constant c such that $\|x_k\|_H \leq c$ for all k.*

Proof. Suppose to the contrary that $\{x_k\}$ is unbounded. Without loss of generality, suppose $\|x_k\|_H \to \infty$ as $k \to \infty$. We will show that this implies the existence of $z^* \in H$ and a subsequence $\{x_{n_k}\}$ such that $(x_{n_k}, z^*)_H \to \infty$ as $n_k \to \infty$; this contradicts the definition of weak convergence for $\{x_k\}$. We start with

(1) Let z be an arbitrary element of H and $\rho > 0$ an arbitrarily small number. The elements
$$z_k = z + \rho \frac{x_k}{2\|x_k\|_H}$$
lie in the closed ball of radius ρ with center z. Note that $\|z_k - z\|_H = \rho/2$. The sequence $\{z_k\}$ has the property
$$(x_k, z_k)_H = (x_k, z)_H + \frac{\rho}{2}\left(x_k, \frac{x_k}{\|x_k\|_H}\right)_H = (x_k, z)_H + \frac{\rho}{2}\|x_k\|_H \to \infty$$
as $k \to \infty$, since $(x_k, z)_H$ is bounded at least.

(2) We construct the above-mentioned $\{x_{n_k}\}$ and z^*. The element z^* will be the limit of the centers of a sequence of nested closed balls $\{B_k\}$ with radii tending to zero.

(2-1) Fix some $\rho_1 > 0$ and take x_1 as the center of the first ball B_1. By (1), we get k_1 and $z_{k_1} \in B_1$ such that $(x_{k_1}, z_{k_1})_H > 1$. As the inner product is continuous with respect to both arguments and z_{k_1} is an interior point of B_1, there is a closed ball $B_2 \subset B_1$ with center z_{k_1} and radius $\rho_2 \le \rho_1/2$ such that for all $z \in B_2$ we have $(x_{k_1}, z)_H > 1$.

(2-2) Consider the tail of the sequence $\{x_k\}$ consisting of those terms having indices greater than k_1. The norms of this subsequence tend to infinity. So considering for this tail the set of values $(x_k, z)_H$ when $z \in B_2$, we find ourselves in the situation of point (1) of the proof, and so there are $k_2 > k_1$ and $z_{k_2} \in B_2$ such that $(x_{k_2}, z_{k_2})_H > 2$. By continuity of $(x_{k_2}, z)_H$ in z, we find a closed ball B_3 with center z_{k_2} and radius $\rho_3 \le \rho_1/3$ such that $B_3 \subset B_2$ and for all $z \in B_3$ we have $(x_{k_2}, z)_H > 3$. We can continue this indefinitely.

(2-n) On the nth step, similarly, we consider the tail of the sequence $\{x_k\}$ consisting of those term whose indices exceed k_{n-1}; we get a closed ball B_{n+1} enclosed in B_n and an element z_{k_n}, the center of B_{n+1} whose radius ρ_n is no more than ρ_1/n such that for all $z \in B_{n+1}$ we have $(x_{k_n}, z)_H > n$.

The sequence $\{z_{k_n}\}$ of centers of the nested closed balls B_n is a Cauchy sequence and therefore has a limit z^*. Because z^* belongs to each B_n,
$$(x_{k_n}, z^*)_H > n \to \infty \quad \text{as } n \to \infty.$$

This contradicts the condition that $\{x_k\}$ is a weak Cauchy sequence in the Hilbert space H. $\square$

The following is a simple but useful sufficient condition for a weakly convergent sequence to converge strongly to the weak limit.

Theorem 3.27.3. *Suppose $\{x_k\}$ converges weakly to x^* in H and*
$$\lim_{k\to\infty} \|x_k\|_H = \|x^*\|_H \quad \text{as } k \to \infty.$$
Then x^* is a strong limit of $\{x_k\}$.

Proof. Let us start with
$$\alpha_k = \|x_k - x^*\|_H^2 = (x_k - x^*, x_k - x^*)_H$$
$$= (x_k, x_k)_H - (x^*, x_k)_H - (x_k, x^*)_H + (x^*, x^*)_H.$$
As $k \to \infty$, each of the first three addends tends to $(x^*, x^*)_H$. Indeed, $(x_k, x_k)_H \to (x^*, x^*)_H$ by the condition of the theorem. We have $(x_k, x^*)_H \to (x^*, x^*)_H$ by the definition of weak convergence of $\{x_k\}$ to x^*, and similarly for $(x^*, x_k)_H$. This means that $\alpha_k \to 0$, which was to be proved. □

We formulate some properties of weak convergence as exercises.

Exercise 3.27.1. *Let $x_k \rightharpoonup x^*$, and let $\{x_k\}$ be a strong Cauchy sequence. Show that x^* is a strong limit of $\{x_k\}$ as well.*

Exercise 3.27.2. *Show that if a sequence $\{x_k\}$ converges weakly to x^* in H, then $\|x^*\|_H \leq \sup_k \|x_k\|_H$.*

Exercise 3.27.3. *Let B be a continuous linear operator in H and $x_k \rightharpoonup x^*$. Show that $Bx_k \rightharpoonup Bx^*$; that is, a continuous linear operator B is weakly continuous.*

Exercise 3.27.4. *Suppose B is a compact linear operator in H. Show that B maps a sequence $\{x_k\}$ that converges weakly to x^* into a sequence $\{Bx_k\}$ that converges strongly to Bx^*.*

Exercise 3.27.5. *Let B be a compact linear operator in a Hilbert space H, and let $x_k \rightharpoonup x^*$. Show that*
$$\lim_{k\to\infty} (Bx_k, x_k)_H = (Bx^*, x^*)_H. \tag{3.27.3}$$

Definition 3.27.1 is somewhat inconvenient to check. Let us establish a more convenient condition for weak convergence.

Theorem 3.27.4. *Let $e_1, e_2, e_3, \ldots$ be a complete system of elements in H. Then $\{x_k\}$ is a weak Cauchy sequence if it is bounded (i.e., if there* is a constant c_0 such that $\|x_k\|_H \le c_0$ for all k), and for any fixed k the numerical sequence $(x_n, e_k)_H$ is a Cauchy sequence.

Proof. We must show that for any $h \in H$ and any $\varepsilon > 0$ we can find a number N such that
$$|(x_n - x_m, h)_H| < \varepsilon \quad \text{whenever } m, n > N.$$
So we take $\varepsilon > 0$ and fix $h \in H$. As $e_1, e_2, e_3, \ldots$ is a complete system in H, we can find a finite linear combination
$$h_R = \sum_{k=1}^{R} c_k e_k$$
such that
$$\|h - h_R\|_H < \frac{\varepsilon}{3c_0}.$$
Since R is finite and for each e_k we have $(x_n - x_m, e_k)_H \to 0$ as $n, m \to \infty$, there exists N such that for $n, m > N$ we get
$$|(x_n - x_m, h_R)_H| < \frac{\varepsilon}{3}.$$
Now we are prepared to demonstrate the needed inequality for $n, m > N$:
$$\begin{aligned}|(x_n - x_m, h)_H| &= |(x_n - x_m, h - h_R + h_R)_H| \\ &\le |(x_n, h - h_R)_H| + |(x_m, h - h_R)_H| + |(x_n - x_m, h_R)_H| \\ &\le (\|x_n\|_H + \|x_m\|_H) \|h - h_R\|_H + |(x_n - x_m, h_R)_H| \\ &\le 2c_0 \frac{\varepsilon}{3c_0} + \frac{\varepsilon}{3} = \varepsilon.\end{aligned}$$
□

We saw that if $\{x_k\}$ is a weak Cauchy sequence, the conditions of the theorem hold. This means the conditions of the theorem are equivalent to the definition of a weak Cauchy sequence. Note that in this new definition in a separable Hilbert space H, we can use an orthonormal basis of H as a complete set $e_1, e_2, \ldots$.

We know that a ball in an infinite-dimensional Hilbert space is not precompact. Let us introduce

Definition 3.27.3. A set S is weakly precompact in a Hilbert space if any sequence from S contains a weakly Cauchy subsequence.

The following property is used to justify numerical methods.

Theorem 3.27.5. *A ball in a separable Hilbert space H is weakly precompact.*

Proof. Let $e_1, e_2, e_3, \ldots$ be a complete system of elements in H. Any sequence $\{x_k\}$ from the ball is bounded; hence, by Theorem 3.27.4, it suffices to find a subsequence $\{x_{m_n}\}$ such that for any k the sequence $\{(x_{m_n}, e_k)_H\}$ is a numerical Cauchy sequence. Using the diagonal process, we will construct this subsequence as follows.

(1) Consider the numerical sequence $\{(x_k, e_1)_H\}$. By hypothesis it is bounded as

$$|(x_k, e_1)_H| \leq c_0 \|e_1\|_H,$$

and so it contains a numerical Cauchy subsequence $\{(x_{k_1}, e_1)_H\}$. Let us take an element of $\{x_{k_1}\}$ as the first element of the needed subsequence. Denote this element by x_{1_1}.

(2) Consider the numerical sequence $\{(x_{k_1}, e_2)_H\}$. It is bounded and therefore contains a Cauchy subsequence $\{(x_{k_2}, e_2)_H\}$ defined by some indices k_2. From $\{x_{k_2}\}$ take an element whose position in the initial sequence is further than that of x_{1_1}. Denote this element by x_{2_2}.

(3) Similarly, consider $\{(x_{k_2}, e_3)_H\}$. Again, it is a bounded numerical sequence containing a Cauchy subsequence $\{(x_{k_3}, e_3)_H\}$. Select an element x_{3_3} from $\{x_{k_3}\}$ such that its position in the initial sequence exceeds that of the previous element.

We can continue this process indefinitely. The result is a subsequence $\{x_{n_n}\}$ that is a subsequence (starting with some number) of any $\{x_{k_m}\}$. Therefore $\{(x_{n_n}, e_k)_H\}$ is a numerical Cauchy sequence for any fixed k. $\square$

3.28 Completeness of the System of Eigenvectors of a Self-Adjoint, Compact, Strictly Positive Linear Operator

The principal goal of this section is to demonstrate that a self-adjoint, compact, strictly positive linear operator A in an infinite-dimensional separable Hilbert space H has a set of eigenvectors that constitutes an orthonormal basis of H. This permits us to apply the Fourier method to dynamics problems in mechanics. We also obtain an extremal method for finding the eigenvalues of the operator. We start with a simple

Lemma 3.28.1. *Let A be a self-adjoint, compact, strictly positive linear operator in H. On the unit ball $B = \{x \in H \colon \|x\|_H \leq 1\}$, the form $(Ax, x)_H$ attains its maximum. That is, there exists x_0 with $\|x_0\|_H = 1$ such that $(Ax_0, x_0)_H \geq (Ax, x)_H$ for any x having $\|x\|_H \leq 1$.*

Proof. The element x_0, if it exists, must lie on the unit sphere. If it is such that $\|x_0\|_H < 1$, then for $y_0 = x_0/\|x_0\|_H$ we get

$$(Ay_0, y_0)_H = (Ax_0, x_0)_H / \|x_0\|_H^2 > (Ax_0, x_0)_H.$$

Next we prove existence of the maximum point. The set of numbers $(Ax, x)_H$ for all $x \in B$ is bounded from above, so there is a sequence $\{x_k\}$ such that

$$\lim_{k \to \infty} (Ax_k, x_k)_H = \sup_{x \in B} (Ax, x)_H.$$

As $\{x_k\}$ is bounded, by Theorem 3.27.5 it contains a subsequence $\{x_{k_1}\}$ having a weak limit x_0 which, by Exercise 3.27.2, lies within B. But then, by Exercise 3.27.5, we have

$$\lim_{k_1 \to \infty} (Ax_{k_1}, x_{k_1})_H = (Ax_0, x_0)_H.$$

This completes the proof. □

We will show that by using this extremum approach we have obtained an eigenvector of A.

Theorem 3.28.1. *Let A be a self-adjoint, compact, strictly positive linear operator in H, and let x_0 be an element at which $(Ax, x)_H$ attains its maximum on the unit sphere S, $\|x\|_H = 1$, of H. Then x_0 is an eigenvector of A and $\mu_0 = 1/(Ax_0, x_0)_H$ is the smallest eigenvalue of A.*

Proof. Let μ_1 be an eigenvalue of A and x_1 a corresponding eigenvector. Then $x_1 = \mu_1 A x_1$ and $(x_1, x_1)_H = \mu_1 (Ax_1, x_1)_H$. Division by $\|x_1\|_H^2$ gives

$$\mu_1 = \frac{1}{(Ay_1, y_1)_H}, \quad y_1 = \frac{x_1}{\|x_1\|_H}.$$

So $\mu_1 \geq \mu_0$, since μ_0 is the infimum of $1/(Ax, x)_H$ on the unit sphere of H.

Now we show that x_0 is an eigenvector and μ_0 an eigenvalue of A. It is evident that the problem of maximum of $(Ax, x)_H$ on the unit sphere is equivalent to the problem of finding the supremum of the functional

$$F(x) = \left(A \frac{x}{\|x\|_H}, \frac{x}{\|x\|_H}\right)_H$$

for $x \in H$, and so $F(x)$ attains its maximum at the same point x_0 as $(Ax, x)_H$ does on the unit sphere. Thus

$$F(x_0) \geq F(x_0 + th)$$

for any real t and any $h \in H$. Let h be fixed. As a function of t, $F(x_0 + th)$ takes its maximum at $t = 0$ and thus

$$\left.\frac{dF(x_0 + th)}{dt}\right|_{t=0} = 0.$$

A quick calculation gives

$$\mathrm{Re}(Ax_0, h)_H - \left(A\frac{x_0}{\|x_0\|_H}, \frac{x_0}{\|x_0\|_H}\right)_H \mathrm{Re}(x_0, h)_H = 0.$$

Because $\|x_0\|_H = 1$, this takes the form

$$\mathrm{Re}(Ax_0, h)_H - \frac{1}{\mu_0} \mathrm{Re}(x_0, h)_H = 0.$$

This holds for any h. It therefore holds when we change h to ih, getting

$$\mathrm{Im}(Ax_0, h)_H - \frac{1}{\mu_0} \mathrm{Im}(x_0, h)_H = 0.$$

Thus

$$(Ax_0, h)_H - \frac{1}{\mu_0}(x_0, h)_H = 0.$$

As h is arbitrary, it follows that

$$Ax_0 - \frac{1}{\mu_0} x_0 = 0.$$

This completes the proof. □

Theorem 3.28.2. *Let A be a self-adjoint, compact, strictly positive linear operator in H. Then A has a countable set of eigenvalues*

$$\mu_0 \leq \mu_1 \leq \mu_2 \leq \cdots$$

having no finite accumulation point ($\mu_k \to \infty$ as $k \to \infty$). To each μ_k there corresponds no more than a finite number of mutually orthogonal eigenvectors which together constitute an orthonormal basis of H.

Proof. *Constructing eigenvectors.* We have constructed the first eigenvalue μ_0 and its corresponding eigenvector x_0. By Theorem 3.26.4, two eigenvectors corresponding to different eigenvalues are orthogonal. Clearly, from the set of eigenvectors that correspond to the same eigenvalue, we can select mutually orthogonal ones. This prompts us to seek the next eigenvector orthogonal to x_0; in general, each eigenvector sought should be orthogonal to all those previously found. We do this as follows. The set of vectors of the form αx_0 with a scalar α is a subspace X_1 of H. By Theorem 1.16.2 we decompose H into the direct orthogonal sum

$$H = X_1 \dotplus H_1$$

where any element of H_1 is orthogonal to x_0. The subspace H_1 is a Hilbert space itself, so we can apply Theorem 3.28.1 and find $\max(Ax,x)_H = 1/\mu_1$ on the unit sphere S_1 of H_1 and a vector $x_1 \in S_1$ at which $(Ax,x)_H$ attains this maximum. As in the proof of Theorem 3.28.1, this element satisfies

$$(Ax_1, h_1)_H - \frac{1}{\mu_1}(x_1, h_1)_H = 0 \tag{3.28.1}$$

where h_1 is an arbitrary element of H_1. But this equality holds when we change h_1 to x_0. Indeed, both terms on the left-hand side are zero: $(x_1, x_0)_H = 0$ as $x_1 \in H_1$ and so is orthogonal to x_0, and $(Ax_1, x_0)_H = (x_1, Ax_0)_H = (1/\mu_0)(x_1, x_0)_H = 0$. Thus (3.28.1) holds for any $h_1 \in H$ and we see that

$$\mu_1 A x_1 = x_1.$$

Hence μ_1 is an eigenvalue of A and x_1 is a corresponding eigenvector. By construction, the maximum of the form $(Ax,x)_H$ is taken on the unit sphere S_1. Meanwhile, S_1 is a part of S, which is the unit sphere of H. Thus, in terms of eigenvalues, we have $\mu_1 \geq \mu_0$.

Let us describe a step-by-step procedure for finding the eigenpairs. Suppose we have found the first k eigenvectors, which are mutually orthogonal, and the corresponding eigenvalues $\mu_0 \leq \mu_1 \leq \cdots \leq \mu_{k-1}$. We introduce the subspace X_k spanned by the eigenvectors $x_0, \ldots, x_{k-1}$ and its orthogonal complement H_k in H. On the unit sphere S_k of the subspace H_k, we seek the maximum of $(Ax,x)_H$. This is the same as to find the supremum of

$$F(x) = \left(A\frac{x}{\|x\|_H}, \frac{x}{\|x\|_H}\right)_H$$

in $x \in H_k$, $x \neq 0$. This maximum is attained at a vector $x_k \in S_k$. Denoting this maximum of $\max_{S_k}(Ax,x)_H$ by $1/\mu_k$, we get

$$(Ax_k, h_k)_H - \frac{1}{\mu_k}(x_k, h_k)_H = 0, \tag{3.28.2}$$

which holds for any $h_k \in H_k$. It still holds when we change h_k to any x_r, $r < k$, since both terms on the left-hand side equal zero as above. So (3.28.2) holds for any $h_k \in H$, and thus

$$\mu_k A x_k = x_k.$$

We have constructed the next eigenpair μ_k, x_k of A. This procedure gives us an orthonormal system $x_1, x_2, x_3, \ldots$. The process cannot be disrupted. A disruption could occur only if $\max_{S_k}(Ax, x)_H = 0$; however, this could happen only if $H_k = \{0\}$, i.e., if H were finite-dimensional, a possibility we have excluded.

Let us show that $x_1, x_2, x_3, \ldots$ is a basis of H. The above procedure results in an infinite orthonormal system $x_1, x_2, x_3, \ldots$. By Theorem 3.26.4, to an eigenvalue there corresponds no more than a finite number of eigenvectors from $x_1, x_2, x_3, \ldots$, and so the number of eigenvalues $\mu_1, \mu_2, \mu_3, \ldots$ is infinite as well. By the same theorem the set $\mu_1, \mu_2, \mu_3, \ldots$ has no finite accumulation point, so

$$\mu_k \to \infty \quad \text{as } k \to \infty. \tag{3.28.3}$$

To demonstrate that $x_1, x_2, x_3, \ldots$ is an orthonormal basis of H, we must show that for any $x \in H$ the Fourier representation

$$x = \sum_{k=0}^{\infty} (x, x_k)_H x_k$$

holds. Take any $x \in H$. On page 248 we saw that the sequence

$$s_n = \sum_{k=0}^{n} (x, x_k)_H x_k$$

converges strongly. Our task is to demonstrate that the limit is x. Consider $z_n = x - s_n$. Its limit exists and is z. It suffices to show that $z = 0$. So suppose to the contrary that $z \neq 0$. Next, z_n is an element of H_{n+1} as $(z_n, x_r)_H = (x, x_r)_H - (x, x_r)_H = 0$ for $r \leq n$. By the maximum construction of the eigenvalues we have

$$\frac{(Az_n, z_n)_H}{\|z_n\|_H^2} \leq \frac{1}{\mu_{n+1}} \to 0 \quad \text{as } n \to \infty.$$

Because $\{z_n\}$ converges strongly to $z \neq 0$, we get

$$\frac{(Az, z)_H}{\|z\|_H^2} = 0.$$

This implies $(Az, z)_H = 0$ and so, by strict positiveness of A, $z = 0$. This contradiction completes the proof. □

We recall that Theorem 3.28.2 can be applied to the problem of oscillations of all the elastic objects we have considered.

Further notes on applications of spectral properties

Certain spectral properties of a self-adjoint, compact, strictly positive linear operator A in a Hilbert space are presented in Theorems 3.27.5 and 3.28.2. Here we mention additional facts and a few literature references.

(1) For any $x \in H$ there is a representation

$$Ax = \sum_{k=0}^{\infty} \frac{(x, x_k)_H}{\mu_k} x_k.$$

The series is strongly convergent and defines a sequence of *finite-dimensional operators* A_n given by

$$A_n x = \sum_{k=0}^{n} \frac{(x, x_k)_H}{\mu_k} x_k,$$

This sequence $\{A_n\}$ converges in the operator norm to A [Lebedev and Vorovich (2002)].

(2) *Courant's minimax method.* R. Courant proposed a method for finding eigenvalues of A without having to find the previous eigenvalues. Courant's formula is

$$\frac{1}{\mu_{n+1}} = \inf_{Q_n} \sup_{x \in S_n} (Ax, x)_H,$$

where Q_n is an arbitrary n-dimensional subspace of H and S_n is the unit sphere in the orthogonal complement of Q_n with respect to H. The infimum is taken over all subspaces Q_n. The proof can be found in [Lebedev and Vorovich (2002)]. A consequence of Courant's principle is that the eigenfrequencies of all the types of elastic bodies we have considered can only increase if we impose additional geometric constraints such as equality to zero on some set or additional elastic supports [Lebedev and Vorovich (2002)].

(3) In [Lebedev and Cloud (2003)] there is a discussion of the separation of variables method for dynamical problems involving elastic bodies. It is based on the spectral properties of A discussed above.

3.29 Other Standard Models of Elasticity

An important part of the theory of elasticity is devoted to approximate theories of objects whose shapes are closely related to those of planes and surfaces. Such objects are called *plates* and *shells*. The theory of these objects reduces to the theory of deformation of surfaces having special elastic properties. As the initial theory is three-dimensional, such approximations cannot be precise (except in special cases) and are not unique. But linear shell theories provide engineers with good descriptions of real elastic objects. They can be studied using more or less the approach taken in this book. To characterize their energy spaces is more challenging, however. A nice introduction to the mathematical theory of plates, with an extensive discussion of the finite element method, is contained in [Destuynder and Salaun (1996)]. The presentation in [Ciarlet (1998), (2000), (2005)] is more serious. Finally, there are numerous open problems in the various nonlinear theories of elasticity and elastic shells. One of the few classes of nonlinear problems that have been studied is treated in [Vorovich (1999)].

We will merely sketch some facts in the theory of plates. A plate is a structure that looks like a tabletop. In the theory of plate bending based on the Kirchhoff hypotheses, which are similar to those of beam theory, the deformation of the plate is described in terms of the displacement of its midplane. The midplane occupies domain S with Cartesian coordinates x_1, x_2. The plate has a relatively small thickness $2h$ and occupies the volume $S \times [-h, h]$. Kirchhoff's assumptions specify that after deformation the normal to the midplane remains straight and normal to its deformed state; moreover, ε_{13} and ε_{23} are small and taken to be zero. A further assumption that $\sigma_3 = 0$ permits one to exclude ε_{33} from all the relations of three-dimensional linear elasticity. Kirchhoff's plate theory is not unique in applications.

Kirchhoff's assumptions facilitate a theory of bending described by just one function: the deflection w of the midplane. The stress characteristics of the plate there are given in terms of couples

$$M_{ij} = \int_{-h}^{h} x_3 \sigma_{ij} \, dx_3 \qquad (i, j = 1, 2)$$

whereas strains are defined by

$$\rho_{ij} = -\frac{\partial^2 w}{\partial x_i \partial x_j}.$$

These are related to M_{ij} by an analogue of Hooke's law:

$$M_{ij} = D_{ijkl} \rho_{kl}$$

(summation convention in force). The rigidity tensor $\{D_{ijkl}\}$ must define a positive strain energy due to deformation, and we assume it satisfies the inequality

$$D_{ijkl}\rho_{ij}\rho_{kl} \geq c_0 \rho_{mn}\rho_{mn}$$

with a constant $c_0 > 0$ that does not depend on the choice of ρ_{ij}.

The strain energy of the plate is given by

$$\frac{1}{2}\int_S D_{ijkl}\rho_{kl}\rho_{ij}\, dx_1\, dx_2.$$

The VWP equation takes the form

$$\int_S D_{ijkl}\rho_{kl}\,\delta\rho_{ij}\, dx_1\, dx_2 - \int_S F\delta w\, dx_1\, dx_2 - \int_{\partial S} f\delta w\, ds = 0,$$

where δw is a virtual displacement, F is the distributed force over S, and f is given on the boundary ∂S of S.

For an isotropic plate, a variational procedure applied to the VWP equation produces the biharmonic equation

$$D\nabla^4 w + F = 0, \qquad (x_1, x_2) \in S$$

and two natural boundary conditions.

For a clamped boundary, we obtain an analogue to Dirichlet's problem with the conditions

$$w|_{\partial S} = 0, \qquad \left.\frac{\partial w}{\partial n}\right|_{\partial S} = 0.$$

So here we can get two boundary value problems: equilibrium of the plate with clamped edge, and equilibrium of a free plate.

An energy inner product for these problems has the form

$$(w_1, w_2)_P = \frac{1}{2}\int_S D_{ijkl}\rho_{kl}(w_1)\rho_{ij}(w_2)\, dx_1\, dx_2.$$

One may then introduce the energy space and show that the energy norm is equivalent to the norm on $W^{2,2}(S)$. Hence it is possible to use Sobolev's imbedding result to see that the energy space in imbedded into $C(S)$. Finally, one may formulate a generalized setup for the problem and establish conditions on the external forces that guarantee existence and uniqueness of the generalized solution.

Additional investigation is needed regarding the self-balance conditions for the equilibrium of a free plate.

Similarly, one can consider equilibrium problems in the linear theory of shells (see, e.g., [Ciarlet, 1998 & 2000]).

Appendix A

Hints for Selected Exercises

1.3.1. $0 = d(A,A) \leq d(A,B) + d(B,A) = d(A,B) + d(A,B) = 2d(A,B)$.

1.3.2. For $p \geq 1$ the function $d_p(A,B)$ satisfies the triangle inequality, which takes the specific form

$$(|b_1-a_1|^p+|b_2-a_2|^p)^{1/p} \leq (|c_1-a_1|^p+|c_2-a_2|^p)^{1/p}+(|b_1-c_1|^p+|b_2-c_2|^p)^{1/p}.$$

This is a special case of Minkowski's inequality (1.3.3), which also holds for this range of p. Minkowski's inequality follows from Hölder's inequality (1.12.6). This latter result, one of the central inequalities of analysis, is a consequence of the elementary inequality

$$ab \leq \frac{a^p}{p} + \frac{b^q}{q} \quad \left(\frac{1}{p}+\frac{1}{q}=1\right)$$

which, in turn, can be proved by establishing that

$$t^{1/p} \leq \frac{t}{p} + 1 - \frac{1}{p} \tag{*}$$

holds for all $t \geq 0$ — again, where $p \geq 1$ — and then putting $t = a^p b^{-q}$. Equality holds in (*) if and only if $t = 1$.

Use differentiation to show that the inequality sign in (*) reverses for $0 < p < 1$. This ultimately leads to a sign reversal in Minkowski's inequality for this same range of p, and shows that the triangle inequality fails to hold for $d_p(A,B)$ whenever $0 < p < 1$.

1.4.1. Consider the set of coefficients of a trigonometric polynomial as the components of an abstract vector $(a_1, \ldots, a_n, b_0, b_1, \ldots, b_n)$. Any of the metrics we have introduced for ordinary vectors would be appropriate.

1.5.1. $0 = \|x - x\| \leq \|x\| + \|-x\| = 2\|x\|$.

1.5.2. The two inequalities
$$\|x\| = \|(x-y)+y\| \le \|x-y\| + \|y\|,$$
$$\|y\| = \|(y-x)+x\| \le \|x-y\| + \|x\|,$$
show that both $\|x\| - \|y\|$ and $\|y\| - \|x\|$ are less than or equal to $\|x-y\|$.

1.5.3. We cannot impose this "norm" on the set of all functions continuous on $[0,1]$. Indeed, it does not always take a finite value — consider $f(x) = 1$ for example. We could introduce a normed space by restricting ourselves to "the set of continuous functions for which the above norm is finite," however.

1.5.4. Partition $[a,b]$ uniformly with segment width Δ. The expressions can be approximated by Riemann sums
$$\sum_k f(x_k) g(x_k) \Delta \quad \text{and} \quad \left(\sum_k f^2(x_k) \Delta \right)^{1/2},$$
where $x_k = a + k\Delta$. Regarding the set $(f(a), f(a+\Delta), f(a+2\Delta), \ldots)$ as the components of a fictitious vector, we see that the expressions (up to a factor) are the dot product and the Euclidean norm. So it remains only to produce the limit passage for the partition.

1.5.5.
$$\begin{aligned}
(f - (f,e)e, e) &= (f,e) - ((f,e)e, e) \\
&= (f,e) - (f,e)(e,e) \\
&= (f,e) - (f,e) \\
&= 0.
\end{aligned}$$

1.5.7.
$$\begin{aligned}
\|x+y\|^2 + \|x-y\|^2 &= (x+y, x+y) + (x-y, x-y) \\
&= (x, x+y) + (y, x+y) + (x, x-y) - (y, x-y) \\
&= (x,x) + (x,y) + (y,x) + (y,y) \\
&\quad + (x,x) - (x,y) - (y,x) + (y,y) \\
&= 2(x,x) + 2(y,y) \\
&= 2\|x\|^2 + 2\|y\|^2.
\end{aligned}$$

1.8.1. (a)
$$\frac{d^2 \mathbf{f}(t)}{dt^2} = \frac{d^2 f^k(t)}{dt^2} \mathbf{e}_k.$$

(b)
$$\frac{d^2 \mathbf{f}(t)}{dt^2} = f^k(t) \frac{d^2 \mathbf{e}_k(t)}{dt^2} + 2 \frac{df^k(t)}{dt} \frac{d\mathbf{e}_k(t)}{dt} + \frac{d^2 f^k(t)}{dt^2} \mathbf{e}_k(t).$$

1.9.1. The vector equation
$$\mathbf{e}_1 \sum_{i=1}^n m_i \xi_i + \mathbf{e}_2 \sum_{i=1}^n m_i \eta_i + \mathbf{e}_3 \sum_{i=1}^n m_i \zeta_i = \mathbf{0}$$
implies the three scalar equations
$$\sum_{i=1}^n m_i \xi_i = 0, \quad \sum_{i=1}^n m_i \eta_i = 0, \quad \sum_{i=1}^n m_i \zeta_i = 0,$$
by linear independence.

1.9.2. Equation (1.9.18) yields
$$\frac{d}{dt} \sum_{i=1}^n (\mathbf{r}_i \times m_i \mathbf{v}_i) = \mathbf{0},$$
hence
$$\sum_{i=1}^n (\mathbf{r}_i \times m_i \mathbf{v}_i) = \mathbf{C}$$
as required.

1.10.1. 5.

1.11.1. We verify the norm axioms.

(1) Obviously $\|f\|_{C(V)} \geq 0$, and from the equality $\|f\|_{C(V)} = 0$ it follows that $f = 0$ on V.

(2) We have
$$\|\alpha f\|_{C(V)} = \max_{\mathbf{x} \in V} |\alpha f(\mathbf{x})| = |\alpha| \max_{\mathbf{x} \in V} |f(\mathbf{x})| = |\alpha| \|f\|_{C(V)}.$$

(3) We have
$$|f(\mathbf{x}) + g(\mathbf{x})| \leq |f(\mathbf{x})| + |g(\mathbf{x})| \leq \max_{\mathbf{x} \in V} |f(\mathbf{x})| + \max_{\mathbf{x} \in V} |g(\mathbf{x})|,$$
so
$$\max_{\mathbf{x} \in V} |f(\mathbf{x}) + g(\mathbf{x})| \leq \max_{\mathbf{x} \in V} |f(\mathbf{x})| + \max_{\mathbf{x} \in V} |g(\mathbf{x})|.$$

1.11.2. Similar to Exercise 1.11.1.

1.11.3. From the definition of $\|A\|$ it follows that
$$\|A\| = \sup_{x \neq 0} \frac{\|Ax\|}{\|x\|}.$$
The chain of equalities and inequalities
$$\|A\| = \sup_{x \neq 0} \frac{\|Ax\|}{\|x\|} = \sup_{x \neq 0} \left\|A\left(\frac{x}{\|x\|}\right)\right\| = \sup_{\|x\|=1} \|Ax\|$$
$$\leq \sup_{\|x\| \leq 1} \|Ax\| \leq \sup_{0 < \|x\| \leq 1} \frac{\|Ax\|}{\|x\|} \leq \sup_{0 < \|x\|} \frac{\|Ax\|}{\|x\|} = \|A\|$$
demonstrates that any of several expressions can be used to represent $\|A\|$.

1.12.1. The first two norm axioms hold trivially. That the triangle inequality holds for (1.12.3) can be seen as follows. For each $i = 1, \ldots, n$ we have
$$|x_i + y_i| \leq |x_i| + |y_i| \leq \max_{1 \leq i \leq n} |x_i| + \max_{1 \leq i \leq n} |y_i|.$$
Therefore,
$$\max_{1 \leq i \leq n} |x_i + y_i| \leq \max_{1 \leq i \leq n} |x_i| + \max_{1 \leq i \leq n} |y_i|.$$
Use Minkowski's inequality to verify the triangle inequality for (1.12.5).

1.12.2. By the triangle inequality we have
$$\|A\mathbf{x}\|_r = \left(\sum_{i=1}^n \left|\sum_{j=1}^n a_{ij} x_j\right|^r\right)^{1/r} \leq \left(\sum_{i=1}^n \left[\left(\sum_{j=1}^n |a_{ij} x_j|\right)^p\right]^{r/p}\right)^{1/r}.$$
Use of Hölder's inequality in the form given on page 42 then gives us
$$\|A\mathbf{x}\|_r \leq \left(\sum_{i=1}^n \left[\left(\sum_{j=1}^n |a_{ij}|^q\right)^{p/q} \left(\sum_{j=1}^n |x_j|^p\right)\right]^{r/p}\right)^{1/r}$$
$$= \left(\sum_{i=1}^n \left[\left(\sum_{j=1}^n |a_{ij}|^q\right)^{p/q}\right]^{r/p}\right)^{1/r} \|\mathbf{x}\|_p.$$
So
$$\|A\| = \left(\sum_{i=1}^n \left(\sum_{j=1}^n |a_{ij}|^q\right)^{r/q}\right)^{1/r}$$

where $q = p/(p-1)$.

1.13.1. $d(x_m, x_n) \leq d(x_m, x) + d(x_n, x)$.

1.13.2. Let $c \in [-1, 1]$. If f is continuous and $f(c) \neq 0$, then $|f(x)|$ is nonnegative and $|f(c)| > 0$. So $|f(x)|$ is positive in some neighborhood of point c and we have $\int_{-1}^{1} |f(x)|\, dx > 0$. This shows that $\|f\|_1 = 0$ implies $f = 0$. The remaining axioms follow from the properties of the definite integral.

1.13.3. Take $f_n(x)$ equal to zero everywhere except on $[-1/n, 1/n]$, where it should equal $1 - n|x|$. Then the norm in $C(-1, 1)$ is 1 for each n, and the L^1 norm tends to 0.

1.15.1. Take a representative $\{f_n\}$ from F and consider the sequence $\{K_n\}$ given by

$$K_n = \|f_n\|_p.$$

This is a numerical Cauchy sequence:

$$|K_m - K_n| = \big|\,\|f_m\|_p - \|f_n\|_p\,\big|$$
$$\leq \|f_m - f_n\|_p \to 0 \quad \text{as } m, n \to \infty.$$

Hence by completeness of $\mathbb{R}$ there exists a number

$$K = \lim_{n \to \infty} K_n = \lim_{n \to \infty} \left(\int_\Omega |f_n(\mathbf{x})|^p \, d\Omega \right)^{1/p}.$$

K is independent of the choice of representative sequence. If $\{\tilde{f}_n\}$ is another representative of F, i.e., if

$$\|f_n - \tilde{f}_n\|_p \to 0 \quad \text{as } n \to \infty,$$

then we can set

$$\tilde{K} = \lim_{n \to \infty} \tilde{K}_n = \lim_{n \to \infty} \|\tilde{f}_n\|_p$$

but subsequently find that

$$|K - \tilde{K}| = \big|\lim_{n \to \infty} \|f_n\|_p - \lim_{n \to \infty} \|\tilde{f}_n\|_p\big|$$
$$= \lim_{n \to \infty} \big|\,\|f_n\|_p - \|\tilde{f}_n\|_p\,\big|$$
$$\leq \lim_{n \to \infty} \|f_n - \tilde{f}_n\|_p = 0.$$

This shows that $\tilde{K} = K$. The uniquely determined number

$$K^p = \lim_{n\to\infty} \int_\Omega |f_n(\mathbf{x})|^p \, d\Omega$$

is the right-hand side of (1.15.9).

1.15.2. The statement $L^p(\Omega) \subseteq L^r(\Omega)$ is trivial when $r = p$, so assume $1 \leq r < p$. Then $p/r > 1$, and we can take q such that

$$\frac{1}{q} + \frac{1}{p/r} = \frac{1}{q} + \frac{r}{p} = 1.$$

Hölder's inequality gives

$$\left| \int_\Omega 1 \cdot |F(\mathbf{x})|^r \, d\Omega \right| \leq \left(\int_\Omega 1^q \, d\Omega \right)^{1/q} \left(\int_\Omega |F(\mathbf{x})|^p \, d\Omega \right)^{r/p}$$

$$= (\operatorname{mes}\Omega)^{1-\frac{r}{p}} \left(\int_\Omega |F(\mathbf{x})|^p \, d\Omega \right)^{r/p}, \quad \operatorname{mes}\Omega = \int_\Omega 1 \, d\Omega.$$

This yields the desired inequality with $C_{p,r} = (\operatorname{mes}\Omega)^{\frac{1}{r}-\frac{1}{p}}$.

1.16.1. The "if" part is obvious. We address the "only if" part. For sufficiently large x the term ax^2 dominates bx, showing that we must have $a \geq 0$. If $a = 0$, then bx will take both positive and negative values unless $b = 0$. For the case $a > 0$, we write

$$f(x) = ax^2 + bx = a\left(x + \frac{b}{2a}\right)^2 - \frac{b^2}{4a}.$$

The minimum is taken at $x = -b/2a$ and is $-b^2/4a$. This implies that b must be zero.

1.16.2. We first show that $M^\perp$ is a subspace. Take any two elements $x, y \in M^\perp$ and any scalars α_1 and α_2. For all $m \in M$ we have

$$(\alpha_1 x + \alpha_2 y, m) = \alpha_1(x, m) + \alpha_2(y, m) = 0,$$

so $\alpha_1 x + \alpha_2 y \in M^\perp$ as required. We proceed to show that $M^\perp$ is closed. Let $\{x_n\}$ be any convergent sequence in $M^\perp$, with $x_n \to x$. Take an arbitrary $v \in M$. Then $(x_n, v) = 0$ for each n. By continuity of the inner product we have

$$(x, v) = \lim_{n\to\infty}(x_n, v) = 0,$$

hence $x \in M^\perp$.

1.22.1. We have

$$\sum_{k=1}^{n} x_k \frac{\partial g}{\partial x_k} = x\frac{\partial g}{\partial x} + y\frac{\partial g}{\partial y}$$
$$= x(2x + 2y) + y(2x + 6y)$$
$$= 2(x^2 + 2xy + 3y^2)$$
$$= 2g.$$

2.9.1. The symmetry of the functional in w and δw is evident, as is its linearity in each of variables. Next,

$$(w, w)_S = \int_0^l T_0 w_x^2 \, dx \geq 0$$

and in the case when

$$(w, w)_S = \int_0^l T_0 w_x^2 \, dx = 0$$

we get $w_x = 0$. So $w = c = $ constant and, by (2.9.2), $w = 0$.

2.10.1. As it is complete we cannot find any "new" element. Here the set that is dense and each element of which contains a stationary sequence constitutes the whole space. So the completion gives us nothing new except another viewpoint on the same initial space. This new space is quite inconvenient to use, but it is in one-to-one correspondence with the initial space. The correspondence preserves the norm and algebraic operations, so so it does not matter which space we work with.

2.16.2. A translation of the coordinate origin does not change the form of the inequality.

2.16.4. Use Theorem 2.16.2.

2.16.5. We apply integration by parts to the first term in (2.16.25). The formulas are

$$\int_\Omega g h_x \, dx \, dy = -\int_\Omega g_x h \, dx \, dy + \oint_\Gamma g h \, n_x \, ds,$$
$$\int_\Omega g h_y \, dx \, dy = -\int_\Omega g_y h \, dx \, dy + \oint_\Gamma g h \, n_y \, ds,$$

where s parameterizes Γ, and n_x and n_y are the direction cosines of the outward normal to Ω. Equation (2.16.25) becomes

$$-\int_\Omega [a(u_{xx}+u_{yy})+f]v\,dx\,dy + \oint_\Gamma [a(u_x n_x + u_y n_y) - \varphi]v\,ds = 0.$$

Now the usual methods of the calculus of variations yield the Euler equation

$$a(u_{xx}+u_{yy}) + f = 0$$

and the natural boundary condition

$$a(u_x n_x + u_y n_y) - \varphi = 0.$$

3.2.1. Because there is no lumped force at the point x, equation (3.2.11) shows that N is continuous there. So we get $ESu''|_{x-0} = ESu''|_{x+0}$. If ES has a jump at x, then so must u'' in order to preserve continuity of $N = ESu''$. Continuity of N is also preserved in the special case when $u''(x) = 0$; in this case there is no jump.

3.2.2. Setting the first variation of $\mathcal{E}(u)$ equal to zero, we get

$$\int_0^l ES \frac{du}{dx} \frac{d(\delta u)}{dx}\,dx - \int_0^l t(x)\,\delta u(x)\,dx - \sum_{k=0}^n F_k\,\delta u(x_k) = 0.$$

Integrating by parts, we obtain

$$-\sum_{k=0}^{n-1} \int_{x_{k-1}}^{x_k} \left[\frac{d}{dx}\left(ES\frac{du}{dx}\right) + t(x)\right]\delta u\,dx$$

$$+ \sum_{k=0}^{n-1} ES\frac{du}{dx}\delta u \bigg|_{x=x_{k-1}+0}^{x=x_{k-1}-0} - \sum_{k=0}^n F_k\,\delta u(x_k) = 0.$$

First we should select a subset of δu that differ from zero only on (x_{k-1}, x_k). This yields the Euler equation

$$\frac{d}{dx}\left(ES\frac{du}{dx}\right) + t(x) = 0$$

on each of the intervals (x_{k-1}, x_k), $k = 1, \ldots, n-1$. Next, considering the integral equation on the segment $[0, l]$ we see that for any δu, the integral term is zero because of the above equalities. Considering successively the non-integrated terms we get the following. Taking δu nonzero at node $x_0 = 0$ and zero at other nodes x_k,

$$ES\frac{du}{dx}\bigg|_{x=0} = -F_0.$$

Then taking δu nonzero only at node $x_n = l$ we get

$$ES\frac{du}{dx}\bigg|_{x=l} = F_n.$$

These are natural boundary conditions at the extremes. Taking δu such that it is nonzero only at node x_k, we get jump conditions at x_k:

$$ES\frac{du}{dx}\bigg|_{x_k+0} - ES\frac{du}{dx}\bigg|_{x_k-0} = -F_k.$$

We should call these natural conditions as well, since the equilibrium equation is not valid at the points x_k. These points now constitute the boundary for the domain where it is valid. Finally, the self-balance condition

$$\int_0^l t(x)\,dx + \sum_{k=0}^n F_k = 0$$

is obtained when we substitute to the equation for first variation $\delta u = 1$.

3.2.3. For the attached spring, we have $v = u(c)$. Add $\frac{1}{2}Ku^2(c)$ to the energy expression of Exercise 3.2.2. When deriving the Euler equation, include the node c in the partition and obtain $n+1$ nodes between which the equilibrium equation of Exercise 3.2.2 holds. The conditions at the points x_k ($k = 0, \ldots, n$) remain the same. At point c we get the condition

$$ES\frac{du}{dx}\bigg|_{c+0} - ES\frac{du}{dx}\bigg|_{c-0} = Ku(c),$$

which can be formally obtained from the equation at node x_k by putting $F_k = -Ku(c)$. The self-balance condition disappears, as the spring plays a role similar to that of a geometric constraint.

3.7.1. Given an initial basis $(\mathbf{e}_1, \mathbf{e}_2, \mathbf{e}_3)$, we can construct the reciprocal basis $(\mathbf{e}^1, \mathbf{e}^2, \mathbf{e}^3)$. We construct $\mathbf{e}^1$ by writing

$$\mathbf{e}^1 = \alpha(\mathbf{e}_2 \times \mathbf{e}_3)$$

and then determining the constant α so that $\mathbf{e}_1 \cdot \mathbf{e}^1 = 1$:

$$\alpha[\mathbf{e}_1 \cdot (\mathbf{e}_2 \times \mathbf{e}_3)] = 1.$$

Denoting the scalar quantity $\mathbf{e}_1 \cdot (\mathbf{e}_2 \times \mathbf{e}_3)$ by V (this is the volume of the parallelepiped spanned by the initial basis), we get $\alpha = 1/V$ and

$$\mathbf{e}^1 = \frac{1}{V}(\mathbf{e}_2 \times \mathbf{e}_3).$$

The vectors $\mathbf{e}^2$ and $\mathbf{e}^3$ are constructed similarly.

3.7.2. The proof follows from the definition of a dual basis and the equalities that define the dual basis: $\mathbf{e}^k \cdot \mathbf{e}_m = \delta^k_m$.

3.7.3. In order to show that the vectors $\mathbf{e}^i = g^{ij}\mathbf{e}_j$ constitute the dual basis, we dot-multiply the equality by $\mathbf{e}_k$. Since the matrices (g^{ij}) and (g_{jk}) are mutually inverse, we obtain

$$\mathbf{e}^i \cdot \mathbf{e}_k = g^{ij}\mathbf{e}_j \cdot \mathbf{e}_k = g^{ij}g_{jk} = \delta^i_k.$$

By definition, $\mathbf{e}^1, \mathbf{e}^2, \mathbf{e}^3$ constitute the dual basis.

3.7.4. Take the identity of Exercise 3.7.3 and dot multiply with $\mathbf{e}^k$:

$$\mathbf{e}^i \cdot \mathbf{e}^k = g^{ij}\mathbf{e}_j \cdot \mathbf{e}^k = g^{ij}\delta^k_j = g^{ik}.$$

3.12.3. The formulation of this exercise is typical for books on mechanics; all conditions needed to make the statement precise were omitted. Let us state them explicitly here. (1) The tensor $\boldsymbol{\sigma}$ must be continuously differentiable; moreover, the coordinates in space should be such that all its components are continuously differentiable. (2) The external force $\mathbf{F}$ must be a continuous vector function. (3) The boundary should have some smoothness, and the boundary force condition holds only at points of smoothness of the boundary and continuity of $\mathbf{f}$. To prove the statement, we start with (3.12.5) and change $\boldsymbol{\sigma} \cdot\cdot\, \delta\boldsymbol{\varepsilon}$ to $\boldsymbol{\sigma} \cdot\cdot\, (\boldsymbol{\nabla}\delta\mathbf{u})$:

$$\int_V \mathbf{F} \cdot \delta\mathbf{u}\, dV + \int_S \mathbf{f} \cdot \delta\mathbf{u}\, dS - \int_V \boldsymbol{\sigma} \cdot\cdot\, (\boldsymbol{\nabla}\delta\mathbf{u})\, dV = 0.$$

Using the third formula of Exercise 3.9.3 and the symmetry of the tensor $\boldsymbol{\sigma}$, we integrate by parts in the last integral:

$$\int_V \mathbf{F} \cdot \delta\mathbf{u}\, dV + \int_S \mathbf{f} \cdot \delta\mathbf{u}\, dS + \int_V \boldsymbol{\nabla} \cdot \boldsymbol{\sigma} \cdot \delta\mathbf{u}\, dV - \int_S \mathbf{n} \cdot \boldsymbol{\sigma} \cdot \delta\mathbf{u}\, dS = 0.$$

Hence

$$\int_V (\boldsymbol{\nabla} \cdot \boldsymbol{\sigma} + \mathbf{F}) \cdot \delta\mathbf{u}\, dV + \int_S (-\mathbf{n} \cdot \boldsymbol{\sigma} + \mathbf{f}) \cdot \delta\mathbf{u}\, dS = 0.$$

Next we write down this formula only for virtual displacements $\delta\mathbf{u}$ that vanish on the boundary

$$\int_V (\boldsymbol{\nabla} \cdot \boldsymbol{\sigma} + \mathbf{F}) \cdot \delta\mathbf{u}\, dV = 0.$$

By the main lemma of the calculus of variations we find, because of the arbitrariness of $\delta\mathbf{u}$, that $\boldsymbol{\nabla} \cdot \boldsymbol{\sigma} + \mathbf{F} = \mathbf{0}$ in V. Finally, we derive the natural boundary condition. So, returning to the last equation but for all virtual displacements $\delta\mathbf{u}$ and taking into account that the equilibrium equation holds, we get

$$\int_S (-\mathbf{n}\cdot\boldsymbol{\sigma}+\mathbf{f})\cdot\delta\mathbf{u}\,dS = 0.$$

Applying the main lemma to the surface integral, we obtain

$$-\mathbf{n}\cdot\boldsymbol{\sigma}\big|_S + \mathbf{f} = \mathbf{0}.$$

3.18.1. Use $\mathbf{r}\times\mathbf{b}\times\mathbf{r}=\mathbf{0}$.

3.25.1. Verify the conditions of Arzela's theorem.

3.27.1. Suppose $\lim_{k\to\infty} x_k = \tilde{x}^*$. By definition of weak convergence, for any $h \in H$ we have

$$\lim_{k\to\infty}(x_k,h)_H = (x^*,h)_H.$$

On the other hand, from continuity of the inner product we obtain

$$\lim_{k\to\infty}(x_k,h)_H = (\tilde{x}^*,h)_H.$$

Thus $(x^*,h)_H = (\tilde{x}^*,h)_H$. Since h is arbitrary, we have $\tilde{x}^* = x^*$.

3.27.2. By definition of weak convergence we have

$$(x^*,x^*)_H = \lim_{k\to\infty}(x_k,x^*)_H \le \sup_k \|x_k\|_H \, \|x^*\|_H.$$

Canceling $\|x^*\|_H$ we get the result.

3.27.3.

$$\lim_{k\to\infty}(Bx_k,h)_H = \lim_{k\to\infty}(x_k,B^*h)_H = (x^*,B^*h)_H = (Bx^*,h)_H.$$

3.27.4. By the previous exercise, $\{Bx_k\}$ converges weakly to Bx^*. The sequence $\{x_k\}$ is bounded and so $\{Bx_k\}$ contains a Cauchy subsequence $\{x_{k_1}\}$ which, by Exercise 3.27.1, converges to Bx^*. The assumption that there is a subsequence $\{x_{k_2}\}$ for which $\{Bx_{k_2}\}$ has no strong limit Bx^* leads to a contradiction.

3.27.5. As B is compact and $x_k \rightharpoonup x^*$, by Exercise 3.27.4 we have

$$\lim_{k\to\infty} Bx_k = Bx^*.$$

Write
$$(Bx_k, x_k)_H - (Bx^*, x^*)_H = (Bx_k - Bx^*, x_k)_H + (Bx^*, x_k - x^*)_H.$$
The right-hand side tends to zero as $k \to \infty$. Indeed, $\|x_k\|_H \leq c$ and so
$$|(Bx_k - Bx^*, x_k)_H| \leq \|Bx_k - Bx^*\|_H \|x_k\|_H \to 0.$$
Moreover, $(Bx^*, x_k - x^*)_H \to 0$ because $x_k \rightharpoonup x^*$.

Bibliography

Adams, R.A., Fournier, J.J.F. (2003). *Sobolev Spaces, 2nd ed.* (Academic Press).
Ciarlet, P.G. (1994). *Mathematical Elasticity, Vol. I: Three-Dimensional Elasticity* (North Holland).
Ciarlet, P.G. (1997). *Mathematical Elasticity, Vol. II: Theory of Plates* (Elsevier).
Ciarlet, P.G. (1998). *Introduction to Linear Shell Theory* (Gauthier-Villars & Elsevier).
Ciarlet, P.G. (2000). *Mathematical Elasticity, Vol. III: Theory of Shells* (Elsevier).
Ciarlet, P.G. (2005). *An Introduction to Differential Geometry, with Applications to Elasticity* (Springer).
Destuynder P., and Salaun, M. (1996). *Mathematical Analysis of Thin Plate Models* (Springer).
Hutson, V., Pym, J.S., and Cloud, M.J. (2005) *Applications of Functional Analysis and Operator Theory*, second edition (Elsevier).
Lebedev, L.P., and Vorovich, I.I. (2002). *Functional Analysis in Mechanics* (Springer).
Lebedev, L.P., and Cloud, M.J. (2004). *Approximating Perfection: A Mathematician's Journey Into the World of Mechanics* (Princeton University Press).
Lebedev, L.P., and Cloud, M.J. (2003). *Calculus of Variations and Functional Analysis: With Optimal Control and Applications in Mechanics* (World Scientific).
Lebedev, L.P., and Cloud, M.J. (2002). *Tensor Analysis* (World Scientific).
Sobolev, S.L. (1991). *Some Applications of Functional Analysis in Mathematical Physics*, 3rd ed. (American Mathematical Society).
Gere, J.M., and Timoshenko, S.P. (1997). *Mechanics of Materials*, 4th ed. (PWS).
Timoshenko, S.P. (1959). *Theory of Plates and Shells* (McGraw-Hill).
Vorovich, I.I. (1999) *Nonlinear Theory of Shallow Shells* (Springer).

Index

action, 83
active force, 68
almost everywhere, 56, 137
angular frequency, 249
angular momentum, 28
   conservation of, 29
anisotropic material, 219
Arzelà's theorem, 257

ball, 6
   closed, 6
   open, 6
Banach space, 47
Banach, S., 20, 47
bar, 158
   equilibrium equation, 163
   total energy functional, 166
   virtual work principle, 166
basis, 10
beam, 168
   bending of, 168
   energy norm, 176
   energy space, 176
   equilibrium equation, 171
   existence-uniqueness theorem, 178
   generalized solution, 177
   total energy functional, 174
Bessel's inequality, 248
Betti's duality theorem, 226
Bubnov–Galerkin method, 240
bulk modulus, 225

Cauchy deformation measure, 213
Cauchy problem, 34
Cauchy sequence, 45
   weak, 267
Cauchy stress tensor, 196
Cauchy's lemma, 184
Cauchy–Green strain tensor, 212
Cauchy–Schwarz inequality, 14
center of mass, 22
Christoffel symbols, 204
classical mechanics, 2, 4
classical setup, 105
closed ball, 6
compact operator, 260
compact set, 35, 256
compatibility condition, 164
complete metric space, 47
complete system, 239
completeness, 239
completion, 50, 118
completion theorem, 50
   inner product space, 54
   normed space, 53
cone property, 140
configuration space, 33
conservation of energy, 87
conservative system, 87
constitutive relation, 200, 218
continuity, 37
   sequential, 38
continuum mechanics
   central principles, 90

contraction operator, 153
contraction principle, 153
contravariant components, 192
convergence, 37, 45
   strong, 267
   weak, 267
coordinate origin, 7
corner point, 151
Cosserat mechanics, 183
countability, 243
couple, 18
   principal, 22
Courant's method, 277
covariant components, 192
covariant derivative, 204
curvilinear coordinates, 202

d'Alembert's principle, 25, 27, 30
decomposition theorem, 62
deformable body, 89
deformation method, 164
degree of freedom, 33
denseness, 49
diagonal process, 272
dimension, 10
dimensionless form, 17
direct sum, 60
Dirichlet conditions, 75
displacement vector, 209
distributions, 17
double-dot product, 195
dual basis, 192
dyad, 189

eigenfrequency, 253
eigensolution, 253
eigenvalue, 253
eigenvector, 251
elastic body
   energy norm, 228
   energy space, 228, 234
   equilibrium equation, 221
   existence-uniqueness theorem, 232, 235
   generalized solution, 228, 234
   virtual work principle, 216

elastic force, 91
elasticity, 2
energy, 63
   conservation of, 87
   kinetic, 64
   potential, 87
   strain, 92, 129, 225
   total, 92
equicontinuity, 257
equilibrium
   conditions for, 22
equivalence class, 55
   stationary, 49
equivalent norms, 43
Euler equation, 77
Euler–Lagrange equation, 79
Eulerian description, 207
external forces, 27

factor space, 123
finite ε-net, 256
first problem of elasticity, 222
first variation, 76
fixed point, 153
force(s), 4, 16
   active, 68
   elastic, 91
   external, 27
   in equilibrium, 22
   inertia, 25
   internal, 27
   line of action of, 18
   moment of, 18
   resultant of, 16
   self-balanced, 103
   tension, 96
forced vibrations, 250
Fourier coefficient, 247
Fourier series, 10
Fredholm alternative, 262
Fredholm integral equation, 59
free vector, 19
Friedrichs inequality, 134
function(s)
   generalized, 17
   sufficiently smooth, 90

unit step, 46
functional, 36

Gâteaux derivative, 75
generalized derivative, 120, 137
generalized functions, 17
generalized orthogonality, 254
generalized solution, 107, 120
gradient operator, 205
Gram determinant, 237
Gram–Schmidt formulas, 246
Green's formula, 206

Hölder's inequality
   for integrals, 58
   for sums, 42
Hamilton's principle, 84
Hausdorff criterion, 256
Hilbert space, 47
holonomic constraints, 86
holonomic system, 69
homogeneity of space, 5, 11
Hooke's law, 65, 81, 92, 96, 99, 161, 218

imbedding, 135
imbedding operator, 117
imbedding theorem, 116, 117, 140, 141
incomplete metric space, 47
inequality
   Cauchy–Schwarz, 14
   Friedrichs, 134
   Hölder, 42, 58
   Korn, 229
   Minkowski, 7
   modulus, 56
   Poincaré, 139
   Schwarz, 118, 134
   triangle, 5, 11
inertial force, 25
inertial frame, 5
initial value problem, 34
inner product, 13
inner product space, 11, 13
   completion of, 54

internal forces, 27, 89
isometric correspondence, 7
isometry, 49
isothermal problem, 218
isotropic material, 219
isotropic space, 5
isotropy, 219

kernel, 59, 113
kinetic energy, 64
kinetic potential, 74
Kirchhoff uniqueness theorem, 226
Korn's inequality, 229
Kronecker delta, 220

Lagrange's equations, 71, 73
Lagrangian, 74
Lagrangian description, 207
Lax–Milgram theorem, 128
Lebesgue integral, 55, 56
Lebesgue norm, 55
limit, 45
limit passage, 187
line of action, 18
linear independence, 10, 237
linear momentum, 28
   conservation of, 29
linearity, 38

main lemma, 76
mass, 3
   center of, 22
mass point, 3
   equation of motion, 64
   total energy principle, 64
   virtual work principle, 67
material point, 3
   equation of motion, 64
   total energy principle, 64
   virtual work principle, 67
mechanics
   classical, 2, 4
   theoretical, 89
membrane, 128
   energy norm, 133
   energy space, 133

equilibrium equation, 132
existence-uniqueness theorem, 135
total energy functional, 130
virtual work principle, 143
method of elastic solutions, 152
metric, 5
  axioms of, 5
  taxicab, 6
metric coefficients, 193
metric space(s), 4, 6
  complete, 47
  completion of, 50
  incomplete, 47
  isometric, 49
metric tensor, 195
minimizing vector, 61
minimum energy principle, 81
Minkowski's inequality, 7, 281
mixed basis, 194
mixed components, 192
modulus inequality, 56
moment, 18
momentum
  angular, 28
  linear, 28
multi-index notation, 138

natural boundary condition, 77
natural norm, 14
Navier–Cauchy equations, 221
neighborhood, 6
Neumann condition, 103
norm(s), 11
  axioms of, 11
  equivalent, 43
  Lebesgue, 55
  natural, 14
normed space, 11
  completion of, 53

open ball, 6
operator(s), 36
  compact, 260
  continuous, 37
  contraction, 153
  domain of, 36

finite dimensional, 277
gradient, 205
linear, 38
matrix, 36
norm of, 39
range of, 36
self-adjoint, 252
strictly positive, 253
orthogonal complement, 62
orthogonal decomposition, 61
orthogonality, 13, 60
orthonormal system, 245
oscillation mode, 253

parallelogram equality, 15
Parseval's equality, 248
particle, 3
  equation of motion, 64
  total energy principle, 64
  virtual work principle, 67
plasticity, 218
plate, 278
Plateau's problem, 129
Poincaré's inequality, 139
Poisson's ratio, 219
position vector, 7
positiveness, 225
potential energy, 87
power of continuum, 244
precompact set, 256
pressure, 182
principal axes, 198
principal couple, 22
principal direction, 212
principal strains, 212
principal stresses, 198
principle
  d'Alembert, 25, 27, 30
  Hamilton's variational, 84
  minimum total energy, 81, 93, 122
  solidification, 90
  virtual work, 67
pure shear deformation, 211

radius vector, 23
rational numbers, 244

reference frame, 4
representative Cauchy sequence, 49
resonance, 249
resultant, 16
Riesz representation theorem, 113
rigid body, 3, 31
rigid motion, 210
Ritz approximation, 236
Ritz method, 238
Ritz, W., 236

Saint Venant's equations, 223
Schwarz inequality, 118, 134
second problem of elasticity, 222
self-adjoint operator, 252
separability, 244
sequence(s)
    Cauchy, 45
    convergent, 37, 45
    equivalent Cauchy, 48
    limit of, 45
    representative Cauchy, 49
sequential continuity, 38
set(s)
    compact, 35, 256
    dense, 49
    precompact, 256
shear modulus, 220, 225
shell, 278
sliding vector, 18
Sobolev space, 138
Sobolev, S.L., 136
solidification principle, 90
space
    Banach, 47
    Hilbert, 47
    metric, 4
    normed, 11
    Sobolev, 138
    vector, 8
spectrum, 252
spring, 91
    total energy, 92
    virtual work principle, 93
star-shaped domain, 139
statically determinate, 160

statically indeterminate, 161
stationary equivalence class, 49
stationary value, 84
strain, 91, 161
strain energy, 225
strain tensor, 208
    of small deformation, 209
    Cauchy–Green, 212
stress, 159
strictly positive operator, 253
string, 95, 100, 122
    energy space, 116, 124
    existence-uniqueness theorem, 122, 125
    generalized solution, 120, 125
    virtual work principle, 109
strong convergence, 267
sufficiently smooth function, 90
summation convention, 24

taxicab metric, 6
tension, 96
tensor(s), 189
    Cauchy stress, 196
    coaxial, 220
    first invariant, 221
    metric, 195
    of elastic moduli, 195
    strain, 208
    zero, 194
theorem
    Arzelà, 257
    Banach's contraction, 153
    Betti, 226
    completion, 50
    decomposition, 62
    imbedding, 117
    Riesz representation, 113
    uniqueness, 226
    Weierstrass, 46
theoretical mechanics, 89
theory of distributions, 107
theory of elasticity, 2
trajectory, 33
triangle inequality, 5, 11
trigonometric polynomial, 9

uniqueness theorem, 226
unit step function, 46

variation, 93
variational methods, 235
variational principle, 84
vector space, 8
  axioms of, 9
  basis of, 10
  dimension of, 10
  infinite dimensional, 10
  norm on, 11
vector(s)
  displacement, 209
  free, 19
  minimizing, 61
  position, 7
  radius, 23
  sliding, 18
  zero, 9
vibration, 249
virtual displacement, 68
virtual work, 67
virtual work principle, 67, 214, 224

weak Cauchy sequence, 267
weak convergence, 267
  criterion for, 270
weak solution, 179
Weierstrass theorem, 46
Winkler's foundation, 145
work, 63
  of internal forces, 107, 217
  virtual, 67

Young's modulus, 81, 162

zero vector, 9